AF775182

CRISPR-Cas Systems

Rodolphe Barrangou · John van der Oost
Editors

CRISPR-Cas Systems

RNA-Mediated Adaptive Immunity in Bacteria and Archaea

Springer

Editors
Rodolphe Barrangou
DuPont Nutrition and Health
Madison, WI
USA

John van der Oost
Laboratory of Microbiology
Wageningen University
Wageningen
The Netherlands

ISBN 978-3-642-34656-9 ISBN 978-3-642-34657-6 (eBook)
DOI 10.1007/978-3-642-34657-6
Springer Heidelberg New York Dordrecht London

Library of Congress Control Number: 2012953370

© Springer-Verlag Berlin Heidelberg 2013
This work is subject to copyright. All rights are reserved by the Publisher, whether the whole or part of the material is concerned, specifically the rights of translation, reprinting, reuse of illustrations, recitation, broadcasting, reproduction on microfilms or in any other physical way, and transmission or information storage and retrieval, electronic adaptation, computer software, or by similar or dissimilar methodology now known or hereafter developed. Exempted from this legal reservation are brief excerpts in connection with reviews or scholarly analysis or material supplied specifically for the purpose of being entered and executed on a computer system, for exclusive use by the purchaser of the work. Duplication of this publication or parts thereof is permitted only under the provisions of the Copyright Law of the Publisher's location, in its current version, and permission for use must always be obtained from Springer. Permissions for use may be obtained through RightsLink at the Copyright Clearance Center. Violations are liable to prosecution under the respective Copyright Law.
The use of general descriptive names, registered names, trademarks, service marks, etc. in this publication does not imply, even in the absence of a specific statement, that such names are exempt from the relevant protective laws and regulations and therefore free for general use.
While the advice and information in this book are believed to be true and accurate at the date of publication, neither the authors nor the editors nor the publisher can accept any legal responsibility for any errors or omissions that may be made. The publisher makes no warranty, express or implied, with respect to the material contained herein.

Printed on acid-free paper

Springer is part of Springer Science+Business Media (www.springer.com)

We dedicate this book to the CRISPR community—past, present and future

Preface

Clustered regularly interspaced short palindromic repeats (CRISPR), together with associated sequences (*cas* genes and Cas proteins) form the CRISPR-Cas adaptive immune system, which is present in most archaea and many bacteria. This relatively novel family of repeats was first discovered in 1987, characterized in 2002, implicated in immunity in 2005, and shown to provide acquired resistance against bacterial viruses in 2007. Since, it has been implicated in providing adaptive immunity against bacteriophages, archaeal viruses, and plasmids in numerous organisms. The development of several functional model systems in the recent past has paved the way for thorough scientific investigations of these unique and intriguing defense systems.

Notwithstanding extensive sequence diversity and gene content polymorphism, CRISPR-Cas systems have recently been categorized into three types, based on phylogeny and molecular mechanism of action. This has set the stage for a revision of the nomenclature, and collective agreement on terminology, representation standards, and definition of the various stages that render CRISPR-mediated immunity. Mechanistically, CRISPR-Cas systems drive immunity through three major steps: (1) *acquisition*, where immunization occurs by uptake of foreign DNA sequence and integration as new CRISPR spacers; (2) *expression*, where Cas proteins are produced and CRISPR-encoded transcripts are processed into small interfering CRISPR RNAs (crRNAs); (3) *interference*, where crRNA-Cas ribonucleoprotein complexes mediate homologous target recognition and specific cleavage. The ability of this idiosyncratic system to integrate short DNA sequences from invasive elements into the chromosome renders adaptive immunity inheritable.

This book provides a unique perspective into the historical events and key discoveries that have unraveled the functions of CRISPR-Cas systems and the roles they play in bacterial and archaeal biology and evolution. Once the occurrence, diversity, function, and evolution of CRISPR are established, each CRISPR-Cas type is specifically characterized. Their roles in various biological processes (not restricted to defense) are discussed, and applications are outlined. Their impact on microbial populations and evolutions are outlined, thus setting the stage for a deeper understanding of CRISPR-Cas systems.

Although there are mechanistic commonalities between CRISPR-mediated immunity and RNAi, notably small non-coding RNA-mediated cleavage of complementary target nucleic acid sequences (generally DNA, but one sub-type eliminates RNA) by a ribonucleoprotein complex, there are fundamental differences in the molecular processes that drive these two phenomena. Functionally, in addition to providing adaptive immunity against exogenous viral and plasmid dsDNA, at least some CRISPR-Cas systems appear to play a role in host-regulatory processes. Several applications have been established, notably build up of phage resistance, and exploiting hypervariability for typing and epidemiological surveys. Moreover, the ability to re-program the cleavage machinery has opened new avenues for customized DNA restriction, nicking, genome engineering and editing.

Some of the contributors have been intimated with CRISPR sequences for many years, and provided their personal perspective on this fast-evolving and exciting field. Likewise, several authors have been very active members of the CRISPR research community and have had the privilege to participate in the annual CRISPR meetings hosted at UC Berkeley since 2008, and at Wageningen University in 2010. The material presented here illustrates the frenetic pace at which the field has evolved over the last five years, and the breadth and scope of topics discussed reflect the scientifically diverse community which has come together, covering foci including molecular studies, genetic analyses, mathematical modeling, evolution, functional implementation, epidemiology, metagenomics, and ecology. The variety of entry ways into the field, and diversity of the vantage points of the various groups involved illustrates the relevance of the topic. Current implementation of CRISPR-Cas systems to develop phage resistance in dairy starter cultures has already shown that CRISPR can be leveraged industrially. Current analyses of CRISPR polymorphism in pathogenic species will determine how relevant these loci may be for epidemiological surveys, clinical analyses, and food safety.

We would like to acknowledge all the authors, our colleagues, collaborators, and CRISPR meeting participants for all their contributions to the field, colorful opinions, and insightful conversations.

Looking back, the significant advances in studying the CRISPR mechanism of action have set the stage for applications and areas of investigation, and establish a solid basis for future studies that will investigate the outstanding mysteries and questions that remain unanswered. We are hopeful that the need for proper bioinformatics tools will be addressed, and that NCBI will integrate CRISPR-related resources. Doubtless, we predict that the community camaraderie and scientific diversity will pave the way for a bright future of CRISPR as a field. Certainly, the visibility of the field as measured by ever-increasing quantity, spectacular quality, and impressive citation rates of CRISPR-related publications warrant a bright future. This may just be the beginning…

Contents

Chapter 1
Discovery and Seminal Developments in the CRISPR Field

Francisco J. M. Mojica and Roger A. Garrett

Abstract In the late 1980s and early 1990s, arrays of regularly spaced repeats were detected in both bacterial and archaeal genomes. They are currently known as Clustered Regularly Interspaced Short Palindromic Repeats or CRISPR. Advances in our understanding of their biological significance and potential applications for biotechnology have followed a two-phased development. Initial studies were few and mainly descriptive of arrays of interspaced repeats in bacteria and archaea and of physically linked conserved genes that were inferred to be co-functional. Moreover, before their function was revealed, repeat-spacer arrays of *Mycobacterium* spp were employed as novel markers for bacterial genotyping. The second phase began in 2005, with the discovery of a link between CRISPR arrays and host protection against invading genetic elements. This finding fuelled a plethora of biochemical and genetic studies directed at characterising the mechanistic details of this novel and complex genetic barrier. First, this led to the finding that the repeats, spacers, CRISPR-associated (Cas) proteins and partially conserved leader regions flanking one end of the CRISPR array, constitute the essential functional components. Subsequently, three primary functional steps were defined: (1) acquisition (also termed adaptation): uptake of new spacers at or near the leader sequence, (2) expression: generation of CRISPR transcripts from within the leader region and their processing into small mature CRISPR RNAs (crRNAs) carrying all or most of the spacer sequence and (3) interference: involving protein-crRNA

F. J. M. Mojica (✉)
Departamento de Fisiología, Genética y Microbiología,
Universidad de Alicante, 03080 Alicante, Spain
e-mail: fmojica@ua.es

R. A. Garrett
Department of Biology Biocenter, Archaea Centre,
University of Copenhagen, 2200N Copenhagen, Denmark
e-mail: garrett@bio.ku.dk

R. Barrangou and J. van der Oost (eds.), *CRISPR-Cas Systems*,
DOI: 10.1007/978-3-642-34657-6_1, © Springer-Verlag Berlin Heidelberg 2013

complexes targeting and cleaving foreign genetic elements. Only now can we begin to comprehend the complex functional interactions and diversity of CRISPR-based systems, and the implications of their adaptive nature. Here, we describe the early developments in the CRISPR field and relate them to our current understanding of how these novel, complex and diverse systems function.

Contents

1.1 Introduction

Clusters of regularly interspaced short palindromic repeats (CRISPR) were first observed in bacteria in the late 1980s and, in the mid-1990s, in the genomes of various archaeal lineages. Their apparent broad distribution suggested that they might play a common and fundamental cellular role in both archaeal and bacterial domains, although their prevalence in extremophilic organisms suggested that they might in some way facilitate adaptation to extreme environments (Mojica et al. 2000; Jansen et al. 2002a). The [repeat-spacer]$_n$ arrays were distinct in structure and sequence from other known functional repeat-based systems and consequently, there were no precedents for inferring their biological function(s). The first evidence documenting their functionality arose from studies on haloarchaea, where transcripts from repeat loci were detected in wild-type isolates and growth defects were observed in cells transformed with artificial plasmids carrying repeat-spacer arrays (Mojica et al. 1993, 1995). At that time, a role for the repeats in replicon partitioning was proposed. In retrospect, CRISPR-mediated autoimmunity could explain the observed phenotypes, in particular chromosome loss. An important development followed with the identification of CRISPR-associated (*cas*) genes, which indicated that repeat-spacer arrays were part of more complex CRISPR-Cas protein systems (Jansen et al. 2002a). Although initial bioinformatical analyses of Cas protein sequences predicted their involvement in nucleic acid metabolism (Jansen et al. 2002a; Makarova et al. 2002; Haft et al. 2005), the first insight into their actual functions arose from the discovery of the origin of the spacers. Detailed sequence studies indicated that spacers and the identical sequences (later called protospacers) within the DNA of invading genetic elements were incompatible and this led to the hypothesis that CRISPR-Cas systems can generate immunity against foreign DNA (Bolotin et al. 2005; Mojica et al. 2005; Pourcel et al. 2005). Experimental evidence supporting this hypothesis followed for the firmicutes *Streptococcus thermophilus* when infected with phages (Barrangou et al. 2007) and for *Staphylococcus epidermidis* based on plasmid conjugation assays (Marraffini and Sontheimer 2008). Thus, CRISPR-Cas systems constitute a unique form of adaptive and heritable genetic barrier.

In brief, during infection by a genetic element, new spacers with sequences identical to those of an invading genetic element are inserted into a genomic CRISPR array of the challenged cell (acquisition). Then, small processed CRISPR RNAs (crRNAs) deriving from the CRISPR array and carrying most or all of the spacer sequence, form a complex with Cas proteins (expression). These ribonucleoprotein complexes subsequently target a complementary sequence in the invading nucleic acid and cleave within the matching sequence (interference).

CRISPR systems have recently been classified into three main types I, II and III, with a few subtypes that differ primarily in their processing and interference mechanistic details (Makarova et al. 2011). DNA targeting has been reported for type I, II and III-A systems (Brouns et al. 2008; Garneau et al. 2010; Marraffini and Sontheimer 2008) while RNA has been shown to be a primary target of some type III-B systems (Hale et al. 2009, 2012; Zhang et al. 2012). This current nomenclature is used throughout this book.

Apart from functioning as a defence system, the gene silencing mechanisms employed by CRISPR systems can, potentially, influence other cellular pathways. Different studies on the genetic and biochemical characterisation of Cas proteins and on the biological activity of diverse CRISPR-Cas systems provide support for this proposition (Viswanathan et al. 2007; Aklujkar and Lovley 2010; Babu et al. 2011; Cady and O'Toole 2011).

1.2 Early Breakthroughs

1.2.1 Regularly Spaced Repeats

The first report of a CRISPR repeat array arose from sequencing of an *Escherichia coli* K12 chromosomal fragment containing the *iap* gene which encodes an alkaline phosphatase isozyme conversion aminopeptidase (Ishino et al. 1987). Nakata and colleagues found "five highly homologous sequences of 29 nucleotides" that "were arranged as direct repeats with 32 nucleotides as spacing" (Ishino et al. 1987). The *iap* proximal repeat, centred 35 nucleotides downstream from the coding sequence, was less conserved than the others. Given the apparent dyad symmetry of this degenerate repeat, the authors suggested that its capacity to produce a stable stem-loop structure could provide a mechanism for transcriptional termination. A 7 bp inverted repeat was also present in the conserved repeats. The entire array was later sequenced and shown to contain 14 repeats. An additional cassette with 7 repeat units was also identified 24 kb downstream from the *iap* gene. Similar repeats were detected by Southern blot hybridisation in other *E. coli* strains and in the closely related *Shigella dysenteriae* and *Salmonella enterica*, although none were found in the proteobacteria *Klebsiella pneumoniae* and *Pseudomonas aeruginosa* (Nakata et al. 1989).

1.2.2 Direct Repeats in Mycobacteria

In 1991, a hot-spot region for integration of insertion sequence (IS) elements was identified in genomes of several strains of the *Mycobacterium tuberculosis* complex (MTC) in a region harbouring 36 bp direct repeats (DRs) interspaced by "35–41 bp of spacer DNA" (Hermans et al. 1991). The DR region showed polymorphism among *M. tuberculosis* strains, mainly due to the absence or presence of repeat-spacer units (also named DVRs after Direct Variable Repeats), but the relative order of the spacers remained constant. These features led to the development of new methods for strain differentiation by targeting DR loci. The first of these typing strategies, referred to as DVR-PCR, was based on PCR amplification of repeat-spacer multimers (Groenen et al. 1993). This approach produced the spoligotyping (SPacer OLIGOnucleotide TYPING) method. It establishes the presence or absence of spacers by hybridisation of labelled PCR products amplified from DR loci to membranes carrying oligonucleotides corresponding to previously identified spacer sequences (Aranaz et al. 1996; Kamerbeek et al. 1997). Spoligotyping is now used widely for genotyping of MTC strains and is exceptionally useful for epidemiological studies (Driscoll 2009). More recently, CRISPR arrays have been examined for their application as molecular markers for typing of other bacterial species (see Chaps. 2 and 11).

1.2.3 TREPs in Haloarchaea

Short tandem repeats of 30–34 bp interspersed with unique sequences of 35–39 bp were identified in archaea in 1993 (Mojica et al. 1993). A stretch of 15 repeat-spacer units was initially detected in a region of the *Haloferax mediterranei* genome associated with salt-dependent differential transcription. Later, two repeat loci (designated TREPs after Tandem REPeats) were located in the *H. mediterranei* chromosome and another was found in the largest resident megaplasmid (Mojica et al. 1995). Sequences similar to TREPs were also detected in other haloarchaeal strains, including *Haloferax volcanii,* which was also subjected to functional studies.

1.2.4 New Family of Prokaryotic Repeats

Several reports describing arrays of regularly spaced repeats appeared during the late 1990s, primarily as a consequence of the rapid progress in the development of genome sequencing technologies. The first description of complete genome repeat arrays appeared for the euryarchaeon *Methanocaldococcus jannaschii* (Bult et al. 1996), where the presence of a long partly conserved sequence adjoining one end of several short repeat (SR) arrays, later termed the leader (Jansen et al. 2002a), was also identified. Similarly, repeat arrays were soon found in other archaeal genomes, notably *Archaeoglobus fulgidus* (Klenk et al. 1997), *Methanothermobacter thermoautotrophicum* (Smith et al. 1997), *Sulfolobus solfataricus* (Sensen et al. 1998), *Pyrococcus horikoshii* (Kawarabayasi et al. 1998), *Aeropyrum pernix* (Kawarabayasi et al. 1999) and the *Sulfolobus* conjugative plasmid pNOB8 (She et al. 1998). Repeat arrays were also detected in the bacterium *Thermotoga maritima* (Nelson et al. 1999), and partial sequences were found in *Streptococcus pyogenes* (Hoe et al. 1999) and in cyanobacterial strains (Masepohl et al. 1996).

The long tandemly repeated repetitive (LTRR) sequences described for the cyanobacteria were very similar to DNA repeats found in mitochondrial plasmids of the bean *Vicia faba* (Flamand et al. 1992; Mojica et al. 2000), in both sequence and structure (partially palindromic repeats separated by 19–34 bp). This was a remarkable finding consistent with the proposed cyanobacterial ancestry of chloroplasts and suggests widespread horizontal gene transfer occurring amongst plant organelles (Hao et al. 2010). The presence of LTRR-like elements in mitochondria was intriguing but their assignment to CRISPR arrays was controversial (Jansen et al. 2002a). Although the repeat sequences were highly conserved, primarily at the internal inverted repeats (Mojica et al. 2000), the spacing was irregular (Flamand et al. 1992). Moreover, *cas* gene homologs have not been detected in the eukaryal domain (Makarova et al. 2006). The absence of elements required for CRISPR-mediated interference, notably *bona fide* spacers and appropriate *cas* genes, suggests that the LTRR-like eukaryal arrays may play a different role.

Asunto: **Re: Acronym**
Fecha: Wed, 21 Nov 2001 16:39:06 +0100
De: "Ruud Jansen" <R.Jansen@vet.uu.nl>
Empresa: Diergeneeskunde
A: "Francisco J. Martínez Mojica" <fmojica@ua.es>

Dear Francis

What a great acronym is CRISPR.
I feel that every letter that was removed in the alternatives made it
less crispy so I prefer the snappy CRISPR over SRSR and SPIDR.
Also not unimportant is the fact that in MedLine CRISPR is a
unique entry, which is not true for some of the other shorter
acronyms.

Fig. 1.1 Text extracted from an e-mail sent by Jansen to Mojica concerning the naming of regularly spaced repeats that led to the proposal of the CRISPR acronym (reproduced with permission)

Nevertheless, conservation of the internal repeats implies that the secondary structure is functionally important.

Comparison of the above-mentioned repeat arrays with those identified in other complete and partial bacterial and archaeal genomes from diverse phylogenetic taxa, led to the hypothesis that they constitute a new family of DNA repeats, and they were referred to as Short Regularly Spaced Repeats or SRSR (Mojica et al. 2000). SRSRs were defined as short sequences, typically containing inverted repeats, and generally arranged in clusters of repeated units regularly interspaced with unique sequences (Mojica et al. 2000). Leader sequences, first observed in the *M. jannaschii* genome (Bult et al. 1996), were associated with many repeat arrays.

Later, Jansen and Mojica proposed replacing SRSR with the more explicit, elegant and now widely adopted acronym CRISPR, for Clustered Regularly Interspaced Short Palindromic Repeats (Fig. 1.1) which served to remove the confusion arising from an abundance of disparate acronyms that had accumulated in the literature including DR, TREP, LTRR, SR, LCTR (She et al. 2001) and SPIDR (Jansen et al. 2002b). However, we now know that many CRISPR repeats do not exhibit dyad symmetry that could potentially generate stable “hairpin-loop” structures in transcripts and, moreover, true palindromes rarely occur in CRISPR repeats.

1.2.5 Distribution of CRISPR Arrays Amongst Archaea and Bacteria

Early searches for CRISPR arrays revealed their widespread distribution and prevalence in extremophiles, particularly in the archaeal domain where the number, diversity and extent of the arrays is, on average, much higher than in bacteria (Mojica et al. 2000; Jansen et al. 2002a, b, see Chaps. 2 and 5). At present, CRISPR arrays have been detected in more than 85 % of sequenced archaeal genomes

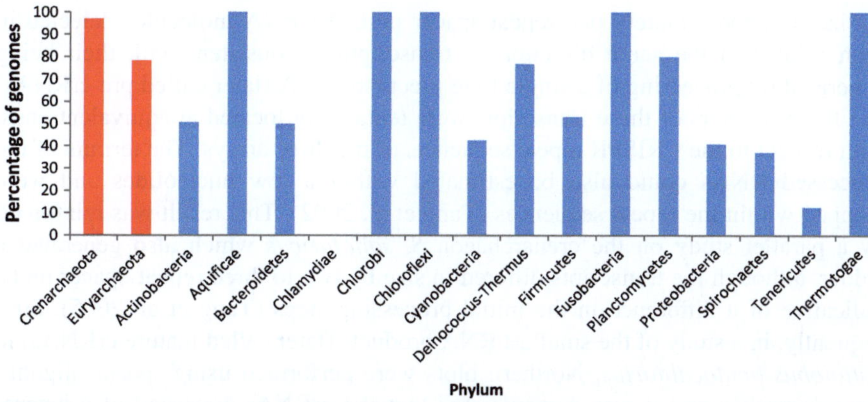

Fig. 1.2 Percentage of sequenced genomes carrying CRISPR arrays, within the main phylogenetic groups of archaea (*red*) and bacteria (*blue*). Only phyla for which more than five genomes have been sequenced are included. Data derive from the CRISPI (Rousseau et al. 2009) and CRISPRdb databases (Grissa et al. 2007)

(Fig. 1.2) and in about 49 % of available bacterial genomes based on data present in publicly available CRISPR databases (CRISPRdb at http://crispr.u-psud.fr/crispr/—Grissa et al. 2007 and CRISPI at http://crispi.genouest.org/—Rousseau et al. 2009). The prevalence in archaea is further supported by the observation that isolates from all major archaeal phyla carry CRISPR arrays and, moreover, CRISPR-deficient archaea are closely related to strains which either harbour complete CRISPR-Cas systems or carry remnants of CRISPR loci. For bacteria, CRISPR arrays have been found in all the major phyla except Chlamydiae, Gemmatimonadetes, Streptophyta and Synergistetes. However, only onc genome has been analysed for each of the latter three phyla and, therefore, the only major bacterial phylum that currently appears to be devoid of CRISPR systems is Chlamydiae, for which over 30 complete genomes have been analysed. This absence could be related to the minimal genome sizes of about 1 Mbp and/or their exceptional lifestyle as obligate intracellular parasites.

1.2.6 Discovery of Processed CRISPR Transcripts

Northern blot hybridisation using a repeat probe against total cellular RNA of *H. mediterranei*, separated by low resolution gel electrophoresis, revealed smears with blurred bands suggesting that processed RNAs were produced from the repeat arrays (Mojica et al. 1993). The nature of the CRISPR transcripts was revealed by sequencing of a cDNA library of RNAs from the euryarchaeon *A. fulgidus* (Tang et al. 2002). It showed that numerous RNAs were produced from each of three CRISPR loci. Moreover, Northern blot hybridisation revealed ladders of regularly spaced discrete bands when probed for the repeat sequence. These bands differed in size by approximately one repeat-spacer unit. All RNA molecules detected in each array had the same direction of transcription, consistent with their being generated by processing of a single long precursor RNA (later called pre-crRNAs) and the promoters of these transcripts were tentatively located at equivalent ends, with respect to the CRISPR repeat sequence, of the three arrays. The termini of the processed RNAs could also be estimated within a few nucleotides and were mapped within the repeat sequences (Tang et al. 2002). This result was reinforced by a parallel study on the crenarchaeon *S. solfataricus* which also generated a ladder, although the transcripts differed in size by two to three repeat-spacer units, indicative of a difference in the initial processing steps (Tang et al. 2005). Subsequently, in a study of the smallest RNA products (later called mature crRNAs) in *Sulfolobus acidocaldarius*, Northern blots were performed using spacer oligonucleotide probes and it was demonstrated that the crRNAs consisted of a heterogeneous mixture of bands falling in the size range 36–45 nt containing mainly spacer sequence (Lillestøl et al. 2006). Specific cleavage by a Cas endonuclease (Cas6) within each repeat of precursors referred to as pre-crRNAs, was subsequently reported for *E. coli* (Brouns et al. 2008) and *Pyrococcus furiosus* (Carte et al. 2008; Hale et al. 2008, see Chap. 6).

Transcripts are also produced from reverse strands of some CRISPR loci. They have been reported in different *Sulfolobus* species, where they can generate discrete bands in Northern blots (Lillestøl et al. 2006, 2009; Deng et al. 2012) and in *Thermus thermophilus* (Agari et al. 2009) and *Pelobacter carbinolicus* (Aklujkar and Lovley 2010). Their significance remains unclear and it has been speculated that they arise primarily from transcriptional initiation signals randomly taken up in spacers or, occasionally, from within a repeat or from repeat-spacer junctions (Deng et al. 2012). In *P. furiosus*, an antisense crRNA was recently shown to be targeted and cleaved by the corresponding crRNA-protein complex (Hale et al. 2012). This suggests that these RNAs may be actively targeted and cleaved in cells which carry type III-B CRISPR systems. However, in *P. carbinolicus,* a spacer at the leader distal end, also referred to as the trailer end, of a CRISPR array was shown to be transcribed equally on both strands and to match perfectly to the gene of the host encoded histidyl-tRNA transferase. The authors provide evidence for the CRISPR system influencing the maintenance and evolution of histidine-rich housekeeping proteins in the *Geobacteriaceae* (Aklujkar and Lovley 2010, see Chap. 10).

1.2.7 Identification of CRISPR-associated Proteins

Four *cas* (CRISPR-associated) genes, *cas1–cas4*, were originally identified in the immediate vicinity of many CRISPR loci of archaea and bacteria, with no preference for a particular side of the repeat array or for their direction of transcription; sometimes complex assemblies of CRISPR loci and diverse *cas* gene cassettes were seen. Importantly, these genes were absent from genomes lacking CRISPR loci. This physical link between repeat arrays and *cas* genes led to the proposal that they were co-functional (Jansen et al. 2002a). These authors also noted that when multiple CRISPR arrays with the same repeat sequence were located in a genome, *cas* genes were associated with only one of them. Further, they noted that if CRISPR arrays with different repeat sequences were present, each CRISPR system carried its own set of *cas* genes. Later work showed that although these CRISPR-associated gene cassettes were common for many organisms, there are exceptions. For example, amongst the crenarchaeal Sulfolobales CRISPR-*cas* gene cassettes sometimes constitute autonomous units within a cell even when CRISPR arrays exhibit closely similar repeat and homologous leader sequences (Shah and Garrett 2011).

In initial bioinformatical analyses of the Cas1–4 proteins, although no functions were predicted for Cas1 and Cas2, the Cas3 and Cas4 protein families were proposed to be homologous to DNA-helicases and exonucleases, respectively (Jansen et al. 2002a). Moreover, Makarova et al. (2002) classified five families of Repair Associated Mysterious Proteins, or RAMPs, that were later linked to CRISPR systems and renamed as Repeat Associated Mysterious Proteins (Haft et al. 2005). Subsequently, a total of 45 Cas protein families were defined and the diverse CRISPR-Cas systems were classified into 8 main subtypes according to the content and organisation of their *cas* gene cassettes (Haft et al. 2005). At present, over 50 different CRISPR-associated gene families have been identified which are generally, but not invariably, located close to CRISPR arrays (Makarova et al. 2006, 2011). The core Cas proteins involved in acquisition, expression and interference are listed in Table 1.2 and considered further in Sect. 1.4.1 and Chaps. 5–9.

The diverse RAMP proteins contain an RNA recognition motif (RRM) and are implicated in RNA binding and/or ribonuclease activity (Makarova et al. 2002, 2006, 2011, see Chap. 3). They were classified into two major groups containing Csm (CRISPR-Cas subtype Mtube; Haft et al. 2005) and Cmr (CRISPR RAMP module; Haft et al. 2005) proteins which make up interference complexes but they also include Cas6 family proteins responsible for pre-crRNA processing.

The recent reclassification of CRISPR-Cas systems into three main types (I, II and III), which include different subtypes, is mainly based on the presence of particular signature *cas* genes (Makarova et al. 2011). During this process, some Cas protein families that had previously been classified as specific for different CRISPR subtypes, and which exhibited highly divergent sequences (Haft et al. 2005; Makarova et al. 2006), were clustered into larger families of putatively homologous proteins and reclassified as core proteins. These include the newly named Cas7 and Cas8 proteins (Table 3.2; see Chap. 3).

The first and only protein shown to bind specifically to CRISPR DNA repeats is a *Sulfolobus* protein, later named Cbp1, that is not linked to CRISPR loci or *cas* genes and is, therefore, likely to have other cellular functions (Peng et al. 2003). This protein was originally thought to be involved in CRISPR DNA packaging but a later study of the effects of Cbp1 knockout and overexpression on pre-crRNA yields led to the inference that it affected transcription and could facilitate uninterrupted transcription of large CRISPR loci (Deng et al. 2012).

1.3 CRISPR-Cas Function Revealed

1.3.1 Early Hypotheses

After their discovery, CRISPR arrays were initially attributed a variety of functions. The observation that *E. coli* repeats occurred in intergenic regions led to the suggestion that they had a role in modulating gene expression (Nakata et al. 1989). In a similar vein, it was speculated that the DR sequences of *Mycobacterium* might influence expression of neighbouring genes by providing binding sites for regulatory proteins (Hermans et al. 1991). However, for the haloarchaeon *H. volcanii*, involvement of repeat arrays in regulation of neighbouring genes was considered unlikely given the large sizes of the arrays and, therefore, the first experimental studies were undertaken to determine their function(s). Recombinant plasmids containing CRISPR arrays (TREPs) were transformed into *H. volcanii* cells and the effects on cellular processes were examined (Mojica et al. 1995). Similar levels of transformation efficiency were observed for a suicide vector carrying TREPs and control constructs, indicating that a major role in recombination was unlikely. However, cell cultures with an otherwise stable TREP-containing plasmid, showed two-fold lower cell viability and a doubling of the recombinant plasmid copy number relative to cells carrying the same vector lacking repeat arrays. Moreover, an altered chromosomal distribution was frequently observed amongst dividing daughter cells transformed with the TREP-containing plasmids and, therefore, a role in replicon partitioning was proposed. In another study, a change in DNA replication kinetics was observed in the chromosome of *S. acidocaldarius* on passing through a CRISPR-rich region but it was speculated that this might be a retarding effect of the repeat binding protein Cbp1 assembled along CRISPR loci (Lundgren et al. 2004).

Additional CRISPR functions have been proposed relating to chromosomal rearrangements for strains of the hyperthermophilic bacteria *Thermotoga* (Deboy et al. 2006), the actinobacterium *Streptomyces kanamiceticus* (Yanai et al. 2006) and *E. coli* (Rhiele et al. 2001). However, the observed effects involved the physical linkage of CRISPR loci with inverted or translocated chromosomal regions and, probably, the CRISPR loci provided sites for homologous recombination via common leaders and/or repeats, rather than the CRISPR-Cas systems actively facilitating these events.

1.3.2 The Link to Invading Genetic Elements

Two reports appeared almost simultaneously in 2005 proposing the origin of CRISPR spacers. One study was based on analyses of isolates of *Streptococcus pyogenes* and *Yersina pestis* (Pourcel et al. 2005) and the other included strains of these two bacteria, as well as over 30 additional bacterial species and 16 archaea

(Mojica et al. 2005). BLAST searches for sequence matches revealed that 2 % of analysed spacers showed close similarities to sequences located in non-CRISPR loci, mostly within viral DNAs (Mojica et al. 2005; Pourcel et al. 2005) but also within plasmids and in chromosomal sequences apparently unrelated to transmissible genetic elements (Mojica et al. 2005). Moreover, the best sequence matches were generally located in genetic elements that could, potentially, invade hosts carrying CRISPR arrays with the matching spacer, or other closely related strains. This suggested that spacers derived from fragments of the invading genetic elements which later were called protospacers (Deveau et al. 2008).

In addition, a literature survey revealed that viruses and plasmids harbouring the protospacers did not infect hosts carrying matching CRISPR spacers (Mojica et al. 2005). Moreover, it was shown that when such viruses were integrated as proviruses in genomes carrying matching spacers, the corresponding protospacer was either altered in sequence or absent. All these lines of evidence provided the basis for the hypothesis that CRISPR arrays are constituents of an adaptable defence system that confers specific immunity against invading DNA elements by a novel defence mechanism (Mojica et al. 2005; Pourcel et al. 2005). It was proposed that targeting of the genetic element is achieved via the base pairing potential of spacer sequences within CRISPR transcripts, similarly to eukaryotic interference RNA systems (Mojica et al. 2005).

These were genuinely surprising and very exciting findings. However, as with many groundbreaking ideas, gaining acceptance through publication proved to be a long and arduous process. Both papers were first submitted at about the same time in 2003 and resubmitted to three and four different journals, respectively, before they were finally accepted for publication. In general, referees' reports revealed a high degree of scepticism with comments such as "I cannot believe it" (an acquired defence mechanism) "but I cannot see the flaw" (Pourcel and Vergnaud, personal communication). In a variation of this, another referee (coauthor RAG) while positive about the idea, felt that initially there were too few short matching sequences to convince the broader scientific community of such a novel and radically different immune system. Eventually, the revised papers were published, with crucial support from the Journal Editors, within a few weeks of each other in 2005.

Following these reports, Bolotin et al. (2005) confirmed the main results (i.e., that 2 % of CRISPR spacers from 14 bacterial species were similar to other genomic sequences mostly located within phage genes) and included a wider analysis of *Streptococcus* spp. spacers. They proposed further that the CRISPR-based system provides immunity against foreign DNA expression via anti-sense crRNAs. In support of the defence hypothesis, the authors documented a negative correlation between the size of CRISPR loci in *S. thermophilus* strains and the number of phages that could infect them. This inverse relationship may have reflected that spacers of the larger CRISPR arrays matched to multiple phages.

Bolotin et al. (2005) also identified a short conserved sequence adjacent to the protospacer end that becomes leader distal in a type II CRISPR array of

S. thermophilus. This polarity was reinforced for additional type II CRISPR-Cas systems (Deveau et al. 2008; Horvath et al. 2008). Later it was demonstrated that a similar short motif occurs at the opposite end of the protospacer of type I systems that becomes leader proximal in CRISPR arrays (Lillestøl et al. 2009; Mojica et al. 2009; Semenova et al. 2009). These diverse motifs were collectively named PAM for Protospacer Adjacent Motif after related signature sequences had been identified for a variety of different organisms and disparate CRISPR systems (Mojica et al. 2009). Since then PAMs have been shown to be essential for CRISPR-Cas activity, at least for type I and II systems (Barrangou et al. 2007; Deveau et al. 2008; Garneau et al. 2010; Gudbergsdottir et al. 2011; Semenova et al. 2011, see Sect. 1.4.7)

The involvement of a CRISPR-Cas system in immunity was demonstrated experimentally for *S. thermophilus* in 2007 when cells carrying a type II CRISPR system became resistant to phages after acquiring new spacers with sequences identical to protospacers on the viral DNA (Barrangou et al. 2007). The new spacers were preferentially inserted at the leader end of the repeat arrays, but occasional internal additions coincident with spacer deletions were also reported (Barrangou et al. 2007; Deveau et al. 2008). Remarkably, phage-insensitive mutants could be selected during these experiments, implying that a reciprocal adaptation between CRISPR and targeted virus populations may occur.

In this context, the first comprehensive environmental study of CRISPRs, in acidophilic biofilms carrying bacteria and archaea, provided strong evidence for a dynamic interplay between viruses and CRISPR arrays, with older spacers being lost, and newer ones acquired, thereby enabling hosts to adapt rapidly to infection by earlier infecting viruses with mutated viral genome sequences as well as to previously unencountered viruses (Andersson and Banfield 2008; Tyson and Banfield 2008, see Chap. 10).

1.3.3 Functional Diversity of CRISPR Systems

Until 2005 progress in the CRISPR field was mainly descriptive as indicated in Table 1.1. However, the discovery of the origin of spacers and their proposed involvement in an immune-like system targeting invading genetic elements of both bacteria and archaea, was a major scientific breakthrough that has radically transformed microbiological research. It stimulated an enormous burst in research and publication activity (Fig. 1.3) much of it initially directed at characterising the molecular mechanisms of the CRISPR-based immune systems.

Initially, there was considerable uncertainty as to the targeting mechanisms of the CRISPR systems. As mentioned above, the small processed RNAs carrying mainly spacer sequences were the likely targeting agents. Early work on *Sulfolobus* CRISPR loci provided examples of many putatively significant spacer matches to viral and plasmid species that were located on either DNA strand of both genes and intergenic regions. This strongly suggested that dsDNA was at least one target (Lillestøl et al. 2006; Shah et al. 2009). Subsequently foreign DNA, as opposed to

Table 1.1 Chronology of seminal developments in the CRISPR field

Years	Contribution	Reference
1987	Discovery of the *iap*-associated repeats of *E. coli*	Ishino et al.
1991	Discovery of DRs in mycobacteria	Hermans et al.
1993	Discovery of TREPs in haloarchaea	Mojica et al.
1993	Evidence for TREPs transcription	Mojica et al.
1993	Development of the first typing method based on the repeats	Groenen et al.
1995	Evidence of TREPs activity	Mojica et al.
1998	Repeat array found in an archaeal conjugative plasmid	She et al.
2000	Recognition and description of the SRSR family of repeats	Mojica et al.
2002	Renaming of repeats as CRISPR	Jansen et al.
2002	Identification of core *cas* genes	Jansen et al.
2002	Characterization of CRISPR transcripts and of regular processing within repeats	Tang et al.
2002	Identification of RAMP proteins carrying RNA recognition motifs	Makarova et al.
2003	First experimental identification of a protein interacting with CRISPR DNA repeats	Peng et al.
2005	Unveiling the origin of spacers and a proposal of a universal defence function for CRISPR	Mojica et al. Pourcel et al.
2005	Identification of a conserved PAM motif associated with protospacers	Bolotin et al.
2005	Classification of 45 Cas protein families	Haft et al.
2006	Identification of small spacer-containing crRNAs	Lillestøl et al.
2006	Demonstration that putative protospacers can be located within genes, intergenically, and on either DNA strand	Lillestøl et al.
2006	Characterisation of antisense crRNAs	Lillestøl et al.
2006	Evidence for horizontal transfer of CRISPR systems	Godde et al.
2007	Repeats-based classification of CRISPR-Cas systems	Kunin et al.
2007	First experimental demonstration of CRISPR interference	Barrangou et al.
2007	First experimental demonstration of acquisition of new spacers, leading to CRISPR adaptation	Barrangou et al.
2008	Demonstration of the rapidly changing spacer contents of CRISPR arrays in environmental biofilms	Tyson et al. Andersson et al.
2008	Direct experimental evidence for DNA targeting by a type III-A system	Marraffini et al.
2008	Demonstration of CRISPR interference against plasmids	Marraffini et al.
2008	Identification of a ribonucleoprotein complex (Cascade) responsible for processing of pre-crRNA to crRNA	Brouns et al.
2008	Characterisation of the mature crRNAs of a type I system and proof of their role to guide Cascade to the target sequence, after which Cas3 is recruited to the trigger interference	Brouns et al.

(continued)

Table 1.1 (continued)

Years	Contribution	Reference
2008	Experimental evidence in support of DNA targeting by a type I system	Brouns et al.
2009	In vitro targeting of RNA by a type III-B system	Hale et al.
2010	In vivo DNA targeting demonstrated for a type II system	Garneau et al.
2010	Unveiling the mechanism of self versus non-self discrimination during interference by a type III-A system	Marraffini et al.
2010	Development of the auto-immune CRISPR concept	Stern et al.
2010	Crystal structure of a complex of Cas6 with a hairpin structured pre-crRNA	Haurwitz et al.
2011	Architecture of type I targeting complexes	Jore et al. Wiedenheft et al.
2011	CryoEM structure determination of type I targeting complex (Cascade)	Jore et al. Wiedenheft et al.
2011	Characterisation of the type II tracrRNA-based processing mechanism	Delcheva et al.
2011	Defining limits of type I DNA targeting specificity	Gudbergsdottir et al. Semenova et al.
2011	Crystal structure of a complex of Cas6 and single stranded pre-crRNA	Wang et al.
2011	Reclassification of Cas proteins and CRISPR systems	Makarova et al.
2012	In vivo cleavage of antisense crRNAs by a type III-B system	Hale et al.
2012	Demonstration of an alternative type III-B interference mechanism	Zhang et al.
2012	EM structural determination of a type III-B interference complex	Zhang et al.
2012	CasA (Cse1) of type I-E system interacts with PAM and is required for target recognition and binding	Sashital et al.
2012	Concerted action of a type I-E interference complex and Cas3	Westra et al.
2012	Induced acquisition in type I-E system by overexpressing Cas1 and Cas2	Yosef et al.
2012	Evidence for a positive feedback between active CRISPR spacers and new spacer uptake in a type I-E system	Swarts et al.
2012	Evidence that prior recognition of protospacers by specific crRNAs stimulates acquisition in a type I-E system	Datsenko et al.
2012	Evidence for a ruler mechanism operating during protospacer excision in a type I-A system	Erdmann and Garrett
2012	Evidence for spacer acquisition throughout a CRISPR array by an alternative mechanism in type I-A system	Erdmann and Garrett

Listed are some of the seminal developments in the CRISPR field. It is not complete and many developments, including, for example, the characterisation of individual Cas proteins are not included for space reasons.

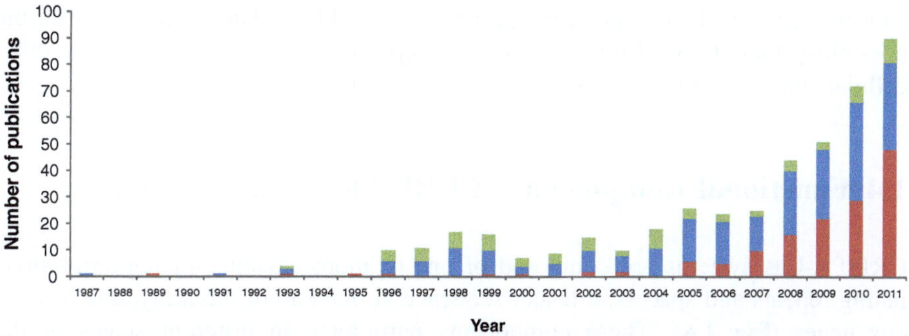

Fig. 1.3 Numbers of papers published between 1987 and 2011 relating to CRISPR systems. To provide a more detailed view of the development of CRISPR publications, reports where CRISPR arrays or associated genes are the main focus (*red*) are distinguished from those where they are of secondary interest (*blue*). Reports addressing the use of DRs for typing of mycobacteria developed independently of CRISPR-based functional studies and are shown separately (*green*). Two marked increases in the number of publications are visible. After 1995, following reports of CRISPR in completed genomes, and after 2005 following the discovery of the origin of CRISPR spacers

mRNA, targeting was demonstrated unambiguously for the type III-A CRISPR-Csm system of *S. epidermidis* (Marraffini and Sontheimer 2008) and indirect evidence for DNA targeting was also provided for the type I CRISPR-Cas system of *E. coli* (Brouns et al. 2008).

In contrast, it was shown by in vitro experiments that the type III-B CRISPR-Cmr system of *P. furiosus* could further process crRNA from the 3′-end by a ruler mechanism yielding two discrete products that could facilitate targeting and cleavage of RNAs carrying the corresponding protospacer sequence (Hale et al. 2008, 2009; Carte et al. 2010). More recently an antisense CRISPR RNA of *P. furiosus* was shown to be targeted and cleaved in the same way in vivo (Hale et al. 2012). Moreover, a different sequence-specific cleavage of targeted RNAs was demonstrated for one of two type III-B systems of *S. solfataricus*, and the structural form of the three dimensional interference complex was also determined (Zhang et al. 2012). This experimental evidence underlines the diversity of the interference modules and the basis for the existence of disparate CRISPR systems. It also provides a rationale for the presence of different systems in one organism, particularly amongst thermophilic bacteria and archaea which can carry combinations of type I, III-A or type III-B CRISPR systems (Agari et al. 2009; Garrett et al. 2011). Moreover, some organisms, for example some strains of *S. thermophilus and S. islandicus*, simultaneously carry all three major types (Horvath and Barrangou 2010; Guo et al. 2011).

Apart from immunity, recent reports appear to implicate CRISPR-Cas systems in other cellular functions including DNA-repair in *E. coli* (Babu et al. 2011), fruiting body development in myxobacteria (Viswanathan et al. 2007), and biofilm formation in *Pseudomonas aeruginosa* (Cady and O'Toole 2011; Zegans et al. 2009). These may arise as indirect effects of CRISPR-based defence activity but since the Cas proteins associated with interference modules are quite heterogeneous and divergent in their sequences (Garrett et al. 2011; Makarova et al. 2011), the possibility that the CRISPR functional complexes could be exploited for other cellular functions remains open (see Chaps. 10 and 11).

1.4 Functional Components of CRISPR-Based Systems

CRISPR-Cas systems are composed of one or more cassettes of regularly alternating repeats and spacers, a leader sequence at one end of the array, and a set of *cas* genes (Fig. 1.4). These components participate in different stages of the CRISPR-Cas pathway, generating CRISPR-RNAs (crRNAs) and Cas proteins. Cas proteins are involved in the uptake of new CRISPR spacers (termed acquisition but also widely referred to as adaptation), the generation of small crRNAs from CRISPR transcripts (crRNA expression and processing) and targeting and cleavage of invading nucleic acids (interference) by protein-crRNA complexes which base pair with complementary protospacer sequences (see Fig. 1.4).

1.4.1 CRISPR-associated (Cas) Proteins

The identification of many different Cas proteins and their recent reclassification was addressed in Sect. 1.2.7 and the main core proteins involved in acquisition, processing and interference are listed in Table 1.2. The most conserved proteins Cas1 and Cas2 have been implicated in the acquisition step, probably together with Cas4 which is often co-transcribed with Cas1 (or its gene is fused with that of Cas1). This high level of protein conservation suggests that this process is universal for all CRISPR-Cas systems (Garrett et al. 2011; Makarova et al. 2011). Although this inference receives some experimental support from the recent flurry of articles on spacer uptake in the type I-E system of *E. coli* under genetically manipulated conditions (Datsenko et al. 2012; Swarts et al. 2012; Yosef et al. 2012) and in the type I-A (and type III-B) systems of *S. solfataricus* using an environmental viral sample (Erdmann and Garrett 2012), there is also evidence of distinct mechanistic differences between these systems regarding the regulation of spacer uptake, the role of the PAM motif and the location of the CRISPR repeat at which insertion occurs.

Cas6 is the primary RNA processing enzyme for all type I and III systems cutting within repeats (see Sect. 1.4.7), and this function is performed by a tracrRNA and the host encoded bacterial-specific RNAse III in type II systems (Deltcheva et al. 2011; see Chap. 5). Further maturation of the crRNA occurs for type III-A and III-B systems that is likely to involve Csm2, Csm3 and/or Csm5 for the former (Hatoum-Aslan et al. 2011).

The interference protein-crRNA modules are the most disparate, where different core Cas proteins are required for each of the type I, II and III-A and type

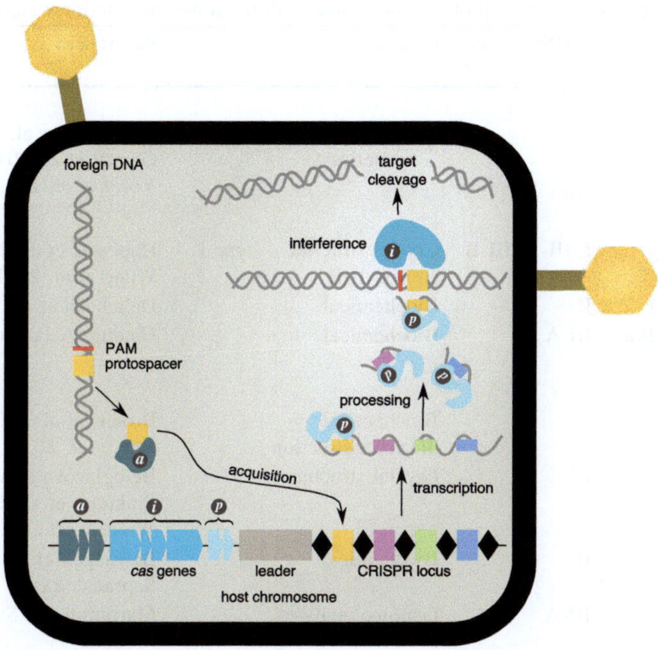

Fig. 1.4 Schematic representation of CRISPR-Cas systems and the main steps of CRISPR-mediated interference. Whereas the acquisition step appears to be universal, differences occur in the processing mechanisms, and particularly in the interference mechanisms. The scheme depicts targeting of double stranded DNA that appears to be a major function of type I, type II and type III-A CRISPR systems. CRISPR-Cas systems are composed of a set of *cas* genes (*blue shaded arrows*), where genes involved in acquisition (*a*), expression (*p*) and interference (*i*) are generally clustered separately. There is invariably at least one array of alternating repeats (*diamonds*) and spacer sequences (*squares*) that is flanked by a leader (*cinnamon bar*) carrying the main promoter of the CRISPR array, located close to the first repeat. In contrast to the high sequence conservation of most repeats in the array, the leader distal repeat often differs from the consensus. New spacers derived from sequences (protospacers) of invading genetic elements, that generally lie adjacent to a short PAM signature sequence (*red rectangle*), are preferentially inserted in a repeat at or near the leader proximal end of the CRISPR cassette. This insertion is accompanied by duplication of the repeat. The mechanism of spacer acquisition remains to be determined, although the universal core Cas proteins Cas1 and Cas2 have been implicated in the process. Spacers yield guide RNAs for the interference stage. Firstly, CRISPR-spacer arrays are transcribed into long pre-crRNAs that are subsequently cleaved by single cut at each repeat with further trimmering occurring at the 5′-end (type II systems) or the 3′-end (type III systems), giving rise to mature mono-spacer crRNAs flanked by partial repeat sequences, at least on one side. Maturation of the crRNA is performed by Cas6 homologs excepting for type II systems where the bacterial host endonuclease RNase III is employed together with a complementary trans-encoded tracrRNA. In the final interference step, disparate and complex crRNA-Cas protein assemblies base-pair to the complementary protospacer sequences of the genetic element leading to cleavage of the target DNA, or RNA in type III-B systems, by an endonuclease which has been identified as Cas3 for type I systems. The presence of a PAM motif adjoining the target sequence is likely to enhance interference efficiency at least for type I and II systems

Table 1.2 Summary of core Cas proteins associated with the different functional steps

Protein	CRISPR-Cas type	Evidence	Reference
Acquisition			
Cas1	All	Genome analyses and genetics	Makarova et al. 2011
Cas2	All		Garrett et al. 2011
Cas4	All		Yosef et al. 2012
Expression			
Cas6	I, III-A, III-B	Crystal structures—type I and IIIB	Haurwitz et al. 2010 Wang et al. 2011
RNAse III (host)	II	Biochemical	Deltcheva et al. 2011
Csm2, Csm3 and/or Csm5	III-A	Biochemical	Hatoum-Aslan et al. 2011
Interference			
Cas5, Cas6, Cas7, Cas8	I	Biochemical 3D reconstruction	Brouns et al. 2008 Jore et al. 2011
Cas3	I	Crystal structure	Beloglazova et al. 2011 Sinkunas et al. 2011 Wiedenheft et al. 2011
Cas9	II	Biochemical	Barrangou et al. 2007 Sapranauskas et al. 2011
Csm1–5	III-A	Genome analyses	Marraffini and Sontheimer 2008
Cmr1–6 or Cmr1–7	III-B	Biochemical Crystal structure	Hale et al. 2009 Zhang et al. 2012

III-B systems (Table 1.2). The most common type I complexes carry Cas5–7 and the Cas3 protein has been demonstrated to be involved in target cleavage (Beloglazova et al. 2011; Sinkunas et al. 2011; Westra et al. 2012). Type II systems use the large Cas9 protein, while type III-A and III-B modules carry Csm1–5 and Cmr1–6 (or 7), respectively (see Table 1.2). Moreover, many non-core Cas proteins are associated with the interference modules which presumably enhance their functional diversity and/or versatility (Garrett et al. 2011; Makarova et al. 2011). In addition a few proteins, including Csa3 and Csm6 have been shown to carry potential transcriptional regulator domains but it remains unclear whether they also perform other functions (Lintner et al. 2011; Makarova et al. 2011).

1.4.2 CRISPR Arrays

Most CRISPR arrays carry from 2 to about 100 repeats, although occasionally larger arrays are found. The largest single array identified to date contains 588 repeats and occurs in the halophilic bacterium *Haliangium ochraceum* DSM 14365 (CRISPRdb database). Moreover, it is separated from two additional arrays with the same repeat sequence by a set of three non-*cas* genes related to integration or transposition. Together, the three CRISPR arrays carry 815 repeats and extend over about 66 kbp.

Some organisms, and especially extremophilic archaea, carry multiple CRISPR loci. An extreme example is *Methanocaldococcus* sp. FS406-22 with 23 genomic arrays (CRISPI and CRISPRdb databases). Although most CRISPRs are located in chromosomal regions, sometimes within proviral-like elements (Sebaihia et al. 2006), repeat arrays are also found on plasmids, an important vehicle of horizontal gene transfer, at least for some bacteria (Godde and Bickerton 2006; Chakraborty et al. 2010). CRISPR arrays have also been located in free viral genomes (Minot et al. 2011; García-Heredia et al. 2012) and in small *Sulfolobus* conjugative plasmids pNOB8 and pKEF9 (She et al. 1998; Greve et al. 2004). For the latter, CRISPR transcripts were shown to be processed and they carry spacers matching to *Sulfolobus* rudiviruses (Lillestøl et al. 2009). Collectively, these results suggest that, by utilising the host Cas proteins, there is a basis for a competitive interplay between viruses and plasmids within a cell. CRISPR locus occurrence and distribution is further discussed in Chaps. 2 and 6.

1.4.3 Leaders

Many CRISPR arrays carry leader sequences. These range from about 100–500 bp. Leaders are oriented specifically with respect to the repeat sequence of the adjoining CRISPR locus and when a degenerate terminal repeat is present it invariably occurs at the leader distal end of the locus. Leaders are non-protein coding and often carry some low complexity sequence regions. Their degree of overall sequence conservation is relatively low and their actual sizes can be quite difficult to estimate. Some organisms, for example *M. jannaschii* (Bult et al. 1996), carry several CRISPR loci with near identical leaders that can be readily identified by sequence comparison (Jansen et al. 2002a; Lillestøl et al. 2009; Mojica et al. 2009). Moreover, leaders of closely related CRISPR-Cas systems often carry partially conserved sequence motifs which may be important functionally (Shah et al. 2009; Lillestøl et al. 2009). Although the degree of conservation of the leader sequence generally follows that of the repeat of the associated CRISPR array, clear deviations from this tendency have been observed suggesting that exchange of leaders between CRISPR arrays carrying different repeats can occur (Mojica et al. 2009; Shah et al. 2009).

Leaders appear to play at least two key functional roles, one demonstrated experimentally and the other proposed. First, the leader carries the main promoter site for transcription of the adjoining CRISPR locus (Lillestøl et al. 2006, 2009; Brouns et al. 2008; Pougach et al. 2010; Pul et al. 2010; Westra et al. 2010; Wurtzel et al. 2010). Moreover, if the leader is absent, at most only low levels of transcription are observed from promoter motifs randomly taken up in spacers or, occasionally, from within a repeat (Lillestøl et al. 2009; Wurtzel et al. 2010; Deng et al. 2012). Second, it adjoins the site where new spacers are primarily incorporated at

(Barrangou et al. 2007) or near (Lillestøl et al. 2006, 2009; Held et al. 2010) the first repeat. Moreover, phylogenetic studies on crenarchaeal CRISPR systems provided evidence for coevolution of leaders, repeats and Cas1 sequences, suggesting that leader plays an important functional role in the acquisition step, possibly facilitating insertion of the new repeat-spacer units by contributing to an assembly site for Cas proteins (Marraffini and Sontheimer 2008; Lillestøl et al. 2009; Shah et al. 2009). Consistent with this idea, some leaderless CRISPR loci that are maintained in cells, appear to be inactive in the acquisition step but if transcribed from a promoter near the leaderless end, or from within the CRISPR array, they may still maintain interference activity (Lillestøl et al. 2006, 2009). The first experimental evidence that DNA elements in the leader are required for acquisition has recently been provided (Yosef et al. 2012).

1.4.4 Repeats

CRISPR repeats in sequenced genomes range from 23 to 55 bp peaking at 24–25, 29–30 and 36–37 bp, where each peak is separated by about half a helical turn of dsDNA (Fig. 1.5). Very few CRISPR repeats are larger than 38 bp and the 55 bp repeats of *Desulfobacca acetoxidans* DSM 11109 are the only ones known to exceed 50 bp (CRISPI and CRISPRdb databases). Although four copies of these repeats are clustered and regularly spaced, highly similar sequences (16 matches with e-values below 10^{-7}) occur throughout the genome that are not linked to *cas* genes which suggests that they may not be cofunctional with CRISPR-based systems. The lower size limit of 23 bp is represented by the repeats of the archaeon

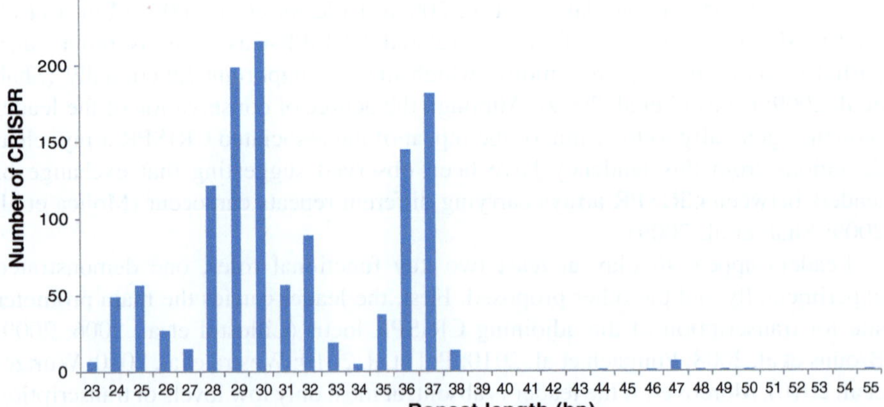

Fig. 1.5 Size distributions are given for consensus CRISPR repeat sequences documented in CRISPRdb (Grissa et al. 2007)

Ferroglobus placidus DSM 10642 (6 arrays with 10–27 units associated with *csa* and *cmr* genes) and other smaller arrays found in distantly related isolates. While most CRISPR arrays carry up to about 100 repeats, they can vary in size from 2 to 588 repeats in *H. ochraceum* DSM 14365, with the latter organism carrying the upper record of a total of 815 repeats (see Sect. 1.4.2).

CRISPR repeats were classified into 12 major and 21 minor groups on the basis of sequence similarity (Kunin et al. 2007). Repeat sequences are generally highly conserved within a given array, although local repeat deviations sometimes occur and the repeat at the distal end from the leader is sometimes degenerate. Repeats present in CRISPR loci of a given organism can be highly divergent and, conversely, similar repeat types are found in distantly related organisms. These observations underpinned the hypothesis that CRISPR-based systems undergo occasional horizontal gene transfer (Godde and Bickerton 2006; Portillo and González 2009; Chakraborty et al. 2010). This view is reinforced by a recent study of 12 *S. islandicus* genomes where strong evidence was provided for exchange of type I CRISPR-Cas systems as well as the type III CRISPR-Csm and CRISPR-Cmr systems (Guo et al. 2011; Shah and Garrett 2011).

A feature of the repeats that was considered to be general when they were originally described is their palindromic character (Mojica et al. 2000). Most bacterial repeat sequences exhibit a partial dyad symmetry, with internal and/or terminal inverted repeats that could potentially produce stable stem-loop structures in RNA transcripts (Kunin et al. 2007). Moreover, recent studies suggest that this secondary structure determines the precise site for processing pre-crRNAs, at least for some organisms (Brouns et al. 2008; Carte et al. 2008; Haurwitz et al. 2010; Hatoum-Aslam et al. 2011). However, in contrast to many bacteria, dyad symmetry is not a dominant feature of many archaeal repeats (Lillestøl et al. 2006) and, moreover, in the crystal structure of the *P. furiosus* Cas6-crRNA complex the RNA is single stranded (Wang et al. 2011—see also Sect 1.4.7).

1.4.5 Spacers

Spacers located within a repeat array are similar in size, varying within a few nucleotides, and carry unique sequences. They can range from 21 to 72 bp but most are either 32 bp or they fall in the range 35–37 bp (Fig. 1.6). A minority of spacers are larger than 51 bp but it remains unclear whether they participate in CRISPR defence. For example, *Synechococcus* sp. PCC 7002, *Chloroflexus aurantiacus* J-10 and the euryarchaeon *Picrophilus torridus* DSM 9790 carry arrays of unevenly spaced repeats; most spacers are about 40 bp long, but a small proportion fall in the range 48–72 bp. Other repeat arrays that carry large spacers may be degenerate or, possibly, have other functions. For example, *Clostridium botulinum* B1 str. Okra contains an array with 2 spacers of 64 bp while other repeat arrays with large spacers show exceptional variations in their repeat sequences. Repeat arrays with spacers below the minimal consensus size of 26 bp

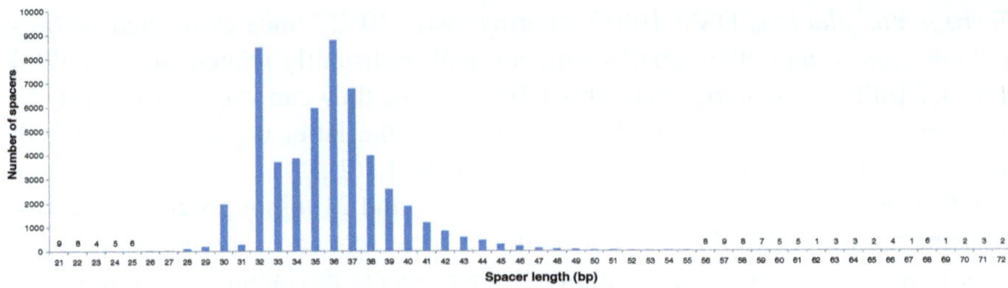

Fig. 1.6 Size distribution of spacers within CRISPR arrays recorded in CRISPRdb (Grissa et al. 2007). Spacer numbers are given when there are less than ten entries

are also anomalous. Thus, present data support the CRISPR spacer size range as lying between 26 and 51 bp for functional arrays.

The spacer sequence content of CRISPR arrays is highly variable and only different strains of a given species tend to carry identical spacers within their arrays. Moreover, these invariant spacers tend to be concentrated towards the trailer end of the array, maintaining their relative order. Closely related strains have often accrued new unique spacers near the leader and this accumulation of new spacers yields a potential chronological record of invading genetic elements. This property also provided the basis for developing new strain typing strategies (see Sect. 1.2.2 and Chap. 2).

Some CRISPR loci constitute dynamic structures. Deletions occur, and evidence has been found for duplication of repeat-spacer units and of recombination events occurring between CRISPR loci carrying similar repeats (Bolotin et al. 2005; Lillestøl et al. 2006; Díez-Villaseñor et al. 2010; Touchon and Rocha 2010). Moreover, challenging CRISPR systems with vector borne-protospacers carrying PAM motifs and maintained under selection, produces deletions in CRISPR loci which include the matching spacer and sometimes whole CRISPR loci (Gudbergsdottir et al. 2011).

Studies of environmentally stable biofilms suggest that there is a dynamic interplay between viruses and the spacer content of CRISPR arrays indicative of CRISPR spacer uptake being activated by new viral infections or by mutations occurring in viral protospacers or their PAM motifs (Andersson and Banfield 2008; Tyson and Banfield 2008). The existence of such mechanisms for the periodic removal of repeat-spacer units, to compensate for the addition of newly added spacers, would seem to be a prerequisite for viability of both CRISPR systems and the host cells (see Chap. 10).

Occasionally, spacers might be added that can target the host genome with possible deleterious consequences. Stern et al. (2010) examined the frequency of occurrence of spacers that exhibited perfect sequence matches within the same genome, for all available genome sequences, as well as their locations, and they concluded that an autoimmune response probably operates whereby the chromosome-matching spacer is deleted or that it undergoes mutation of, for example, its flanking repeat sequences such that the corresponding crRNA is inactivated.

1.4.6 Protospacer Adjacent Motifs

An important sequence element of CRISPR-Cas systems lies adjacent to the protospacers. It constitutes a short signature sequence of 2–5 nt located immediately adjacent or up to two positions from the protospacer and it can exhibit a range of base sequences depending on the CRISPR type and the organism (Bolotin et al. 2005; Deveau et al. 2008; Horvath et al. 2008; Lillestøl et al. 2009; Mojica et al. 2009; Semenova et al. 2009; Gudbergdottir et al. 2011). The sequence element is called the protospacer adjacent motif (PAM) and is located at one end of the protospacer (Mojica et al. 2009). To date, PAMs have only been defined for those systems where the number of identified protospacers on genetic elements is sufficiently high to yield obvious patterns. Moreover, PAMEs (spacer end corresponding to the PAM-adjacent protospacer edge) of most (if not all) CRISPR arrays belonging to type I systems are oriented towards the leader (Lillestøl et al. 2009; Mojica et al. 2009; Semenova et al. 2009), whereas those associated with the bacteria-specific type II CRISPR-Cas systems have the opposite orientation (Bolotin et al. 2005; Barrangou et al. 2007; Deveau et al. 2008; Horvath et al. 2008; Deltcheva et al. 2011). Additionally, it has been shown that the presence of the PAM is essential for interference in both type I CRISPR-Cas systems (Gudbergsdottir et al. 2011; Semenova et al. 2011) and type II systems (Barrangou et al. 2007; Deveau et al. 2008; Horvath et al. 2008; Garneau et al. 2010). More specifically PAM has been implicated in target recognition in the type I-E system of *E. coli* (Sashital et al. 2012; Westra et al. 2012) although, at least for the latter system, protospacer adjacent sequences that differ from the predicted PAM can also produce interference (Brouns et al. 2008; Datsenko et al. 2012; Swarts et al. 2012; Westra et al. 2012). In contrast, experimental evidence suggests that PAM motifs are not essential for interference by type III systems (Hale et al. 2009, 2012; Marraffini and Sontheimer 2008, 2010; Hatoum-Aslan et al. 2011; Manica et al. 2011; Zhang et al. 2012).

1.4.7 CRISPR RNAs

CRISPR loci are generally considered to be transcribed from within the leader to yield long transcripts (pre-crRNAs) probably terminating downstream of the CRISPR locus, although potential transcriptional signals taken up in spacers are likely to generate alternative start-stop sites of varying degrees of effectivity (Lillestøl et al. 2009; Deng et al. 2012; Hale et al. 2012).

The pre-crRNAs are initially processed (primary processing) within the repeats to generate intermediate crRNAs. Experimentally localised primary processing sites lie at the base of putative hairpin-loops that form in some type I (Brouns et al. 2008; Hatoum-Aslam et al. 2011), or at the helix-loop junction in some type III CRISPR systems (Carte et al. 2008; Hatoum-Aslam et al. 2011). Moreover, a crystal structure of a type I complex of a Cas6 homolog (Cas6f, Csy4) and crRNA of *P. aeruginosa* revealed that the hairpin structure was maintained in the complex (Haurwitz et al. 2010). In addition, in the type III-A system of *S. epidermidis,* the capacity to form a hairpin-loop was shown to enhance the yields of mature crRNAs produced (Hatoum-Aslam et al. 2011).

These results suggested that the pre-crRNA secondary or tertiary structure plays a role in determining processing sites although, whatever the detailed cleavage mechanism, repeat sequences of 8 nt (named the 5′-handle) are produced upstream of the spacer in crRNAs of type I and III systems. However, some pre-crRNAs appear to lack the ability to generate stable secondary structures suggesting that processing mechanisms differ. A "deep" RNA (cDNA) sequencing study of *S. solfataricus* P2 showed that all intermediate processed RNAs, from those CRISPR transcripts which lacked significant dyad symmetry in their repeats, carried all or most of the terminal 8 nt of the repeat sequence at their 5′-ends (Wurtzel et al. 2010; Deng et al. 2012). Furthermore, crystallographic studies of a pre-crRNA-Cas6 complex from *P. furiosus* indicated that the RNA was in a single-stranded state prior to cleavage (Wang et al. 2011).

Although no maturation of intermediate crRNAs appears to occur in type I CRISPR systems (Brouns et al. 2008; Gesner et al. 2011; Jore et al. 2011; Sashital et al. 2011; Wiedenheft et al. 2011), for the type III-A and III-B, CRISPR-Csm and CRISPR-Cmr systems, respectively, a further maturation step occurs at the 3′-end of the intermediate crRNA to yield shorter mature crRNAs. For the type III-A system, the RAMP proteins Csm2, Csm3 and/or Csm5 have been shown to be directly or indirectly involved in the final maturation step that employs a sequence-independent ruler mechanism (Hatoum-Aslan et al. 2011). The 3′-ends of two mature crRNA products, lie within the ranges 37–39 and 43–45 nt for type III-A and III-B systems of *S. epidermidis* and *P. furiosus*, respectively (Hale et al. 2009; Hatoum-Aslan et al. 2011), although less discretely sized products were observed for the type III-B system of *S. solfataricus* (Zhang et al. 2012).

The radically different CRISPR-RNA processing of the bacteria-specific type II CRISPR system involves tracrRNA, encoded adjacent to the CRISPR locus, that anneals to the repeat in pre-crRNA molecules and the double helix is cleaved by RNase III (Deltcheva et al. 2011). Trimming from the 5′-end of intermediate crRNAs yields interference-competent RNAs of 39–42 nt which carry a partial spacer with a repeat tag at the 3′-end. The subtype specific Cas protein Csn1 may execute both processing events (see Chap. 5 for details).

In summary, although mature crRNAs can differ significantly in size, they all contain complete or partial spacer sequences and portions of flanking repeats. During interference, the spacer sequence pairs with the complementary protospacer sequence while the residual repeat sequence(s) probably provide attachment sites for assembly of Cas proteins of the disparate interference complexes. Moreover, whereas the 5′-repeat tags of the type I crRNAs overlap with the PAM motif of the protospacer on targeting, the 3′-residual repeat sequence of type II mature crRNAs also overlaps with the PAM motif located at the opposite end of the protospacer. This provides at least a potential molecular basis for both types of PAM motif to influence interference.

1.5 Reflections

After two decades of research, many of the secrets and mysteries of the CRISPR defence systems have now been revealed. Three primary mechanistic steps, acquisition, expression and processing of crRNAs and interference have been defined, although they remain to be elucidated in detail in the various systems. CRISPR arrays have been shown to constitute the functional cores of the disparate CRISPR immune systems. Each step, acquisition, pre-crRNA processing, and interference is associated with a group of defined core Cas proteins.

The most conserved stage is acquisition, consisting of the uptake of foreign DNA as new spacers at or near the leader-end of the CRISPR array and it seems to involve the ubiquitous Cas1 and Cas2 and frequently Cas4. The recent developments on spacer uptake in type I systems of *E. coli* and *S. solfataricus* provide novel insights into their molecular mechanisms which appear to be more diverse that was expected. Expression of the CRISPR RNA from within the leader, and its subsequent cleavage by Cas6 within the repeat sequence, appears to be a universal property of the type I and III systems while the bacterial-specific type II system employs a tracrRNA and the host enzyme RNAse III to cleave within the repeats. The interference step is clearly the most diverse mechanistically and differs radically for the three CRISPR types. Moreover, the associated protein modules are quite disparate and their genes are often linked with those of non-core *cas* genes suggesting that the different modules are functionally versatile.

We still have limited insight into the basis of the functional diversity. Type I, II and III-A systems have been implicated in DNA targeting in vivo while the type III-B system targets RNA in vitro and in vivo but further details of the specificities remain to be determined. Many archaea and bacterial extremophiles carry multiple interference modules and sometimes different modules are coupled together in gene cassettes. Moreover, for archaeal type III-A and III-B systems their gene cassettes are often located independently in genomes, detached from CRISPR loci and other *cas* genes. Interference gene cassettes have also been shown to exchange between different type I CRISPR systems. This reinforces the view that they provide functional diversity but it remains unclear what the range of targeting and cleavage specificity is.

Future work will determine whether defence targets extend beyond dsDNA genetic elements to ssDNA elements, or to ssRNA and dsRNA viruses, or to gene transcripts and non-coding RNAs, and will also provide insights into the extent to which these processes, and the protein components involved, are limited to defence-related functions (see Chap. 10). There are also a plethora of potential applications of particular interest to molecular biologists, the food industry, and medicine and these are considered further in Chap. 11.

Acknowledgments Shiraz A. Shah is thanked for help with Fig. 1.4 and for constructive discussions. F. J. M. M. is supported by the Spanish Ministerio de Ciencia e Innovación (BIO2011-24417). R. A. G. is supported by the Danish Natural Science Research Council.

References

Agari Y, Sakamoto K, Tamakoshi M, Oshima T, Kuramitsu S, Shinkai A (2009) Transcription profile of *Thermus thermophilus* CRISPR systems after phage infection. J Mol Biol 395:270–281

Aklujkar M, Lovley DR (2010) Interference with histidyl-tRNA synthetase by a CRISPR spacer sequence as a factor in the evolution of *Pelobacter carbinolicus*. BMC Evol Biol 10:230

Andersson AF, Banfield JF (2008) Virus population dynamics and acquired virus resistance in natural microbial communities. Science 320:1047–1050

Aranaz A, Liebana E, Mateos A, Dominguez L, Vidal D, Domingo M et al (1996) Spacer oligonucleotide typing of *Mycobacterium bovis* strains from cattle and other animals: a tool for studying epidemiology of tuberculosis. J Clin Microbiol 34:2734–2740

Babu M, Beloglazova N, Flick R, Graham C, Skarina T, Nocek B et al (2011) A dual function of the CRISPR-Cas system in bacterial antivirus immunity and DNA repair. Mol Microbiol 79:484–502

Barrangou R, Fremaux C, Deveau H, Richards M, Boyaval P, Moineau S et al (2007) CRISPR provides acquired resistance against viruses in prokaryotes. Science 315:1709–1712

Beloglazova N, Petit P, Flick R, Brown G, Savchenko A, Yakunin AF (2011) Structure and activity of the Cas3 HD nuclease MJ0384 an effector enzyme of the CRISPR interference. EMBO J 30:4616–4627

Bolotin A, Quinquis B, Sorokin A, Ehrlich SD (2005) Clustered regularly interspaced short palindrome repeats (CRISPRs) have spacers of extrachromosomal origin. Microbiol 151:2551–2561

Brouns SJ, Jore MM, Lundgren M, Westra ER, Slijkhuis RJ, Snijders AP et al (2008) Small CRISPR RNAs guide antiviral defense in prokaryotes. Science 321:960–964

Bult CJ, White O, Olsen GJ, Zhou L, Fleischmann RD, Sutton GG et al (1996) Complete genome sequence of the methanogenic archaeon *Methanococcus jannaschii*. Science 273:1058–1073

Cady KC, O'Toole GA (2011) Non-identity-mediated CRISPR-bacteriophage interaction mediated via the Csy and Cas3 proteins. J Bacteriol 193:3433–3445

Carte J, Wang R, Li H, Terns RM, Terns MP (2008) Cas6 is an endoribonuclease that generates guide RNAs for invader defense in prokaryotes. Genes Dev 22:3489–3496

Carte J, Pfister NT, Compton MM, Terns RM, Terns MP (2010) Binding and cleavage of CRISPR RNA by Cas6. RNA 16:2181–2188

Chakraborty S, Snijders AP, Chakravorty R, Ahmed M, Tarek AM, Hossain MA (2010) Comparative network clustering of direct repeats (DRs) and cas genes confirms the possibility of the horizontal transfer of CRISPR locus among bacteria. Mol Phylogenet Evol 56:878–887

Datsenko KA, Pougach K, Tikhonov A, Wanner BL, Severinov K, Semenova E (2012) Molecular memory of prior infections activates the CRISPR/Cas adaptive bacterial immunity system. Nat Commun 3:945

Deboy RT, Mongodin EF, Emerson JB, Nelson KE (2006) Chromosome evolution in the thermotogales: large-scale inversions and strain diversification of CRISPR sequences. J Bacteriol 188:2364–2374

Deltcheva E, Chylinski K, Sharma CM, Gonzales K, Chao Y, Pirzada ZA et al (2011) CRISPR RNA maturation by trans-encoded small RNA and host factor RNase III. Nature 471:602–607

Deng L, Kenchappa CS, Peng X, She Q, Garrett RA (2012) Modulation of CRISPR locus transcription by the repeat-binding protein Cbp1 in *Sulfolobus*. Nucleic Acids Res 40:2470–2480

Deveau H, Barrangou R, Garneau JE, Labonte J, Fremaux C, Boyaval P et al (2008) Phage response to CRISPR-encoded resistance in *Streptococcus thermophilus*. J Bacteriol 190:1390–1400
Díez-Villaseñor C, Almendros C, García-Martínez J, Mojica FJM (2010) Diversity of CRISPR loci in *Escherichia coli*. Microbiol 153:1351–1361
Driscoll JR (2009) Spoligotyping for molecular epidemiology of the *Mycobacterium tuberculosis* complex. Methods Mol Biol 551:117–128
Erdmann S, Garrett RA (2012) Selective and hyperactive uptake of foreign DNA by adaptive immune systems of an archaeon via two distinct mechanisms. Mol Microbiol 85:1044–1056
Flamand MC, Goblet JP, Duc G, Briquet M, Boutry M (1992) Sequence and transcription analysis of mitochondrial plasmids isolated from cytoplasmic male-sterile lines of *Vicia faba*. Plant Mol Biol 19:913–923
García-Heredia I, Martín-Cuadrado A, Mojica FJM, Santos F, Mira-Obrador A, Antón J, Rodríguez-Valera F (2012) Reconstructing viral genomes from the environment using fosmid clones: the case of haloviruses. PLoS ONE 7:e33802
Garneau JE, Dupuis ME, Villion M, Romero DA, Barrangou R, Boyaval P et al (2010) The CRISPR/cas bacterial immune system cleaves bacteriophage and plasmid DNA. Nature 468:67–71
Garrett RA, Vestergaard G, Shah SA (2011) Archaeal CRISPR-based immune systems: exchangeable functional modules. Trends Microbiol 19:549–556
Gesner EM, Schellenberg MJ, Garside EL, George MM, MacMillan AM (2011) Recognition and maturation of effector RNAs in a CRISPR interference pathway. Nat Struct Mol Biol 18:688–692
Godde JS, Bickerton A (2006) The repetitive DNA elements called CRISPRs and their associated genes: evidence of horizontal transfer among prokaryotes. J Mol Evol 62:718–729
Greve B, Jensen S, Brügger K, Zillig W, Garrett RA (2004) Genomic comparison of archaeal conjugative plasmids from *Sulfolobus*. Archaea 1:231–239
Grissa I, Vergnaud G, Pourcel C (2007) The CRISPRdb database and tools to display CRISPRs and to generate dictionaries of spacers and repeats. BMC Bioinf 8:172
Groenen PM, Bunschoten AE, van Soolingen D, van Embden JD (1993) Nature of DNA polymorphism in the direct repeat cluster of *Mycobacterium tuberculosis*; application for strain differentiation by a novel typing method. Mol Microbiol 10:1057–1065
Gudbergsdottir S, Deng L, Chen Z, Jensen JVK, Jensen LR, She Q, Garrett RA (2011) Dynamic properties of the *Sulfolobus* CRISPR/Cas and CRISPR/Cmr systems when challenged with vector-borne viral and plasmid genes and protospacers. Mol Microbiol 79:35–49
Guo L, Brügger K, Liu C, Shah SA, Zheng H, Zhu Y et al (2011) Genome analyses of icelandic strains of *Sulfolobus islandicus* model organisms for genetic and virus-host interaction studies. J Bacteriol 193:1672–1680
Haft DH, Selengut J, Mongodin EF, Nelson KE (2005) A guild of 45 CRISPR-associated (Cas) protein families and multiple CRISPR/Cas subtypes exist in prokaryotic genomes. PLoS Comput Biol 1:474–483
Hale C, Kleppe K, Terns RM, Terns MP (2008) Prokaryotic silencing (psi)RNAs in *Pyrococcus furiosus*. RNA 14:2572–2579
Hale CR, Zhao P, Olson S, Duff MO, Graveley BR, Wells L et al (2009) RNA-guided RNA cleavage by a CRISPR RNA-Cas protein complex. Cell 139:945–956
Hale CR, Majumdar S, Elmore J, Pfister N, Compton M, Olson S et al (2012) Essential features and rational design of CRISPR RNAs that function with the Cas RAMP module complex to cleave RNAs. Mol Cell 45:292–302
Hao W, Richardson AO, Zheng Y, Palmer JD (2010) Gorgeous mosaic of mitochondrial genes created by horizontal transfer and gene conversion. Proc Natl Acad Sci U S A 107:21576–21581
Hatoum-Aslan A, Maniv I, Marraffini LA (2011) Mature clustered regularly interspaced short palindromic repeats RNA (crRNA) length is measured by a ruler mechanism anchored at the precursor processing site. Proc Natl Acad Sci U S A 108:21218–21222
Haurwitz RE, Jinek M, Wiedenheft B, Zhou K, Doudna JA (2010) Sequence- and structure-specific RNA processing by a CRISPR endonuclease. Science 329:1355–1358

Held NL, Herrera A, Quiroz HC, Whitaker RJ (2010) CRISPR associated diversity within a population of *Sulfolobus islandicus*. PLoS ONE 5:e12988

Hermans PW, van Soolingen D, Bik EM, de Haas PE, Dale JW, van Embden JD (1991) Insertion element IS987 from *Mycobacterium bovis* BCG is located in a hot-spot integration region for insertion elements in *Mycobacterium tuberculosis* complex strains. Infect Immun 59:2695–2705

Hoe N, Nakashima K, Grigsby D, Pan X, Dou SJ, Naidich S et al (1999) Rapid molecular genetic subtyping of serotype M1 group A *Streptococcus* strains. Emerg Infect Dis 5:254–263

Horvath P, Barrangou R (2010) CRISPR/Cas the immune system of bacteria and archaea. Science 327:167–170

Horvath P, Romero DA, Coute-Monvoisin AC, Richards M, Deveau H, Moineau S et al (2008) Diversity, activity and evolution of CRISPR loci in *Streptococcus thermophilus*. J Bacteriol 190:1401–1412

Ishino Y, Shinagawa H, Makino K, Amemura M, Nakata A (1987) Nucleotide-sequence of the Iap gene responsible for alkaline-phosphatase isozyme conversion in *Escherichia coli* and identification of the gene product. J Bacteriol 169:5429–5433

Jansen R, Embden JD, Gaastra W, Schouls LM (2002a) Identification of genes that are associated with DNA repeats in prokaryotes. Mol Microbiol 43:1565–1575

Jansen R, Van Embden JDA, Gaastra W, Schouls LM (2002b) Identification of a novel family of sequence repeats among prokaryotes. OMICS A J Integrat Biol 6:23–33

Jore MM, Lundgren M, van Duijn E, Bultema JB, Westra ER, Waghmare SP et al (2011) Structural basis for CRISPR RNA-guided DNA recognition by cascade. Nat Struct Mol Biol 18:529–536

Kamerbeek J, Schouls L, Kolk A, van Agterveld M, van Soolingen D, Kuijper S et al (1997) Simultaneous detection and strain differentiation of *Mycobacterium tuberculosis* for diagnosis and epidemiology. J Clin Microbiol 35:907–914

Kawarabayasi Y, Sawada M, Horikawa H, Haikawa Y, Hino Y, Yamamoto S et al (1998) Complete sequence and gene organization of the genome of a hyper-thermophilic archaebacterium *Pyrococcus horikoshii* OT3. DNA Res 5:55–76

Kawarabayasi Y, Hino Y, Horikawa H, Yamazaki S, Haikawa Y, Jin-no K et al (1999) Complete genome sequence of an aerobic hyper-thermophilic crenarchaeon *Aeropyrum pernix* K1. DNA Res 6:83–101

Klenk HP, Clayton RA, Tomb JF, White O, Nelson KE, Ketchum KA et al (1997) The complete genome sequence of the hyperthermophilic sulphate-reducing archaeon *Archaeoglobus fulgidus*. Nature 390:364–370

Kunin V, Sorek R, Hugenholtz P (2007) Evolutionary conservation of sequence and secondary structures in CRISPR repeats. Genome Biol 8:R61

Lillestøl RK, Redder P, Garrett RA, Brügger K (2006) A putative viral defence mechanism in archaeal cells. Archaea 2:59–72

Lillestøl RK, Shah SA, Brügger K, Redder P, Phan H, Christiansen J, Garrett RA (2009) CRISPR families of the crenarchaeal genus *Sulfolobus*:bidirectional transcription and dynamic properties. Mol Microbiol 72:259–272

Lintner NG, Frankel KA, Tsutakawa SE, Alsbury DL, Copié V, Young MJ, Tainer JA, Lawrence CM (2011) The structure of the CRISPR-associated protein Csa3 provides insight into the regulation of the CRISPR/Cas system. J Mol Biol 405:939–955

Lundgren M, Andersson A, Chen L, Nilsson P, Bernander R (2004) Three replication origins in *Sulfolobus* species: synchronous initiation of chromosome replication and asynchronous termination. Proc Natl Acad Sci U S A 101:7046–7051

Makarova KS, Aravind L, Grishin NV, Rogozin IB, Koonin EV (2002) A DNA repair system specific for thermophilic Archaea and bacteria predicted by genomic context analysis. Nucleic Acids Res 30:482–496

Makarova KS, Grishin NV, Shabalina SA, Wolf YI, Koonin EV (2006) A putative RNA-interference-based immune system in prokaryotes: computational analysis of the predicted enzymatic machinery functional analogies with eukaryotic RNAi and hypothetical mechanisms of action. Biol Direct 1:7

Makarova KS, Haft DH, Barrangou R, Brouns SJJ, Charpentier E, Horvath P et al (2011) Evolution and classification of the CRISPR-Cas systems. Nature Rev Microbiol 9:467–477
Manica A, Zebec Z, Teichmann D, Schleper C (2011) In vivo activity of CRISPR-mediated virus defence in a hyperthermophilic archaeon. Mol Microbiol 80:481–491
Marraffini LA, Sontheimer EJ (2008) CRISPR interference limits horizontal gene transfer in *Staphylococci* by targeting DNA. Science 322:1843–1845
Marraffini LA, Sontheimer EJ (2010) Self versus non-self discrimination during CRISPR RNA-directed immunity. Nature 463:568–571
Masepohl B, Görlitz K, Böhme H (1996) Long tandemly repeated repetitive (LTRR) sequences in the filamentous cyanobacterium *Anabaena* sp. PCC2120. Biochim Biophys Acta 1307:26–30
Minot S, Sinha R, Chen J, Li H, Keilbaugh SA, Wu GD et al (2011) The human gut virome: inter-individual variation and dynamic response to diet. Genome Res 21:1616–1625
Mojica FJM, Juez G, Rodríguez-Valera F (1993) Transcription at different salinities of *Haloferax mediterranei* sequences adjacent to partially modified *PstI* sites. Mol Microbiol 9:613–621
Mojica FJM, Ferrer C, Juez G, Rodríguez-Valera F (1995) Long stretches of short tandem repeats are present in the largest replicons of the archaea *Haloferax mediterranei* and *Haloferax volcanii* and could be involved in replicon partitioning. Mol Microbiol 17:85–93
Mojica FJM, Díez-Villaseñor C, Soria E, Juez G (2000) Biological significance of a family of regularly spaced repeats in the genomes of archaea, bacteria and mitochondria. Mol Microbiol 36:244–246
Mojica FJM, Díez-Villaseñor C, García-Martínez J, Soria E (2005) Intervening sequences of regularly spaced prokaryotic repeats derive from foreign genetic elements. J Mol Evol 60:174–182
Mojica FJM, Díez-Villaseñor C, García-Martínez J, Almendros C (2009) Short motif sequences determine the targets of the prokaryotic CRISPR defence system. Microbiol 155:733–740
Nakata A, Amemura M, Makino K (1989) Unusual nucleotide arrangement with repeated sequences in the *Escherichia coli* K-12 chromosome. J Bacteriol 171:3553–3556
Nelson KE, Clayton RA, Gill SR, Gwinn ML, Dodson RJ, Haft DH et al (1999) Evidence for lateral gene transfer between archaea and bacteria from genome sequence of *Thermotoga maritima*. Nature 399:323–329
Peng X, Brügger M, Shen B, Chen LM, She QX, Garrett A (2003) Genus-specific protein binding to the large clusters of DNA repeats (short regularly spaced repeats) present in *Sulfolobus* genomes. J Bacteriol 185:2410–2417
Portillo MC, González JM (2009) CRISPR elements in the thermococcales: evidence for associated horizontal gene transfer in *Pyrococcus furiosus*. J Appl Genet 50:421–430
Pougach K, Semenova E, Bogdanova E, Datsenko KA, Djordjevic M, Wanner BL, Severinov K (2010) Transcription processing and function of CRISPR cassettes in *Escherichia coli*. Mol Microbiol 77:1367–1379
Pourcel C, Salvignol G, Vergnaud G (2005) CRISPR elements in *Yersinia pestis* acquire new repeats by preferential uptake of bacteriophage DNA and provide additional tools for evolutionary studies. Microbiol 151:653–663
Pul U, Wurm R, Arslan Z, Geissen R, Hofmann N, Wagner R (2010) Identification and characterization of *E. coli* CRISPR-cas promoters and their silencing by H-NS. Mol Microbiol 75:1495–1512
Riehle MM, Bennett AF, Long AD (2001) Genetic architecture of thermal adaptation in *Escherichia coli*. Proc Natl Acad Sci U S A 98:525–530
Rousseau C, Nicolas J, Gonnet M (2009) CRISPI: a CRISPR interactive database. Bioinformatics 25:3317–3318
Sapranauskas R, Gasiunas G, Fremaux C, Barrangou R, Horvath P, Siksnys V (2011) The *Streptococcus thermophilus* CRISPR/Cas system provides immunity in *Escherichia coli*. Nucleic Acids Res 39:9275–9282
Sashital DG, Jinek M, Doudna JA (2011) An RNA-induced conformational change required for CRISPR RNA cleavage by the endoribonuclease Cse3. Nat Struct Mol Biol 18:680–687

Sashital DG, Wiedenheft B, Doudna JA (2012) Mechanism of foreign DNA selection in a bacterial adaptive immune system. Mol Cell 46:606–615

Sebaihia M, Wren BW, Mullany P, Fairweather NF, Minton N, Stabler R et al (2006) The multidrug-resistant human pathogen *Clostridium difficile* has a highly mobile mosaic genome. Nature Genet 38:779–786

Semenova E, Nagornykh M, Pyatnitskiy M, Artamonova II, Severinov K (2009) Analysis of CRISPR system function in plant pathogen *Xanthomonas oryzae*. FEMS Microbiol Lett 296:110–116

Semenova E, Jore MM, Datsenko KA, Semenova A, Westra ER, Wanner B et al (2011) Interference by clustered regularly interspaced short palindromic repeat (CRISPR) RNA is governed by a seed sequence. Proc Natl Acad Sci U S A 108:10098–10103

Sensen CW, Charlebois RL, Chow C, Clausen IG, Curtis B, Doolittle WF et al (1998) Completing the sequence of the *Sulfolobus solfataricus* P2 genome. Extremophiles 2:305–312

Shah SA, Garrett RA (2011) CRISPR/Cas and Cmr modules mobility and evolution of adaptive immune systems. Res Microbiol 162:27–38

Shah SA, Hansen NR, Garrett RA (2009) Distribution of CRISPR spacer matches in viruses and plasmids of crenarchaeal acidothermophiles and implications for their inhibitory mechanism. Biochem Soc Trans 37:23–28

She Q, Phan H, Garrett RA, Albers SV, Stedman KM, Zillig W (1998) Genetic profile of pNOB8 from *Sulfolobus*: the first conjugative plasmid from an archaeon. Extremophiles 2:417–425

She Q, Singh RK, Confalonieri F, Zivanovic Y, Allard G, Awayez MJ et al (2001) The complete genome of the crenarchaeon *Sulfolobus solfataricus* P2. Proc Natl Acad Sci U S A 98:7835–7840

Sinkunas T, Gasiunas G, Fremaux C, Barrangou R, Horvath P, Siksnys V (2011) Cas3 is a single-stranded DNA nuclease and ATP-dependent helicase in the CRISPR/Cas immune system. EMBO J 30:1335–1342

Smith DR, Doucette-Stamm LA, Deloughery C, Lee H, Dubois J, Aldredge T et al (1997) Complete genome sequence of *Methanobacterium thermoautotrophicum* deltaH: functional analysis and comparative genomics. J Bacteriol 179:7135–7155

Stern A, Keren L, Wurtzel O, Amitai G, Sorek R (2010) Self-targeting by CRISPR: gene regulation or autoimmunity? Trends Genet 26:335–340

Swarts DC, Mosterd C, van Passel MWJ, Brouns SJJ (2012) CRISPR interference directs strand specific spacer acquisition. PLoS ONE 7:e35888

Tang TH, Bachellerie JP, Rozhdestvensky T, Bortolin ML, Huber H, Drungowski M et al (2002) Identification of 86 candidates for small non-messenger RNAs from the archaeon *Archaeoglobus fulgidus*. Proc Natl Acad Sci U S A 99:7536–7541

Tang TH, Polacek N, Zywicki M, Huber H, Brügger K, Garrett RA et al (2005) Identification of novel non-coding RNAs as potential antisense regulators in the archaeon *Sulfolobus solfataricus*. Mol Microbiol 55:469–481

Touchon M, Rocha EP (2010) The small slow and specialized CRISPR and anti-CRISPR of *Escherichia* and *Salmonella*. PLoS ONE 5:e11126

Tyson GW, Banfield JF (2008) Rapidly evolving CRISPRs implicated in acquired resistance of microorganisms to viruses. Environ Microbiol 10:200–207

Viswanathan P, Murphy K, Julien B, Garza AG, Kroos L (2007) Regulation of dev an operon that includes genes essential for *Myxococcus xanthus* development and CRISPR-associated genes and repeats. J Bacteriol 189:3738–3750

Wang R, Preamplume G, Terns MP, Terns RM, Li H (2011) Interaction of the Cas6 riboendonuclease with CRISPR RNAs: recognition and cleavage. Structure 19:257–264

Westra ER, Pul U, Heidrich N, Jore MM, Lundgren M, Stratmann T et al (2010) H-NS-mediated repression of CRISPR-based immunity in *Escherichia coli* K12 can be relieved by the transcription activator LeuO. Mol Microbiol 77:1380–1393

Westra ER, van Erp PB, Künne T, Wong SP, Staals RH, Seegers CL et al (2012) CRISPR immunity relies on the consecutive binding and degradation of negatively supercoiled invader DNA by cascade and Cas3. Mol Cell 46:595–605

Wiedenheft B, van Duijn E, Bultema JB, Waghmare SP, Zhou K, Barendregt A et al (2011) RNA-guided complex from a bacterial immune system enhances target recognition through seed sequence interactions. Proc Natl Acad Sci U S A 108:10092–10097

Wurtzel O, Sapra R, Chen F, Zhu Y, Simmons BA, Sorek R (2010) A single-base resolution map of an archaeal transcriptome. Genome Res 20:133–141

Yanai K, Murakami T, Bibb M (2006) Amplification of the entire kanamycin biosynthetic gene cluster during empirical strain improvement of *Streptomyces kanamyceticus*. Proc Natl Acad Sci U S A 103:9661–9666

Yosef I, Goren MG, Qimron U (2012) Proteins and DNA elements essential for the CRISPR adaptation process in *Escherichia coli*. Nucleic Acids Res 40:5569–5576

Zegans ME, Wagner JC, Cady KC, Murphy DM, Hammond JH, O'Toole GA (2009) Interaction between bacteriophage DMS3 and host CRISPR region inhibits group behaviors of *Pseudomonas aeruginosa*. J Bacteriol 191:210–219

Zhang J, Rouillon C, Kerou M, Reeks J, Brügger K, Graham S et al (2012) Structure and mechanism of the Cmr complex of CRISPR-mediated antiviral immunity. Mol Cell 45: 303–313

Chapter 2
Occurrence, Diversity of CRISPR-Cas Systems and Genotyping Implications

Christine Pourcel and Christine Drevet

Abstract This chapter describes the overall variability of CRISPR-Cas systems as observed in publicly available genomes and how this can be used to draw hypotheses on phylogenetic relationships between species and between strains of a given species. The fact that spacers are added sequentially at the leader end and that a given spacer is rarely acquired twice or duplicated are key elements for building hierarchical relationships between strains. Presence/absence of a given CRISPR locus and variability in the number of direct repeats and spacers in that locus between strains have been frequently reported, providing in some cases phylogenic information, but this polymorphism was extensively used for genotyping in only a few instances. The observation that not all strains possess a CRISPR locus in a given species precludes its use as a general typing tool. However, in some species, the degree of variability is a powerful marker of the species diversity and evolution. Through examples found among 1,434 published genomes of bacteria and archaea, different features of the CRISPR-Cas systems diversity will be highlighted.

Contents

C. Pourcel (✉) · C. Drevet
Institut de Génétique et Microbiologie, Université Paris-Sud, Bât 400, 91405 Orsay cedex, France
e-mail: christine.pourcel@u-psud.fr

C. Drevet
e-mail: christine.drevet@igmors.u-psud.fr

R. Barrangou and J. van der Oost (eds.), *CRISPR-Cas Systems*,
DOI: 10.1007/978-3-642-34657-6_2, © Springer-Verlag Berlin Heidelberg 2013

2.1 Introduction

CRISPRs are remarkable structures found in bacterial and archaeal genomes and known to interact with a set of genes called *cas* (Haft et al. 2005; Horvath and Barrangou 2010; Makarova et al. 2011a). Functional analysis of CRISPR-Cas systems in different species showed that it can play a number of roles, including defense against foreign genetic elements, regulation of lysogeny, regulation of biofilm formation, and others (Barrangou et al. 2007; Edgar and Qimron 2010; Zegans et al. 2009), as discussed in Chap. 10. A CRISPR locus is typically made of a succession of direct repeats (repeats) separated by spacers. *Cas* gene products interact with various CRISPR sequences and the target sequences to mediate the interference pathway (van der Oost et al. 2009). Spacers provide the specificity of the defense mechanism and mostly originate from phages or plasmids (Bolotin et al. 2005; Mojica et al. 2005; Pourcel et al. 2005). The occurrence of self-targeting spacers in some CRISPRs (1 in every 250 spacers in average) might lead to autoimmunity or be a part of a regulatory mechanism (Cui et al. 2008; Stern et al. 2010).

Investigation of publicly available genome sequences shows that CRISPRs are present in about 48 % of bacteria and 80 % of archaea, mostly on chromosomes but also on plasmids (Grissa et al. 2007a). *Cas* genes are found in the majority of CRISPR-containing genomes and when several CRISPRs of the same CRISPR-Cas system are present in a single genome, a single set of *cas* genes is generally clustered with one of the CRISPRs. Little is known on the mechanisms that drive multiplication of CRISPRs within a genome and acquisition and loss of spacers. New spacers are acquired by insertion next to the leader and are lost by internal deletion (Pourcel et al. 2005; Lillestol et al. 2006). It was proposed that when a CRISPR locus reaches a certain length, spacers must be lost and the older ones are preferably and more frequently lost first (Tyson and Banfield 2008). Although this may be true for certain CRISPRs in which the total number of spacer seems limited, in some extreme cases, several hundreds of spacers have been observed. Thus, the equilibrium between acquisition and loss appears to be highly different from one system to the other and this must be related to the ecology of the organism, its reliance on CRISPR-mediated immunity, and the pressure applied by foreign elements. A large body of information indicates that horizontal transfer of CRISPR and *cas* genes takes place between strains and between occasionally distant species and genera (Godde and Bickerton 2006; Horvath et al. 2008; Chakraborty et al. 2010; Shah and Garrett 2011). As a consequence, not all strains of a given species systematically possess the same sets of CRISPRs and *cas* genes. Owing to the huge amount of diversity observed in some CRISPR-Cas systems, examination of their elements (repeats, spacers, flanking sequences, and associated genes) provides important phylogenetic information (Grissa et al. 2008a).

With the advent of new sequencing technologies, more and more genomes are made available including multiple strains of a given species. Next-generation sequencing methods are well adapted to the investigation of CRISPRs and facilitate metagenomic analysis, which is an interesting source of sequences for both microorganisms and the viruses that infect them. In this evolving context, bioinformatic tools are needed to confidently and reliably identify and characterize CRISPRs and their elements (Grissa et al. 2009).

2.2 Assessing the Overall Diversity of CRISPRs

2.2.1 Bioinformatic Tools

2.2.1.1 Identification of CRISPRs

One important concern when trying to identify CRISPRs in a genome sequence is the exact definition of these structures. This is challenging, given the extensive sequence diversity of the CRISPR repeats, and the relative paucity of CRISPR-Cas systems that have been thoroughly characterized and shown to be active in the laboratory. A few specific programs have been developed for this purpose, the most used being CRISPRfinder (Grissa et al. 2007b), PILER-CR (Edgar 2007), and CRT (Bland et al. 2007). Additional programs were employed in different studies such as Pygram (Durand et al. 2006), LUNA (Lillestol et al. 2006), or Dotter (Sonnhammer and Durbin 1995). All programs perform as expected on typical CRISPR structures showing one or more of the following characteristics: five or more spacers, *cas* genes located nearby, homogeneous spacer length, and perfectly conserved repeats. Unfortunately, many CRISPRs do not typically meet these criteria. In this chapter, we will essentially discuss the performance of CRISPRfinder [which relies on the REPuter program (Kurtz et al. 2001)] which is at the basis of several other tools aimed at comparing and classifying CRISPRs.

CRISPR finder is based on specific features that were common to the well-characterized CRISPRs at the time of the program implementation and includes a tolerance margin: 23–55 bp repeats interspaced by sequences of 25–60 bp (spacer) and with spacer lengths between 0.6 and 2.5 the repeat length (Grissa et al. 2007b). The repeats are remarkably conserved in the majority of CRISPRs including those with a very large number of repeat-spacer units, but in some structures they show a high degree of heterogeneity such as in *Streptococcus sanguinis* SK36 (Genbank

ID CP000387), in several *Clostridium sp.* strains, or in *Amycolatopsis mediterranei*, for example. The parameters for defining the consensus repeat were chosen in order to cope with these unusual structures. CRISPRfinder returns all compatible structures and classifies them into "confirmed" (more than three units) and "questionable" (1–3 units) CRISPRs. In CRISPRdb, additional filters have been added to validate or exclude some structures, including a comparison of short CRISPRs' repeats to previously identified repeats and restriction on the spacer allowed length when the corresponding repeat has no classical flanking nucleotides such as GTTT or GAAC (Grissa et al. 2007a). Manual curation and further characterization often alleviate the issues inherent to false positives and occasionally false negatives. However, some of the CRISPR-like structures may correspond to other types of genetic elements such as, for example, portions of genes encoding proteins with repeated amino acid segments. Conversely, some of the shortest CRISPR-like structures containing one or two spacers may be true CRISPRs and they need to be evaluated using additional parameters. Therefore, a critical inspection of the results must still be made to discard sequences that are not true CRISPRs and validate short candidates. The presence of *cas* genes in the vicinity, or the identification of the source of one spacer or more [the proto-spacer (Deveau et al. 2008)] are probably the best criteria to fully validate a CRISPR structure. An indirect strong proof is the presence of an identical repeat in a fully validated CRISPR. Some tools are available for this purpose after running CRISPRfinder: spacers BLAST at NCBI to identify proto-spacers, search for *cas* genes using BLASTX, and search for CRISPRs with a significantly similar repeat in the database.

2.2.1.2 Tools to Analyze CRISPR Loci and Components

When comparing two strains with several CRISPR-Cas systems and/or several CRISPRs with the same repeat, it is necessary to individually identify each locus before listing the spacers. CRISPRcompar (Grissa et al. 2008b) has been developed to help in this classification by comparing sequences flanking CRISPRs that have similar repeats. The program is set to consider as similar, two loci with strictly identical repeats and flanking sequences showing 90 % identity over 200 base pairs. In some species, the accumulation of mutations may hide the common origin of two loci. Another important challenge concerns the identification and numbering of spacers especially when hundreds of unique sequences are to be compared and classified. Graphical representation of spacers have been used which can help to visually assess similarities between alleles but which will show limits when processing very large amounts of sequences (Horvath et al. 2008). CRISPRtionary was specifically developed to produce a catalog of spacers for a given CRISPR locus from submitted alleles, to number them and show their order in each allele (Grissa et al. 2008b). A detailed procedure to use these tools has been described (Grissa et al. 2009). Work is now in progress to provide a database of spacers that can be queried online.

2.2.1.3 CRISPR Databases

The challenge in building a database of CRISPRs is to faithfully identify these loci in order to be both exhaustive (reduce false negatives) and correct (eliminate false positives). This relies on the efficiency and quality of the program used to detect CRISPRs in a sequenced genome but also on manual curation since there is no perfect solution due to the diversity of CRISPR structures. At present two databases exist, CRISPRdb [http://crispr.u-psud.fr/ (Grissa et al. 2007a)] and CRISPI [http://crispi.genouest.org/ (Rousseau et al. 2009)], respectively listing 48.4 % (880/1,817) and 47.3 % (755/1,594) of bacterial genomes, and 83.7 % (105/123) and 80 % (96/120) of archeal genomes as possessing a CRISPR. Although these percentages are very similar, the identified CRISPR-like structures are often different, due to the parameters used to define the repeat sequence and also the repeat and spacer lengths.

CRISPRdb is a repertoire of the characteristics and locations of CRISPRs identified by the CRISPRfinder program in published bacteria and archaea genome sequences (chromosomes and associated plasmids) recovered from the RefSeq database released at the NCBI ftp site (ftp://ftp.ncbi.nih.gov/genomes/). Each sequence is submitted to the CRISPRfinder program and the resulting data is further analyzed by making use of the data previously stored in the database, in particular to validate some of the questionable CRISPRs. In addition, a manual curation step is performed after an initial automatic import to eliminate structures that possess typical characteristics of CRISPRs but correspond to tandem repeats.

CRISPI makes use of PYGRAM to identify CRISPRs, and apparently does not apply restrictions to the repeat and spacer lengths. Consequently, many CRISPRs present in CRISPI are not labeled as CRISPRs in CRISPRdb. For example, at the time of this publication, the CRISPI highlights displayed a CRISPR structure in *Xylella fastidiosa* M23 with five 8 bp-long repeats (70 % identity between repeats) and spacers 446–574 bp-long (the longest observed spacer in this database). This genome does not possess any *cas* gene, and it is very likely that the aforementioned structure is not a CRISPR. In the same highlights, the CRISPR with the longest repeats (five 92 bp-long repeats and four 27–45 bp-long spacers) found in *Shewanella putrefaciens* CN-32 corresponded to the tRNA-Asn locus. Many other differences exist between the CRISPRs identified by the two programs, in the number of CRISPRs, but also the sequence of repeats in a given locus. For example in *Cyanothece* sp. ATCC 51142, CRISPRfinder extracts a 37 bp repeat from CRISPR NC 01056-4, whereas Pygram finds a 45 bp repeat. CRISPRdb identifies 5 "confirmed" CRISPRs in *Sorangium cellulosum* "So ce 56", CRISPI finds 7 CRISPRs, including one which is composed of four 17 bp"repeats" separated by 4 bp-long "spacers". Our current knowledge of CRISPR-Cas systems clearly indicates that such structures cannot be legitimate CRISPR candidates since a four-nucleotide spacer cannot provide any specificity to the interference mechanism.

The parameters used in CRISPRdb to label a CRISPR-like structure as "confirmed" most probably accommodate the vast majority of existing CRISPRs. The smallest repeat recorded to date corresponds to the lower limit of 23 bp and was found only once in the archaeon *Ferroglobus placidus* DSM 10642 (four CRISPRs which contain 26, 21, 18, and 9 spacers, respectively). In a few instances, the repeat of CRISPRs containing a single spacer was wrongly estimated to be 23 bp-long but this could be corrected by comparison with longer CRISPR alleles in other strains of the same species. The largest repeat identified to date is 50 bp-long in *Weeksella virosa* (1 CRISPR harboring 20 spacers). This suggests that the higher limit of the program (set at 55 bp) is acceptable. Last, among the numerous "questionable" CRISPRs usually possessing one or two spacers, some might be indeed real CRISPRs and may be confirmed later when new genomes containing larger CRISPRs with identical repeats will be processed.

A recent addition to CRISPRdb is the possibility to view annotated *cas* genes in genomes harboring a CRISPR and to perform a BlastX analysis using a local database of Cas proteins extracted from the Uniprot database (http://www.uniprot.org/). Similarly, CRISPI displays a detailed list of CRISPR-associated genes with in the vicinity of each CRISPR.

CRISPRdb provides a list of repeats and spacers from published sequenced genomes. However, there is still a need for databases containing all the spacers that have been identified to date, including sequenced alleles as part of intra-species diversity studies and spacers extracted from metagenomes. Specific databases are being constructed to record spacers of a given species and variations in CRISPR alleles. This is the case of the SpolDB4 database dedicated to *Mycobacterium tuberculosis*, and of the *Salmonella enterica* and *Legionella pneumophila* databases held at the Pasteur Institute in Guadeloupe http://www.pasteur-guadeloupe.fr:8081/SITVITDemo/ or in Paris http://www.pasteur.fr/recherche/genopole/PF8/crispr/CRISPRDB.html. Of note, the program iSpacer has been created by Aaron White (http://epilityblog.com/blog/) to compare large spacer libraries to the NCBI sequence database in order to search for proto-spacers. It was used to analyze a collection of spacers from *Pseudomonas aeruginosa* CRISPRs (Cady et al. 2011).

2.2.2 Diversity of CRISPRs Found in Published Genomes

As of June 2012, more than 1,800 bacterial and 120 archeal genomes have been publicly released allowing an assessment of the CRISPR diversity. New genomes are released on a continuous basis. Notwithstanding the current bias(es) in the currently sequenced bacterial and archaeal genomes, some key observations can be made which help in tracing the origin of CRISPR-Cas systems.

2.2.2.1 Repeat and Spacer Features

The available data in CRISPRdb were submitted to global analysis in order to investigate and characterize CRISPRs variability. The largest CRISPR was observed in *Haliangium ochraceum* DSM 14365 with 588 repeats. One set of *cas* genes and two other CRISPRs were found in this bacterium (with 190 and 37 repeats), spanning 75 kb. A second group of *cas* genes was found at another location in this genome but no CRISPR seems present in the vicinity of these genes. Interestingly, archaea and thermophilic bacteria as well as others living in extreme habitats frequently have 2 CRISPR-Cas systems and a large number of CRISPRs. The six members of the genus *Caldicellulosiruptor,* which contains the most thermophilic bacteria, have multiple CRISPRs and very large sets of *cas* genes (33 *cas* genes clustered at a single locus in *C. kristjansoni*). This suggests that CRISPR-Cas systems are an essential element for survival in these organisms. The largest number of CRISPRs is observed in the bacterium *Thermomonospora curvata* strain DSM 43183 with 15 CRISPRs and the archeon *Methanocaldococcus* sp. strain FS406-22 with 23 CRISPRs. *Thermincola* sp JR (Genbank CP002028) possesses three CRISPR-Cas systems made of three clusters of 2, 1, and 4 CRISPRs with different repeats and a different set of *cas* genes at each locus. *Truepera radiovictrix* DSM17093 possesses 4 different CRISPR-Cas systems for 9 CRISPRs at 7 different genetic loci. Some genera and/or species with multiple genome sequences available seem to completely lack CRISPRs such as *Chlamydia* sp. and *Chlamydophila* sp. or *Streptococcus pneumoniae*.

When analyzing the distribution of repeats into size groups, clear differences are seen between archaea and bacteria. Both show two main peaks at 29–30 bp and 36–37 bp but the smaller class of 24–25 bp is seen essentially in archaea, whereas the large repeats of 44 bp and more are only seen in bacteria (Fig. 2.1a). The diagram in Fig. 2.1b shows the length difference between repeats and spacers (average of spacers lengths) in relation to the repeat length. It suggests that a selective pressure exists for total repeat-spacer sizes of 60–75 bp. The diagram in Fig. 2.1c shows a tendency for archaeal CRISPRs to possess the larger number of repeats. When submitting the repeats list to UCLUST de novo clustering with option–id 0.50 at http://drive5.com/usearch (Edgar 2010), 185 clusters were found. 42 % of the sequences cluster into 10 groups containing more than 20 similar repeats (max 79). An analysis by Kunin et al. of 561 repeats from 195 genomes based on their folding score led to the definition of 33 clusters, 12 of which contained 10 or more members (Kunin et al. 2007). In our analysis, 13 % of the repeats belong to small groups (containing less than 5 sequences) including 7 % of single sequences. Among the latter, there are CRISPRs with a high number of spacers. Most of the repeats within a particular group show similar repeat length but there are exceptions. Indeed, internal insertions and deletions (INDELs) are frequently observed in the alignment between short and long repeats within a group. Of note, there is no repeat between 39 and 43 bp (Fig. 2.1a). The 44–50 bp repeats are clustered in a specific group. When the similarity is low, one side of the repeat is often better conserved.

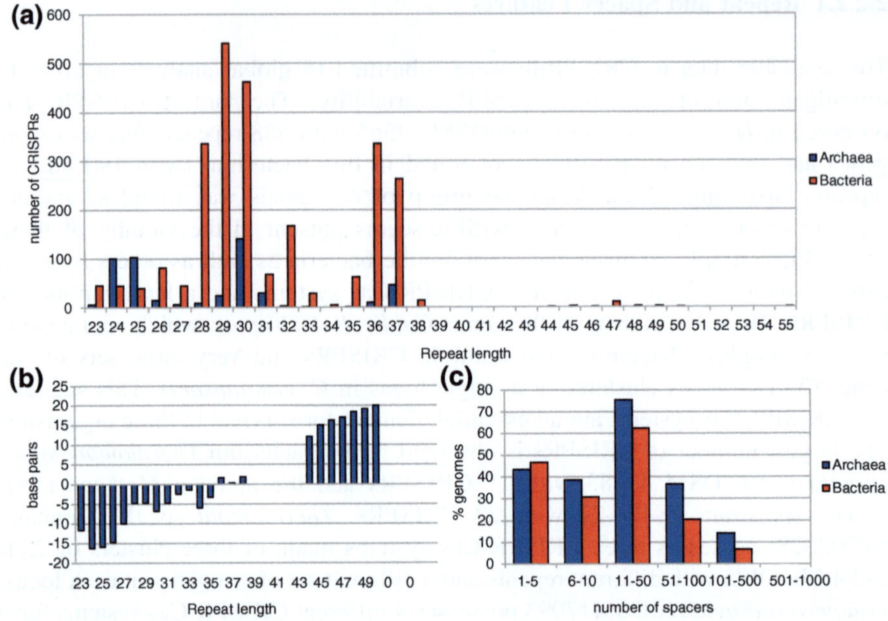

Fig. 2.1 Characteristics and distribution of repeats and spacers as observed in CRISPRdb. **a** Repeat length variability, **b** difference between repeats and average spacers length, **c** number of spacers

Works by Carte et al. (2008), Brouns et al. (2008), and Deltcheva et al. (2011) have shown that repeat sequences are targets for the cleavage by endoribonucleases. The large diversity of repeat sequences suggests that only part of the sequence is recognized by the Cas machinery and that the secondary structure is essential, in agreement with previous observations (Mojica et al. 2000; Jansen et al. 2002). Kunin et al. (2007) showed that among their 12 larger clusters, some but not all repeats were able to form stem-loop structures. To test whether a limited number of short sequences could be recognized in repeats, we performed a search for motifs using MEME (http://meme.sdsc.edu/meme/) with default parameters (zero or one motif per sequence, 3 maximum number of motifs to find). We found that 922 out of 1,041 repeats possessed one of three motifs and these motifs were differently localized over the repeat sequence (Fig. 2.2) (Bailey and Elkan 1994). In cluster 3 described by Kunin et al., the repeat possesses motif 1 forming the loop and motif 2 responsible for the formation of the stem. It may be of interest to note that the ten 44 bp and longer repeats (44, 46, 47, 48, 49, and 50 bp-long) show important similarity over 23 bp on one side (containing motif 1) and are associated with the csn1/Cas9 of Type II CRISPR-Cas systems (Fig. 2.3) (Makarova et al. 2011b). In 9 out of 10 cases, the corresponding CRISPRs are present in members of each of the three classes of the phylum *Bacteroidetes*.

The mechanism of acquisition of new spacers has been shown to involve insertion of a new repeat and a new spacer at the leader end (Barrangou et al. 2007;

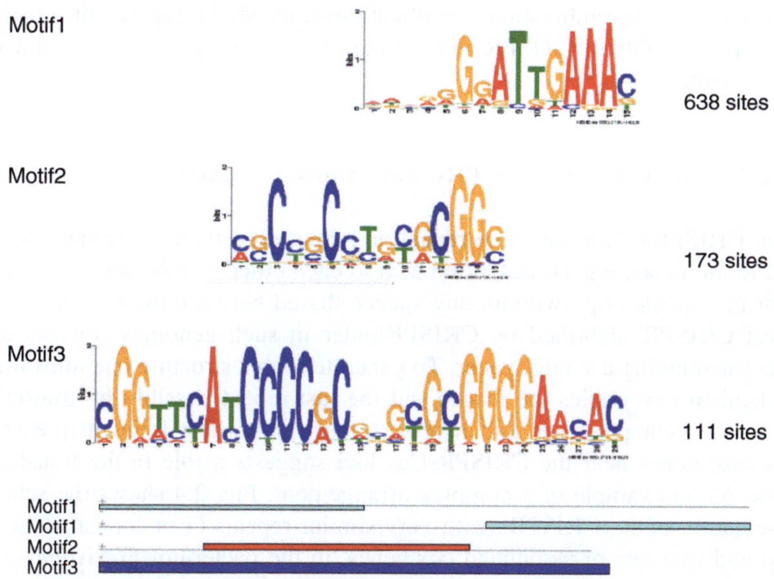

Fig. 2.2 Identification of common motifs in repeats. Three sequence logos produced by MEME at http://meme.sdsc.edu/, Motif1, Motif2, and Motif3, are observed in 638, 173, and 111 repeat types, respectively. At the *bottom* of the figure the diagram shows the most frequent position of the three motifs in the repeat sequences

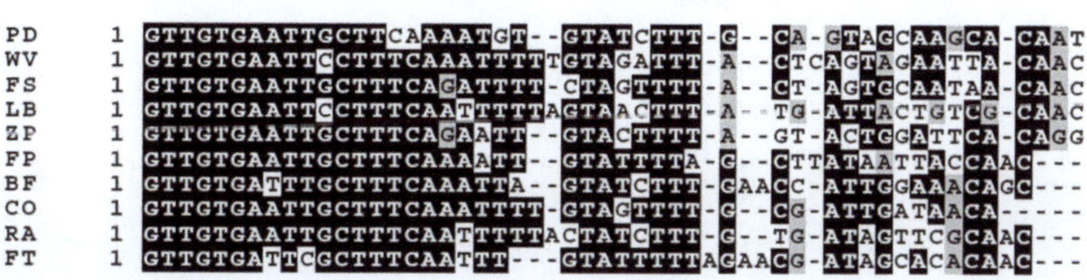

Fig. 2.3 Alignment of the 10 long consensus repeat sequences. PD: *Prevotella denticola*, WV: *Weeksella virosa*, FS: *Fibrobacter succinogenes*, LB: *Leadbetterella byssophila*, ZP: *Zunongwangia profunda*, FP: *Flavobacterium psychrophilum*, BF: *Bacteroides fragilis*, CO: *Capnocytophaga ochracea*, RA: *Riemerella anatipestifer*, and FT: *Fluviicola taffensis*

Deveau et al. 2008). Several spacers can be added during the adaptation process. In the majority of CRISPR structures all the spacers are unique, but duplication of single or groups of spacers can be observed principally in long CRISPRs. For example, in *Spirochaeta caldaria* DSM7334 NC-015732-3, 28 different sequences are observed out of 52 spacers and only 14 are present once. Another example is found in *Myxococcus fulvus* HW-1: adjacent CRISPRs NC-015711-8 and NC-015711-9 with 41 unique sequences out of 101 spacers; 14 spacers are present once while others occur up to 9 times. Although it is possible that some spacers are acquired several times independently, the most probable mechanism for spacer duplication is via recombination or replication slippage. Also, given the challenges inherent to assembly of CRISPR loci, it might be necessary to validate some of the observed patterns.

2.2.2.2 Creation of New CRISPRs and Transfer of the System

Several CRISPRs with the same repeat and conserved flanking leader sequences can be found in some genomes, often next to each other, but the set of *cas* genes is present in a single copy without any spacer shared between these structures. The smallest CRISPR identified by CRISPRfinder in such genomes consists of two repeats surrounding a single spacer. To generate such a structure one must imagine a mechanism that copies the leader and the last repeat possibly by transcription from an adjacent promoter and reverse transcription. The frequent presence of transposase genes near the CRISPR-Cas loci suggests a role in the translocation process. As an example of a complex arrangement, Fig. 2.4 shows the schematic representation of six CRISPRs with very similar repeats (1 or 3 mismatches over 30 bp) and two sets of associated *cas* genes, in the bacterium *Flexistipes sinusarabici* DSM 4947.

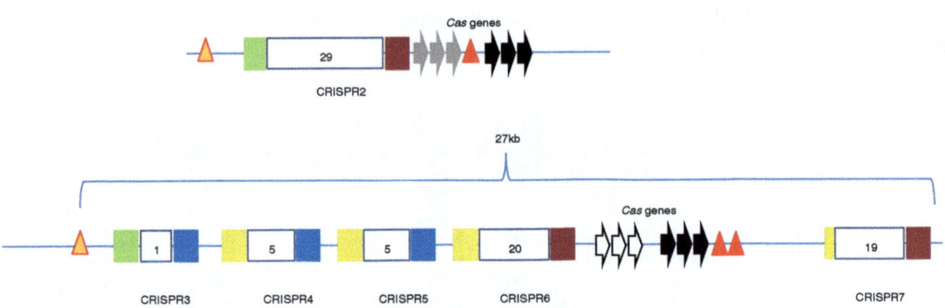

CRISPR_id	Start Position	End Position	Number of spacers	DR consensus	System type	Cas
NC_015672_1	1011494	1014319	43	CTTTAAATCCCAGAATTGGGAACTAAAAC	I	Cas1 to Cas6
NC_015672_2	1428495	1430685	29	GTCGTAATCCCTTCAAATCAGGTCATTAATTCCAAT	III-A	Cas1, Cas2, Cas6, Csm3, Csm5
NC_015672_3	1709462	1709573	1	GTCGTAATCCCTTCTAATCAGGTCATGGATTCCAAT		
NC_015672_4	1709819	1710226	5	GTCGTAATCCCTTCTAATCAGGTCATGGATTCCAAT		
NC_015672_5	1710472	1710885	5	GTCGTAATCCCTTCTAATCAGGTCATGGATTCCAAT		
NC_015672_6	1711131	1712652	20	GTCGTAATCCCTTCTAATCAGGTCATGGATTCCAAT	III-B	Cas1, Cas2, Cas6, Cmr2, Cmr3, Cmr4, Cmr5, Cmr6,
NC_015672_7	1734298	1735756	19	GTCGTAATCCCTTCAAGTCAGGTCATGAATTCCAAT		

Fig. 2.4 Schematic representation of CRISPR-Cas clusters with similar repeats in the bacterium *Flexistipes sinusarabici.* Six CRISPRs are represented as *boxes* showing the number of spacers. Flanking sequences are depicted with *colored boxes. Triangles* represent transposases genes. *Cas* genes are shown by *small white, gray, or black arrows.* Below are shown the position of the 7 CRISPRs found in the genome, the number of spacers, the sequence of the consensus repeat, the *cas* names, and the CRISPR-Cas system type

Analysis of published sequences clearly shows that different genera share similar CRISPR-Cas systems although these genera are not phylogenetically linked when using other genetic markers (Haft et al. 2005; Godde and Bickerton 2006; Chakraborty et al. 2010). A most intriguing observation is the presence of CRISPR-Cas in archae and in bacteria but not in Eukaryota, even the monocellular ones. It has been suggested that CRISPR-Cas systems which are present in the majority of archaea have been transferred to thermophilic bacteria and subsequently spread to other bacterial species. Indeed although the repeats seem to show specific characteristics in archaea, the *cas* gene systems are shared by members of the two domains (Makarova et al. 2011a). Plasmids may be vectors for CRISPRs-Cas systems as some of them have been found to possess complete systems. A total of 121 CRISPRs are carried by 58 plasmids out of 1,269 present in 50 taxons in CRISPRdb. For example, the *L. pneumophila* strain Lens possesses on a plasmid two CRISPRs and a complete set of *cas* genes (subtype I–F) similar to those found in the chromosome of the same strain but also in *P. aeruginosa*, *Yersinia pestis* and *Escherichia coli*.

Most of the *Staphylococcus aureus* sequenced genomes are devoid of CRISPR-Cas except for the livestock-associated ST398 lineage (Golding et al. 2010) and an ST75 early branching lineage (Holt et al. 2011). It is interesting to note that the CRISPR-Cas system is present next to the staphylococcal cassette chromosome (SCC) *mec*V subtype. A similar CRISPR-Cas system is present in some strains of *Staphylococcus epidermidis* (Gill et al. 2005) and *Staphylococcus lugdunensis*.

2.3 The Historical Use of CRISPR Polymorphism for Genotyping

2.3.1 Why and How to Perform Intra-Species Typing

A species may be defined as the sum of numerous strains that are classically differentiated by phenotypic and genetic characteristics, the precision of which depends on the question asked. When performing epidemiological investigations during outbreaks for example, it is critical to be able to trace the source of an infection similarly to forensics investigations in humans. Likewise, it is also important to evaluate the genetic complexity of a species and the speed at which it is evolving. The ultimate genotyping is the determination of the complete genome sequence of an organism. Alternatively, specific genetic polymorphisms can be used to compare strains, such as presence/absence of insertion elements (IS), single nucleotide polymorphisms (SNP), and variable number of tandem repeats (VNTR). The frequency of the genetic changes at these loci and the level of homoplasia (the independent occurrence of identical mutations) influence the informational value of the method and the possibility to use it to infer phylogenetic relationships between strains. In addition, the simplicity and cost of the method is of key importance when numerous samples must be simultaneously processed. Finally in order to be able to compare results between laboratories and to store the data into shared databases, genotypes in the form of a numerical code must be favored over gel/picture-based fingerprints. The characteristics of CRISPRs make them intriguing genetic markers for genotyping and population structure analysis but there is still a lot to be understood with regard to the molecular mechanisms that induce polymorphism in these sequences.

Several techniques have been used to assess the variability at a given CRISPR locus. Each of them will be described in the following paragraphs while discussing species for which a CRISPR-based genotyping scheme has been developed. Sequencing is the most straightforward but is not easily applicable to large alleles. In this case only portions may be amplified and sequenced (perhaps using primers designed in selected spacers). Because the evolution of an active CRISPR occurs via insertion of new spacers at the leader end, sequencing of this portion can be particularly informative. In contrast, sequencing the opposite end, which can contain spacers conserved across various strains, can be useful to cluster related phylogenetic group of strains. Hybridization to spacer-derived oligonucleotides, called "spoligotyping" had been used for some bacterial species but this will only investigate the presence of a pre-established selection of known spacers. Finally it is possible to rapidly differentiate alleles by high-resolution DNA melt curve analysis.

2.3.2 Intra-Species CRISPR Variations

2.3.2.1 *Mycobacterium tuberculosis* and *Mycobacterium canettii*

The first use of CRISPR polymorphism for diagnosis and genotyping was described in the *M. tuberculosis* complex (MTBC) which encompasses different species including *M. tuberculosis*, *M. africanum*, *M. bovis,* and the *M. cannettii* taxon (Kamerbeek et al. 1997). Groenen et al. (1993) were the first to analyze an MTBC CRISPR locus they called "DR" which showed polymorphism between different strains. They initially applied a PCR-based method called direct variable repeat PCR (DVR-PCR) derived from the minisatellite variant repeat PCR technique (MVR-PCR) (Jeffreys et al. 1991). This method, which is not suitable for routine use and high-throughput genotyping was replaced by a very elegant PCR and hybridization-based method called spacer oligotyping or "spoligotyping" (Kamerbeek et al. 1997). Oligonucleotides corresponding to 37 spacers present in the genome of *M. tuberculosis* strain H37Rv and 6 spacers of *M. bovis* BCG are bound to a membrane which is hybridized to amplification products generated by PCR between two repeats. Later, the addition of 25 new spacers improved the discriminatory power of the technique (van der Zanden et al. 2002).

Spoligotyping provides a pattern which can easily be coded and shared between laboratories. In the spolDB4 version of the international spoligotyping database,

1939 shared types (observed twice or more) were identified among 39,295 strains (Brudey et al. 2006).

Spoligotyping and derived methods such as the microbead-based hybridization assay (Zhang et al. 2010) can only indicate the presence/absence profile of known spacers. Sequencing of many MTBC isolates showed that this CRISPR locus is not acquiring new spacers and that polymorphism is only generated by loss of spacers (van Embden et al. 2000), some of which may be the result of IS element insertions (Groenen et al. 1993; Warren et al. 2002). Because all the members of the MTBC appear to possess a CRISPR locus and since the number of spacers remains low and is not increasing, spoligotyping is perfectly adapted to the analysis of this complex. The situation is different for members of the *M. cannettii* taxon. Indeed the first strains analyzed in detail appeared to possess a CRISPR with the same repeat as in *M. tuberculosis* but with a different set of spacers (van Embden et al. 2000). Later analysis of a larger collection of *M. canettii* isolates showed that many did not possess any CRISPR and others had new set of spacers (Fabre et al. 2004, 2010). This confirms the higher degree of diversity within the *M. canettii* taxon which is believed to be the most probable source species of the whole complex (Fabre et al. 2004).

Overall, spoligotyping has been central in the identification of clades in the MTBC, and it is a useful approach for phylogenetic studies (Filliol et al. 2003) but it has a limited value for evolutionary studies (Comas et al. 2009). Yet it remains the cheapest assay to rapidly classify strains.

2.3.2.2 *Yersinia pestis*

Yersinia pestis is a rather monomorphic species, highly pathogenic, and recently emerged (less than 20,000 years and may be not more than a few thousand years) from the more diverse *Yersinia pseudotuberculosis* species (Morelli et al. 2010; Bos et al. 2011). In the eight currently publicly available genomes, 1–3 CRISPRs (initially called Yp1, Yp2, Yp3, and subsequently renamed Ypa, Ypb, Ypc) have been observed with an identical repeat and with a single set of *cas* genes near one of the loci (Ypest I–F subtype). In 2005, the analysis of CRISPR polymorphism in a large collection of isolates mostly from a single epidemic episode provided key information on the mechanism of acquisition of new spacers and the origin of these spacers while opening the way to a new genotyping approach for epidemiological and phylogenetic studies (Pourcel et al. 2005).

This study was based on the analysis of amplicon size for three CRISPR loci in 182 isolates of which 142 originated from Dalat, Vietnam, during the 1964–1967 epidemic, and sequencing of 109 different alleles. Twenty-six unique spacers were observed for CRISPR Ypa, 14 for CRISPR Ypb, and 5 for CRISPR Ypc (Pourcel et al. 2005). The most variable locus is CRISPR Ypa, which is perhaps linked to the presence of *cas* genes in the immediate vicinity. When alleles were compared it appeared that common spacers were found at one end of the locus near the incomplete last repeat, whereas unique spacers were found at the other end, near the leader sequence. This is clearly shown by comparing the 8 CRISPR1 loci from published genomes, using CRISPRcompar. Spacers 8–12 at the leader end are unique, whereas spacers 1–7 at the trailer are shared by at least two strains. Genotyping by multiple locus VNTR analysis (MLVA) and CRISPR confirmed that strains from the Dalat epidemic in Vietnam with an almost identical MLVA type, could be distinguished by the presence of unique spacers located at the leader end of CRISPR Ypa alleles. The only plausible explanation was that they had been recently added to the CRISPR (Pourcel et al. 2004). This was the first evidence that addition of new spacers in CRISPR was polarized, a distinctive feature of CRISPR locus evolution, which was later confirmed in studies by Lillestol et al. (2006) and Barrangou et al. (2007). This observation is not only essential for understanding the mechanism of spacer acquisition but also to infer phylogenetic relationship between strains. Deletions of spacers on the contrary appear to be randomly distributed.

Since the first report on CRISPR polymorphism in *Y. pestis*, almost four hundred additional isolates have been studied (Cui et al. 2008; Riehm et al. 2012). More than 130 *Y. pestis* spacers have been identified so far among 600 isolates representing almost all known *Y. pestis* foci and including both subspecies *pestis* and *microtus* (Cui et al. 2008; Riehm et al. 2012). Apart from 14 ancestral spacers (6 in CRISPR1, 5 in CRISPR2 and 3 in CRISPR3) a proto-spacer can be found for all the others, majoritarily corresponding to a single prophage sequence, but also to a non-viral region in the chromosome. In most instances 100 % identity between the spacer and the proto-spacer is observed which raises the question of potential autoimmunity. It was suggested that autoimmunity is prevented by mutations in the CRISPR-Cas or in adjacent CRISPR motifs and that it does not constitute a regulatory mechanism (Stern et al. 2010). Interestingly, the proto-spacer-adjacent motif (PAM) (Horvath et al. 2008; Deveau et al. 2008; Mojica et al. 2009) shows a very weak conservation in *Y. pestis* proto-spacers (Cui et al. 2008; Mojica et al. 2009). This however may not be important for self versus non-self discrimination as demonstrated in *S. epidermidis* (Marraffini and Sontheimer 2010). Thus, it is possible that in *Y. pestis*, the CRISPR-Cas system serves another function apart from defense against foreign DNA.

2.3.2.3 *Yersinia pseudotuberculosis*

The three *Y. pestis* CRISPRs can also be found in most *Y. pseudotuberculosis* strains but the diversity of spacers is tremendously higher, reflecting the ancestral nature of the loci in this species and the position of *Y. pestis* within the much larger *Y. pseudotuberculosis* species. Actually, the whole *Y. pestis* species represents a single multilocus sequence type (ST) among 90 other STs uncovered so far in *Y. pseudotuberculosis* (Laukkanen-Ninios et al. 2011). In the initial study by Pourcel et al. (2005) 132 different spacers could be observed in the CRISPR Ypa alleles from 9 *Y. pseudotuberculosis* strains. The sequencing of 20 additional alleles identified 160 new spacers (Pourcel, unpublished results).

2.3.2.4 *Streptococcus pyogenes*

S. pyogenes, also called group A *Streptococcus,* is a species in which phages are the major source of genome diversification, constituting up to 12.4 % of the genome (Beres et al. 2002). Ten out of fifteen strains in available genomes possess one or two CRISPRs. Out of 41 unique spacers, 25 [CRISPR1 (11/17) CRISPR2 (14/24)] match with prophage sequences. Interestingly, a prophage is absent from a strain when a corresponding spacer is present in a CRISPR (Pourcel et al. 2005; Nozawa et al. 2011).

Early on, Hoe et al. (1999) investigated the interest of CRISPR variations for genotyping of *S. pyogenes* by sequencing 31 alleles from serotype M1 strains. Although deletion polymorphism was demonstrated, they showed that the informational value of the assay was lower than sequencing of the streptococcal inhibitor of complement (*sic*) gene. Since then no report of the use of CRISPR for typing of this species has been published.

2.3.2.5 *Campylobacter jejuni*

Campylobacter species, notably *C. jejuni* and *C. coli* are the leading cause of gastroenteritis worldwide. *Campylobacter* populations are characterized by high genetic diversity, weak clonality, and high level of recombination. Many genotyping techniques have been developed, of which multi locus sequence typing (MLST) is currently the leading method since its development by Dingle et al. (2001).

The genome sequence of 5 out of 6 strains contain a single CRISPR at the same locus as shown by CRISPRcompar. In 2003 Schouls et al. genotyped 184 strains with three different techniques, amplified fragment length polymorphism (AFLP), MLST, and sequencing of the CRISPR locus (Schouls et al. 2003a). They showed that 19 out of 184 tested strains did not yield a PCR product and 28 contained a CRISPR locus carrying a single repeat and thus no spacer. In the remaining strains 2–8 repeat-spacer units were found, yet 170 different spacers were detected which represents a high degree of polymorphism. There was a large inter-strain variability and the congruence between MLST, AFLP, and CRISPR typing was good. Because 26 % of strains were not typable by CRISPR sequencing it was concluded that this was not the method of choice for typing, but could be useful rather for subtyping of strains with similar AFLP or MLST profiles. Later, Price et al. (2007) developed a high-resolution DNA melt curve (HRM) analysis of the *C. jejuni* and *C. coli* CRISPR locus. They analyzed the CRISPR locus of 138 isolates containing between 1 and 13 spacers. Sequencing of 32 alleles produced 55 novel unique spacers. The CRISPR HRM genotype was determined for 29 isolates, producing highly reproducible and specific melt profiles. Further 125 isolates were then analyzed, demonstrating the power of the HRM method for discriminating "same" or "different" CRISPR genotypes.

2.3.2.6 *Streptococcus thermophilus* and Other Lactic Acid Bacteria

S. thermophilus is a lactic acid bacterium (LAB) widely used in milk fermentation processes as a starter culture. A deep investigation of CRISPR-Cas systems in 102 LAB genomes revealed the presence of eight distinct families in 46.1 % of strains (Horvath et al. 2009). A large diversity in *cas* genes, repeat and spacers content reflects the lateral origin, and the rapid evolution of CRISPR-Cas systems.

The genetic diversity of *S. thermophilus* has been investigated by different fingerprinting techniques, mostly by Random Amplified Polymorphic DNA (RAPD) and more recently by AFLP but these techniques produce genotypes that cannot be easily compared (Lazzi et al. 2009). Because of the commercial importance of this species it is necessary to generate comparable genotypes to identify genetic signatures that characterize specific strains. In that respect, the use of CRISPR polymorphism could be relevant. *S. thermophilus* genome sequences possess 1–4 CRISPR-Cas systems. CRISPR1 was found in all 124 strains analyzed whereas CRISPR2 was found in 59 out of 65 strains and CRISPR3 in 53 out of 66 strains (Horvath et al. 2008). A total of 39.5 % isolates carried all three loci. CRISPR1 shows the highest spacer diversity, followed by CRISPR3, due to internal deletions of spacers and additions at the leader end. Clustering of strains according to their spacer content can help in reconstructing phylogeny in this species. The authors suggest that the dynamic nature of CRISPR loci is potentially valuable for typing and comparative analysis of strains. Furthermore, the fact that *S. thermophilus* acquires new spacers at a high frequency upon challenge by phage infection allows the selection of multiresistant strains that show new and easily detectable genetic elements (Deveau et al. 2008). CRISPR-Cas systems have been analyzed in other species of LAB, notably in *Lactobacillus* (*Lb. acidophilus*, *Lb. casei*, *Lb. delbrueckii*, *Lb. paracasei*, *Lb. rhamnosus*, *and Lb. salivarius*), and *Bifidobacteria.*

2.3.2.7 *Corynebacterium diphtheriae*

Genotyping of *C. diphtheriae* by different methods and more recently by MLST has revealed a significant intraspecies diversity and the existence of clones although recombination hinders the structure of the population (Bolt et al. 2010). Strain NCTC 13129 which genome has been sequenced, possesses two CRISPRs, one with a 36 bp repeat and 7 spacers and another with a 29 bp repeat and 27 spacers, each locus being associated with a set of *cas* genes. Mokrousov et al. first described a spoligotyping method for this species making use of the polymorphism at the two CRISPR loci called DRA and DRB (Mokrousov et al. 2005, 2007). A reverse hybridization macroarray-based assay similar to the *M. tuberculosis* spoligotyping method was developed to study both DRA and DRB. A total number of 27 spacers (21 from DRB and 6 from DRA) were investigated, allowing to subdivide 156 strains of the 1990s ‘Russian epidemic clone’ into 45 spoligotypes. Later, 20 *C. diphteriae* biotype gravis strains collected in Belarus in 2005 in a suspected epidemic foci and showing the same ribotype were investigated by this method, displaying three different spoligotypes (Mokrousov et al. 2009). This confirmed that spoligotyping provides additional discrimination as compared to MLST. To generalize the method, it would be necessary to sequence alleles from more strains of diverse origin, in order to assess the polymorphism of existing loci and to determine whether there are evidences of spacer acquisition at one end of the loci (Mokrousov 2009).

2.3.2.8 *Escherichia coli*

Two CRISPR-Cas systems can be found in *E. coli,* one I-E (Ecoli) subtype (CRISPR2) and one I-F (Ypest) subtype (CRISPR4) (Haft et al. 2005; Makarova et al. 2011a). The presence and diversity of several CRISPRs belonging to the two systems was investigated by Diez-Villasenor et al. in a total of 100 strains representative of the species (including 28 sequenced genomes and 72 strains of the reference collection ECOR) (Diez-Villasenor et al. 2010). Sequencing of CRISPR2.1 and CRISPR2.3 spacers defined 58 and 52 alleles, respectively. Of 153 spacers analyzed in strains possessing the Type I-Ft CRISPR-Cas system, 100 were unique (47 out of 73 in CRISPR4.1 and 53 out of 80 in CRISPR4.2). Comparison of alleles allowed the clustering of strains possessing common spacers, but in the absence of data from another genotyping technique it was not possible to evaluate the informativity of CRISPR typing.

Another study (Touchon et al. 2011) investigated 263 strains and 27 sequenced genomes. The diversity of several Type I-E (CRISPR2) loci was assessed and compared to the phylogeny derived from MLST. A complete lack of CRISPR was observed in strains of the phylogenetic group B2, a major source of extra intestinal infection. CRISPRs shared common spacers within MLST groups and diversity was observed for example within clonal group C, although it appears that deletion rather than acquisition of new spacers was the source of polymorphism. Because there is no exact correlation between CRISPR arrangement and MLST grouping, probably related to horizontal transfer, CRISPR typing cannot be used as a general typing method for *E. coli,* but it could be useful in association with MLST to differentiate strains from a single clonal group, as illustrated for the C group.

2.3.2.9 *Pseudomonas aeruginosa*

The population structure of *P. aeruginosa* has been described as panmictic/epidemic to reflect the fact that only a few clones can be identified, related to antibiotic resistance or linked to specific clinical conditions such as cystic fibrosis (Romling et al. 2005). According to available genome sequence data, two CRISPR-Cas systems are observed: one Type I-F (Ypest) in the reference strain

UCBPP-PA14 and a Type I-E (Ecoli) in reference strain 2,192. The prevalence of these two subtypes was determined in collections of clinical isolates from different countries [unpublished and (Cady et al. 2011)]. In the work of Cady et al., 122 clinical isolates were investigated by amplification of *csy1* (Type I-F) and *cse3* (Type I-E) using PCR primers derived, respectively from strains PA14 and PA2192. Forty out of 122 isolates putatively harbor Type I-F and 6 % Type I-E. In all instances a single localization was found in the complete genome showing that the locus has not been inserted several times independently. Sequencing of all the loci resulted in 656 unique spacers. Among the spacers that showed 100 % identity to non-CRISPR sequences, 65 independent spacers were identical to lysogenic *P. aeruginosa* bacteriophages. We observed similar percentages of the two subtypes in a collection of 200 French isolates and also found a majority of spacers that match with lysogenic phage DNA. To determine whether CRISPR polymorphism could be used for genotyping, we compared the distribution of isolates possessing CRISPR-Cas systems to the clustering obtained using MLVA. In isolates genetically linked to strain PA14 and in all isolates from clone C found in cystic fibrosis patients (Romling et al. 2005), the Type I-FCRISPR-Cas system was found. The CRISPRs polymorphism in these clones allowed fine subtyping and also some phylogenetic reconstruction.

2.3.2.10 *Salmonella enterica*

Multiple serovars of *Salmonella enterica* subsp. *enterica* are associated with food-borne infection. Molecular techniques with high discriminatory power are necessary to investigate outbreaks. In the 15 available genome sequences 1–3 CRISPRs are detected. In a study of 28 sequenced genomes Fricke et al. (2011) investigated the polymorphism of these structures and observed a considerable variability which in part reflected the phylogeny of the species.

Liu et al. (2011a) described an "MLST" scheme in which they combined the sequence analysis of virulence genes *sseL* and *fimH* with that of two CRISPR loci. This assay was applied to the genotyping of 171 clinical isolates from nine *Salmonella* serovars. CRISPR profiles were converted into a CRISPR type and treated as an allele into the MLST scheme. Outbreak strains/clones could be differentiated by addition of CRISPR sequences as compared to using virulence genes only. Investigation of CRISPR polymorphism provided better discrimination of *Salmonella* serovar Enteritidis than PFGE and showed high epidemiologic concordance for all serovars screened except Muenchen. Later, these authors characterized 168 *Salmonella* serovar Enteritidis isolates using the assay now called CRISPR-MVLST to differentiate it from classical MLST (based on 7 housekeeping genes) leading to 27 sequence types (Liu et al. 2011b).

At present, several teams are investigating the polymorphism of CRISPRs in *Salmonella* and developing new hybridization-based assays for genotyping, which could complement the currently used methods. A database of *S. enterica* spacers is available for Blast at the Pasteur Institute http://www.pasteur.fr/recherche/genopole/PF8/crispr/CRISPRDB.html (Fabre et al. 2012).

The CRISPRs in *S. typhimurium* also show a high level of polymorphism which is being used for genotyping (Fabre et al. 2012).

2.3.2.11 *Erwinia amylovora*

E. amylovora, a phytopathogenic bacterium causing fire blight, has relatively low genetic diversity within the species. Commonly used genotyping methods provide poor discrimination of strains within local infested region (Rezzonico et al. 2011). Three CRISPR loci are present in the genome sequence of strain CFBP1430. A total number of 454 unique spacers were identified from the three CRISPR loci among 37 *E. amylovora* isolates (Rezzonico et al. 2011). The shortest CRISPR locus with 5 spacers was almost invariant. When combining the result for all three loci, 18 genotypes were identified. In this work, MEGA version 4.0 was used to infer phylogenetic relationships based on spacers present in strains (Tamura et al. 2007). McGeeh et al. identified 588 individual spacers among 85 isolates within the three CRISPR arrays present in *E. amylovora* (McGhee and Sundin 2012) and defined 28 distinct genotypes. The shortest locus with 5 spacers was invariant as shown in the study by Rezzonico et al. (2011), whereas variability was observed with the other two loci. CRISPR genotyping enabled the differentiation of strains that were shown, in previous studies, to belong to the same genotype using other methods. Furthermore, cluster analysis revealed the similarities and differences among isolates related to geographic source and host isolation.

2.3.2.12 Other Species

In *Mycoplasma gallisepticum* (Delaney et al. 2012), *S. agalactiae*, (Lopez-Sanchez et al. 2012), and *Microcystis aeruginosa* (Kuno et al. 2012), CRISPR locus variability offers new possibilities to perform population structure analysis. Based on our current understanding of CRISPR implementation for typing purposes, CRISPR polymorphism is being investigated in *L. pneumophila* and *Acinetobacter baumannii* to subdivide strains with similar MLST or MLVA genotype. Indeed some of the major *L. pneumophila* clonal complexes including Paris, Lens, and Corby possess one or two of Type I-E (Ecoli) or Type I-F (Ypest) CRISPR-Cas systems and spacer polymorphism can be observed that provide additional discriminatory power to the current genotyping methods (Ginevra et al. 2012). Likewise, clonal complex AYE in *A. baumannii* CRISPR shows spacer polymorphism which might be useful for subtyping (Hauck et al. PLoS one 2012).

2.3.3 Follow-up of CRISPR Diversity in Complex Microbiomes

The development of metagenome analyses provides increasing information about virus population dynamics and interaction with bacterial population, such as for example in acidophilic microbial biofilms (Tyson and Banfield 2008; Andersson and Banfield 2008), a microbial mat in hotsprings (Heidelberg et al. 2009), the oral cavity of a rat (van der Ploeg 2009), the ocean (Sorokin et al. 2010), the human gut (Minot et al. 2011), or in the rumen microbiome (Berg Miller et al. 2011). The results of these studies showed that CRISPR polymorphism reflect virus encounters, acquisition of new spacers, and locus evolution. As more metagenomic studies get underway, we anticipate that investigating CRISPR polymorphism will provide insights into microbial population structures, and their interplay with predatory viruses.

2.4 Discussion

The diversity of repeats, spacers and *cas* genes is amazing considering that the primary function of the system seems to be resistance against invasive DNA (Horvath and Barrangou 2010).

The presence of CRISPR-Cas immune systems in members of two of the three domains of life questions its origin and evolution. An evolutionary scenario based on the analysis of Cas protein families proposes that the system originated in thermophilic Archaea, and spread horizontally to bacteria, but numerous unanswered questions remain (Makarova et al. 2007, 2011b). It is also possible that such a primitive system existed in the last universal common ancestor LUCA (Glansdorff et al. 2008). Because the CRISPR-Cas systems evolve in response to pressures from invasive DNA and are deeply affected by horizontal transfer, it has been suggested that they function like a *bona fide* Lamarckian mechanism (Koonin and Wolf 2009).

Investigation of intra-species CRISPR-Cas systems polymorphism provides some clues on their evolution while sometimes constituting new genotyping tools. The different examples described above show that CRISPR polymorphism is elevated in some species, and can be exploited for rapid genotyping of even closely related strains, but cannot be the sole source of genetic diversity for bacterial genotyping. In some cases, it provides a rapid means to assign a strain to a phylogenetic group, or to identify a new branch. Analysis of CRISPR diversity in strains of species with a long history of evolution identifies large collections of spacers. On the contrary, inside a clonal complex, it appears that CRISPR variability may provide additional information for genotyping. For recently emerged species such as *M. tuberculosis* or *Y. pestis,* which have the complexity of a clonal complex, CRISPR typing provides important phylogenetic information. In many species in which only a fraction of the strains possess a CRISPR, it may be a valuable marker to identify subgroups of strains. Overall, it is necessary to increase the knowledge of intraspecies diversity to better understand the evolution rate of these structures, both by deletion or gain of spacers. In certain species this will depend on the selection forces applied by invasive DNA. In others, on the contrary, a CRISPR may just be slowly losing its spacers via internal deletion(s). When several CRISPRs are present, the locus next to a cluster of *cas* genes may be more active in terms of spacer acquisition, whereas loss of spacers by deletion may be similar (Pourcel et al. 2005; Horvath et al. 2008). This will have important consequences for phylogenetic studies.

In the future, the analysis of huge amounts of sequencing data from isolated microorganisms or from complex microbiomes will constitute a challenge. New bioinformatics tools will be necessary to identify and classify CRISPR elements and alleles. Efforts to maintain up-to-date databases will be needed in order to provide the community with high quality information.

Acknowledgments We thank Gilles Vergnaud for his helpful comments.

References

Andersson AF, Banfield JF (2008) Virus population dynamics and acquired virus resistance in natural microbial communities. Science 320:1047–1050

Bailey TL, Elkan C (1994) Fitting a mixture model by expectation maximization to dicover motifs in biopolymers. In: Proceedings of the second international conference on intelligent systems for molecular biology. pp 28–36

Barrangou R, Fremaux C, Deveau H, Richards M, Boyaval P, Moineau S, Romero DA, Horvath P (2007) CRISPR provides acquired resistance against viruses in prokaryotes. Science 315:1709–1712

Beres SB, Sylva GL, Barbian KD, Lei B, Hoff JS, Mammarella ND, Liu MY, Smoot JC, Porcella SF, Parkins LD, Campbell DS, Smith TM, McCormick JK, Leung DY, Schlievert PM, Musser JM (2002) Genome sequence of a serotype M3 strain of group A *Streptococcus*: phage-encoded toxins, the high-virulence phenotype, and clone emergence. Proc Natl Acad Sci USA 99:10078–10083

Berg Miller ME, Yeoman CJ, Chia N, Tringe SG, Angly FE, Edwards RA, Flint HJ, Lamed R, Bayer EA,and White BA (2011) Phage-bacteria relationships and CRISPR elements revealed by a metagenomic survey of the rumen microbiome. Environ Microbiol 14:207–227

Bland C, Ramsey TL, Sabree F, Lowe M, Brown K, Kyrpides NC, Hugenholtz P (2007) CRISPR recognition tool (CRT): a tool for automatic detection of clustered regularly interspaced palindromic repeats. BMC Bioinform 8:209

Bolotin A, Quinquis B, Sorokin A, Ehrlich SD (2005) Clustered regularly interspaced short palindrome repeats (CRISPRs) have spacers of extrachromosomal origin. Microbiology 151:2551–2561

Bolt F, Cassiday P, Tondella ML, Dezoysa A, Efstratiou A, Sing A, Zasada A, Bernard K, Guiso N, Badell E, Rosso ML, Baldwin A, Dowson C (2010) Multilocus sequence typing identifies evidence for recombination and two distinct lineages of *Corynebacterium diphtheriae*. J Clin Microbiol 48:4177–4185

Bos KI, Schuenemann VJ, Golding GB, Burbano HA, Waglechner N, Coombes BK, McPhee JB, DeWitte SN, Meyer M, Schmedes S, Wood J, Earn DJ, Herring DA, Bauer P, Poinar HN, Krause J (2011) A draft genome of *Yersinia pestis* from victims of the Black Death. Nature 478:506–510

Brouns SJ, Jore MM, Lundgren M, Westra ER, Slijkhuis RJ, Snijders AP, Dickman MJ, Makarova KS, Koonin EV, van der Oost J (2008) Small CRISPR RNAs guide antiviral defense in prokaryotes. Science 321:960–964

Brudey K, Driscoll JR, Rigouts L, Prodinger WM, Gori A, Al-Hajoj SA, Allix C, Aristimuno L, Arora J, Baumanis V, Binder L, Cafrune P, Cataldi A, Cheong S, Diel R, Ellermeier C, Evans JT, Fauville-Dufaux M, Ferdinand S, Garcia de Viedma D, Garzelli C, Gazzola L, Gomes HM, Guttierez MC, Hawkey PM, van Helden PD, Kadival GV, Kreiswirth BN, Kremer K, Kubin M, Kulkarni SP, Liens B, Lillebaek T, Ho ML, Martin C, Martin C, Mokrousov I, Narvskaia O, Ngeow YF, Naumann L, Niemann S, Parwati I, Rahim Z, Rasolofo-Razanamparany V, Rasolonavalona T, Rossetti ML, Rusch-Gerdes S, Sajduda A, Samper S, Shemyakin IG, Singh UB, Somoskovi A, Skuce RA, van Soolingen D, Streicher EM, Suffys PN, Tortoli E, Tracevska T, Vincent V, Victor TC, Warren RM, Yap SF, Zaman K, Portaels F, Rastogi N, Sola C (2006) Mycobacterium tuberculosis complex genetic diversity: mining the fourth international spoligotyping database (SpolDB4) for classification, population genetics and epidemiology. BMC Microbiol 6:23

Cady KC, White AS, Hammond JH, Abendroth MD, Karthikeyan RS, Lalitha P, Zegans ME, O'Toole GA (2011) Prevalence, conservation and functional analysis of Yersinia and Escherichia CRISPR regions in clinical *Pseudomonas aeruginosa* isolates. Microbiology 157:430–437

Carte J, Wang R, Li H, Terns RM, Terns MP (2008) *Cas*6 is an endoribonuclease that generates guide RNAs for invader defense in prokaryotes. Genes Dev 22:3489–3496

Chakraborty S, Snijders AP, Chakravorty R, Ahmed M, Tarek AM, Hossain MA (2010) Comparative network clustering of direct repeats (DRs) and cas genes confirms the possibility of the horizontal transfer of CRISPR locus among bacteria. Mol Phylogenet Evol 56:878–887

Comas I, Homolka S, Niemann S, Gagneux S (2009) Genotyping of genetically monomorphic bacteria: DNA sequencing in *Mycobacterium tuberculosis* highlights the limitations of current methodologies. PLoS ONE 4:e7815

Cui Y, Li Y, Gorge O, Platonov ME, Yan Y, Guo Z, Pourcel C, Dentovskaya SV, Balakhonov SV, Wang X, Song Y, Anisimov AP, Vergnaud G, Yang R (2008) Insight into microevolution of *Yersinia pestis* by clustered regularly interspaced short palindromic repeats. PLoS ONE 3: e2652

Delaney NF, Balenger S, Bonneaud C, Marx CJ, Hill GE, Ferguson-Noel N, Tsai P, Rodrigo A, Edwards SV (2012) Ultrafast evolution and loss of CRISPRs following a host shift in a novel wildlife pathogen. *Mycoplasma gallisepticum* PLoS Genet 8:e1002511

Deltcheva E, Chylinski K, Sharma CM, Gonzales K, Chao Y, Pirzada ZA, Eckert MR, Vogel J, Charpentier E (2011) CRISPR RNA maturation by trans-encoded small RNA and host factor RNase III. Nature 471:602–607

Deveau H, Barrangou R, Garneau JE, Labonte J, Fremaux C, Boyaval P, Romero DA, Horvath P, Moineau S (2008) Phage response to CRISPR-encoded resistance in *Streptococcus thermophilus*. J Bacteriol 190:1390–1400

Diez-Villasenor C, Almendros C, Garcia-Martinez J, Mojica FJ (2010) Diversity of CRISPR loci in *Escherichia coli*. Microbiology 156:1351–1361

Dingle KE, Colles FM, Wareing DR, Ure R, Fox AJ, Bolton FE, Bootsma HJ, Willems RJ, Urwin R, Maiden MC (2001) Multilocus sequence typing system for *Campylobacter jejuni*. J Clin Microbiol 39:14–23

Durand P, Mahe F, Valin AS, Nicolas J (2006) Browsing repeats in genomes: Pygram and an application to non-coding region analysis. BMC Bioinform 7:477

Edgar RC (2007) PILER-CR: fast and accurate identification of CRISPR repeats. BMC Bioinform 8:18

Edgar RC (2010) Search and clustering orders of magnitude faster than BLAST. Bioinformatics 26:2460–2461

Edgar R, Qimron U (2010) The *Escherichia coli* CRISPR system protects from lambda lysogenization, lysogens, and prophage induction. J Bacteriol 192:6291–6294

Fabre M, Koeck JL, Le Fleche P, Simon F, Herve V, Vergnaud G, Pourcel C (2004) High genetic diversity revealed by variable-number tandem repeat genotyping and analysis of hsp65 gene polymorphism in a large collection of *Mycobacterium canettii* strains indicates that the *M. tuberculosis* complex is a recently emerged clone of *M. canettii*. J Clin Microbiol 42:3248–3255

Fabre M, Hauck Y, Soler C, Koeck JL, van Ingen J, van Soolingen D, Vergnaud G, Pourcel C (2010) Molecular characteristics of "*Mycobacterium canettii*" the smooth *Mycobacterium tuberculosis* bacilli. Infect Genet Evol 10:1165–1173

Fabre L, Zhang J, Guigon G, Le Hello S, Guibert V, Accou-Demartin M, de Romans S, Lim C, Roux C, Passet V, Diancourt L, Guibourdenche M, Issenhuth-Jeanjean S, Achtman M, Brisse S, Sola C, Weill FX (2012) CRISPR typing and subtyping for improved laboratory surveillance of Salmonella infections. PLoS ONE 7:e36995

Filliol I, Driscoll JR, van Soolingen D, Kreiswirth BN, Kremer K, Valetudie G, Dang DA, Barlow R, Banerjee D, Bifani PJ, Brudey K, Cataldi A, Cooksey RC, Cousins DV, Dale JW, Dellagostin OA, Drobniewski F, Engelmann G, Ferdinand S, Gascoyne-Binzi D, Gordon M, Gutierrez MC, Haas WH, Heersma H, Kassa-Kelembho E, Ho ML, Makristathis A, Mammina C, Martin G, Mostrom P, Mokrousov I, Narbonne V, Narvskaya O, Nastasi A, Niobe-Eyangoh SN, Pape JW, Rasolofo-Razanamparany V, Ridell M, Rossetti ML, Stauffer F, Suffys PN, Takiff H, Texier-Maugein J, Vincent V, de Waard JH, Sola C, Rastogi N (2003) Snapshot of moving and expanding clones of *Mycobacterium tuberculosis* and their global distribution assessed by spoligotyping in an international study. J Clin Microbiol 41:1963–1970

Fricke WF, Mammel MK, McDermott PF, Tartera C, White DG, Leclerc JE, Ravel J, Cebula TA (2011) Comparative genomics of 28 *Salmonella enterica* isolates: evidence for CRISPR-mediated adaptive sublineage evolution. J Bacteriol 193:3556–3568

Gill SR, Fouts DE, Archer GL, Mongodin EF, Deboy RT, Ravel J, Paulsen IT, Kolonay JF, Brinkac L, Beanan M, Dodson RJ, Daugherty SC, Madupu R, Angiuoli SV, Durkin AS, Haft DH, Vamathevan J, Khouri H, Utterback T, Lee C, Dimitrov G, Jiang L, Qin H, Weidman J, Tran K, Kang K, Hance IR, Nelson KE, Fraser CM (2005) Insights on evolution of virulence and resistance from the complete genome analysis of an early methicillin-resistant *Staphylococcus aureus* strain and a biofilm-producing methicillin-resistant *Staphylococcus epidermidis* strain. J Bacteriol 187:2426–2438

Ginevra C, Jacotin N, Diancourt L, Guigon G, Arquilliere R, Meugnier H, Descours G, Vandenesch F, Etienne J, Lina G, Caro V, Jarraud S (2012) *Legionella pneumophila* sequence type 1/Paris pulsotype subtyping by spoligotyping. J Clin Microbiol 50:696–701

Glansdorff N, Xu Y, Labedan B (2008) The last universal common ancestor: emergence, constitution and genetic legacy of an elusive forerunner. Biol Direct 3:29

Godde JS, Bickerton A (2006) The repetitive DNA elements called CRISPRs and their associated genes: evidence of horizontal transfer among prokaryotes. J Mol Evol 62:718–729

Golding GR, Bryden L, Levett PN, McDonald RR, Wong A, Wylie J, Graham MR, Tyler S, Van Domselaar G, Simor AE, Gravel D, Mulvey MR (2010) Livestock-associated methicillin-resistant *Staphylococcus aureus* sequence type 398 in humans. Canada Emerg Infect Dis 16:587–594

Grissa I, Vergnaud G, Pourcel C (2007a) The CRISPRdb database and tools to display CRISPRs and to generate dictionaries of spacers and repeats. BMC Bioinform 8:172

Grissa I, Vergnaud G, Pourcel C (2007b) CRISPRFinder: a web tool to identify clustered regularly interspaced short palindromic repeats. Nucleic Acids Res 35:W52–W57

Grissa I, Bouchon P, Pourcel C, Vergnaud G (2008a) On-line resources for bacterial micro-evolution studies using MLVA or CRISPR typing. Biochimie 90:660–668

Grissa I, Vergnaud G, Pourcel C (2008b) CRISPRcompar: a website to compare clustered regularly interspaced short palindromic repeats. Nucleic Acids Res 36:W145–W148

Grissa I, Vergnaud G, Pourcel C (2009) Clustered regularly interspaced short palindromic repeats (CRISPRs) for the genotyping of bacterial pathogens. Methods Mol Biol 551:105–116

Groenen PM, Bunschoten AE, van Soolingen D, van Embden JD (1993) Nature of DNA polymorphism in the direct repeat cluster of *Mycobacterium tuberculosis;* application for strain differentiation by a novel typing method. Mol Microbiol 10:1057–1065

Haft DH, Selengut J, Mongodin EF, Nelson KE (2005) A Guild of 45 CRISPR-Associated (Cas) Protein Families and Multiple CRISPR/Cas Subtypes Exist in Prokaryotic Genomes. PLoS Comput Biol 1:e60

Hauck Y, Soler C, Jault P, Merens A, Gerome P, Nab CM, Trueba F, Bargues L, Thien HV, Vergnaud G , Pourcel C. (2012) Diversity of acinetobacter baumannii in four french military hospitals, as assessed by multiple locus variable number of tandem repeats analysis. PLoS One 7:e44597

Heidelberg JF, Nelson WC, Schoenfeld T, Bhaya D (2009) Germ warfare in a microbial mat community: CRISPRs provide insights into the co-evolution of host and viral genomes. PLoS ONE 4:e4169

Hoe N, Nakashima K, Grigsby D, Pan X, Dou SJ, Naidich S, Garcia M, Kahn E, Bergmire-Sweat D, Musser JM (1999) Rapid molecular genetic subtyping of serotype M1 group A *Streptococcus* strains. Emerg Infect Dis 5:254–263

Holt DC, Holden MT, Tong SY, Castillo-Ramirez S, Clarke L, Quail MA, Currie BJ, Parkhill J, Bentley SD, Feil EJ, Giffard PM (2011) A Very Early-Branching *Staphylococcus aureus* Lineage Lacking the Carotenoid Pigment Staphyloxanthin. Genome Biol Evol 3:881–895

Horvath P, Barrangou R (2010) CRISPR/Cas, the immune system of bacteria and archaea. Science 327:167–170

Horvath P, Romero DA, Coute-Monvoisin AC, Richards M, Deveau H, Moineau S, Boyaval P, Fremaux C, Barrangou R (2008) Diversity, activity, and evolution of CRISPR loci in Streptococcus thermophilus. J Bacteriol 190:1401–1412

Horvath P, Coute-Monvoisin AC, Romero DA, Boyaval P, Fremaux C, Barrangou R (2009) Comparative analysis of CRISPR loci in lactic acid bacteria genomes. Int J Food Microbiol 131:62–70

Jansen R, Embden JD, Gaastra W, Schouls LM (2002) Identification of genes that are associated with DNA repeats in prokaryotes. Mol Microbiol 43:1565–1575

Jeffreys AJ, MacLeod A, Tamaki K, Neil DL, Monckton DG (1991) Minisatellite repeat coding as a digital approach to DNA typing. Nature 354:204–209

Kamerbeek J, Schouls L, Kolk A, van Agterveld M, van Soolingen D, Kuijper S, Bunschoten A, Molhuizen H, Shaw R, Goyal M, van Embden J (1997) Simultaneous detection and strain differentiation of *Mycobacterium tuberculosis* for diagnosis and epidemiology. J Clin Microbiol 35:907–914

Koonin EV, Wolf YI (2009) Is evolution Darwinian or/and Lamarckian? Biol Direct 4:42

Kunin V, Sorek R, Hugenholtz P (2007) Evolutionary conservation of sequence and secondary structures in CRISPR repeats. Genome Biol 8:R61

Kuno S, Yoshida T, Kaneko T, Sako Y (2012) Intricate interactions between the bloom-forming cyanobacterium *Microcystis aeruginosa* and foreign genetic elements, revealed by diversified clustered regularly interspaced short palindromic repeat (CRISPR) signatures. Appl Environ Microbiol 78:5353–5360

Kurtz S, Choudhuri JV, Ohlebusch E, Schleiermacher C, Stoye J, Giegerich R (2001) REPuter: the manifold applications of repeat analysis on a genomic scale. Nucleic Acids Res 29: 4633–4642

Laukkanen-Ninios R, Didelot X, Jolley KA, Morelli G, Sangal V, Kristo P, Brehony C, Imori PF, Fukushima H, Siitonen A, Tseneva G, Voskressenskaya E, Falcao JP, Korkeala H, Maiden MC, Mazzoni C, Carniel E, Skurnik M, Achtman M (2011) Population structure of the *Yersinia pseudotuberculosis* complex according to multilocus sequence typing. Environ Microbiol 13:3114–3127

Lazzi C, Bove CG, Sgarbi E, Gatti M, La Gioia F, Torriani S, Neviani E (2009) Application of AFLP fingerprint analysis for studying the biodiversity of *Streptococcus thermophilus*. J Microbiol Methods 79:48–54

Lillestol RK, Redder P, Garrett RA, Brugger K (2006) A putative viral defence mechanism in archaeal cells. Archaea 2:59–72

Liu F, Barrangou R, Gerner-Smidt P, Ribot EM, Knabel SJ, Dudley EG (2011a) Novel virulence gene and clustered regularly interspaced short palindromic repeat (CRISPR) multilocus sequence typing scheme for subtyping of the major serovars of *Salmonella enterica* subsp. *enterica*. Appl Environ Microbiol 77:1946–1956

Liu F, Kariyawasam S, Jayarao BM, Barrangou R, Gerner-Smidt P, Ribot EM, Knabel SJ, Dudley EG (2011b) Subtyping *Salmonella enterica* serovar *enteritidis* isolates from different sources by using sequence typing based on virulence genes and clustered regularly interspaced short palindromic repeats (CRISPRs). Appl Environ Microbiol 77:4520–4526

Lopez-Sanchez MJ, Sauvage E, Da Cunha V, Clermont D, Ratsima Hariniaina E, Gonzalez-Zorn B, Poyart C, Rosinski-Chupin I,and Glaser P (2012) The highly dynamic CRISPR1 system of *Streptococcus agalactiae* controls the diversity of its mobilome. Mol Microbiol 85:1057–1071

Makarova KS, Sorokin AV, Novichkov PS, Wolf YI, Koonin EV (2007) Clusters of orthologous genes for 41 archaeal genomes and implications for evolutionary genomics of archaea. Biol Direct 2:33

Makarova KS, Haft DH, Barrangou R, Brouns SJ, Charpentier E, Horvath P, Moineau S, Mojica FJ, Wolf YI, Yakunin AF, van der Oost J, Koonin EV (2011a) Evolution and classification of the CRISPR-Cas systems. Nat Rev Microbiol 9:467–477

Makarova KS, Aravind L, Wolf YI, Koonin EV (2011b) Unification of Cas protein families and a simple scenario for the origin and evolution of CRISPR-Cas systems. Biol Direct 6:38

Marraffini LA, Sontheimer EJ (2010) Self versus non-self discrimination during CRISPR RNA-directed immunity. Nature 463:568–571

McGhee GC, Sundin GW (2012) *Erwinia amylovora* CRISPR Elements Provide New Tools for Evaluating Strain Diversity and for Microbial Source Tracking. PLoS ONE 7:e41706

Minot S, Sinha R, Chen J, Li H, Keilbaugh SA, Wu GD, Lewis JD, Bushman FD (2011) The human gut virome: inter-individual variation and dynamic response to diet. Genome Res 21:1616–1625

Mojica FJ, Diez-Villasenor C, Soria E, Juez G (2000) Biological significance of a family of regularly spaced repeats in the genomes of archaea bacteria and mitochondria. Mol Microbiol 36:244–246

Mojica FJ, Diez-Villasenor C, Garcia-Martinez J, Soria E (2005) Intervening sequences of regularly spaced prokaryotic repeats derive from foreign genetic elements. J Mol Evol 60:174–182

Mojica FJ, Diez-Villasenor C, Garcia-Martinez J, Almendros C (2009) Short motif sequences determine the targets of the prokaryotic CRISPR defence system. Microbiology 155:733–740

Mokrousov I (2009) *Corynebacterium diphtheriae*: genome diversity, population structure and genotyping perspectives. Infect Genet Evol 9:1–15

Mokrousov I, Narvskaya O, Limeschenko E, Vyazovaya A (2005) Efficient discrimination within a *Corynebacterium diphtheriae* epidemic clonal group by a novel macroarray-based method. J Clin Microbiol 43:1662–1668

Mokrousov I, Limeschenko E, Vyazovaya A, Narvskaya O (2007) *Corynebacterium diphtheriae* spoligotyping based on combined use of two CRISPR loci. Biotechnol J 2:901–906

Mokrousov I, Vyazovaya A, Kolodkina V, Limeschenko E, Titov L, Narvskaya O (2009) Novel macroarray-based method of Corynebacterium diphtheriae genotyping: evaluation in a field study in Belarus. Eur J Clin Microbiol Infect Dis 28:701–703

Morelli G, Song Y, Mazzoni CJ, Eppinger M, Roumagnac P, Wagner DM, Feldkamp M, Kusecek B, Vogler AJ, Li Y, Cui Y, Thomson NR, Jombart T, Leblois R, Lichtner P, Rahalison L, Petersen JM, Balloux F, Keim P, Wirth T, Ravel J, Yang R, Carniel E, Achtman M (2010) *Yersinia pestis* genome sequencing identifies patterns of global phylogenetic diversity. Nat Genet 42:1140–1143

Nozawa T, Furukawa N, Aikawa C, Watanabe T, Haobam B, Kurokawa K, Maruyama F, Nakagawa I (2011) CRISPR inhibition of prophage acquisition in *Streptococcus pyogenes*. PLoS ONE 6:e19543

Pourcel C, Andre-Mazeaud F, Neubauer H, Ramisse F, Vergnaud G (2004) Tandem repeats analysis for the high resolution phylogenetic analysis of *Yersinia pestis*. BMC Microbiol 4:22

Pourcel C, Salvignol G, Vergnaud G (2005) CRISPR elements in *Yersinia pestis* acquire new repeats by preferential uptake of bacteriophage DNA, and provide additional tools for evolutionary studies. Microbiology 151:653–663

Price EP, Smith H, Huygens F, Giffard PM (2007) High-resolution DNA melt curve analysis of the clustered, regularly interspaced short-palindromic-repeat locus of *Campylobacter jejuni*. Appl Environ Microbiol 73:3431–3436

Rezzonico F, Smits TH, Duffy B (2011) Diversity, evolution, and functionality of clustered regularly interspaced short palindromic repeat (CRISPR) regions in the fire blight pathogen Erwinia amylovora. Appl Environ Microbiol 77:3819–3829

Riehm JM, Vergnaud G, Kiefer D, Damdindorj T, Dashdavaa O, Khurelsukh T, Zoller L, Wolfel R, Le Fleche P, Scholz HC (2012) *Yersinia pestis* lineages in Mongolia. PLoS ONE 7:e30624

Romling U, Kader A, Sriramulu DD, Simm R, Kronvall G (2005) Worldwide distribution of *Pseudomonas aeruginosa* clone C strains in the aquatic environment and cystic fibrosis patients. Environ Microbiol 7:1029–1038

Rousseau C, Gonnet M, Le Romancer M, Nicolas J (2009) CRISPI: a CRISPR interactive database. Bioinformatics 25:3317–3318

Schouls LM, Reulen S, Duim B, Wagenaar JA, Willems RJ, Dingle KE, Colles FM, Van Embden JD (2003) Comparative genotyping of *Campylobacter jejuni* by amplified fragment length polymorphism, multilocus sequence typing, and short repeat sequencing: strain diversity, host range, and recombination. J Clin Microbiol 41:15–26

Shah SA, Garrett RA (2011) CRISPR/Cas and Cmr modules, mobility and evolution of adaptive immune systems. Res Microbiol 162:27–38

Sonnhammer EL, Durbin R (1995) A dot-matrix program with dynamic threshold control suited for genomic DNA and protein sequence analysis. Gene 167:GC1–GC10

Sorokin VA, Gelfand MS, Artamonova II (2010) Evolutionary dynamics of clustered irregularly interspaced short palindromic repeat systems in the ocean metagenome. Appl Environ Microbiol 76:2136–2144

Stern A, Keren L, Wurtzel O, Amitai G, Sorek R (2010) Self-targeting by CRISPR: gene regulation or autoimmunity? Trends Genet 26:335–340

Tamura K, Dudley J, Nei M, Kumar S (2007) MEGA4: Molecular Evolutionary Genetics Analysis (MEGA) software version 4.0. Mol Biol Evol 24:1596–1599

Touchon M, Charpentier S, Clermont O, Rocha EP, Denamur E, Branger C (2011) CRISPR distribution within the *Escherichia coli* species is not suggestive of immunity-associated diversifying selection. J Bacteriol 193:2460–2467

Tyson GW, Banfield JF (2008) Rapidly evolving CRISPRs implicated in acquired resistance of microorganisms to viruses. Environ Microbiol 10:200–207

van der Oost J, Jore MM, Westra ER, Lundgren M, Brouns SJ (2009) CRISPR-based adaptive and heritable immunity in prokaryotes. Trends Biochem Sci 34:401–407

van der Ploeg JR (2009) Analysis of CRISPR in *Streptococcus mutans* suggests frequent occurrence of acquired immunity against infection by M102-like bacteriophages. Microbiology 155:1966–1976

van der Zanden AG, Kremer K, Schouls LM, Caimi K, Cataldi A, Hulleman A, Nagelkerke NJ, van Soolingen D (2002) Improvement of differentiation and interpretability of spoligotyping for *Mycobacterium tuberculosis* complex isolates by introduction of new spacer oligonucleotides. J Clin Microbiol 40:4628–4639

van Embden JD, van Gorkom T, Kremer K, Jansen R, van Der Zeijst BA, Schouls LM (2000) Genetic variation and evolutionary origin of the direct repeat locus of *Mycobacterium tuberculosis* complex bacteria. J Bacteriol 182:2393–2401

Warren RM, Streicher EM, Sampson SL, van der Spuy GD, Richardson M, Nguyen D, Behr MA, Victor TC, van Helden PD (2002) Microevolution of the direct repeat region of *Mycobacterium tuberculosis*: implications for interpretation of spoligotyping data. J Clin Microbiol 40:4457–4465

Zegans ME, Wagner JC, Cady KC, Murphy DM, Hammond JH, O'Toole GA (2009) Interaction between bacteriophage DMS3 and host CRISPR region inhibits group behaviors of *Pseudomonas aeruginosa*. J Bacteriol 191:210–219

Zhang J, Abadia E, Refregier G, Tafaj S, Boschiroli ML, Guillard B, Andremont A, Ruimy R, Sola C (2010) *Mycobacterium tuberculosis* complex CRISPR genotyping: improving efficiency, throughput and discriminative power of 'spoligotyping' with new spacers and a microbead-based hybridization assay. J Med Microbiol 59:285–294

Chapter 3
Evolution and Classification of CRISPR-Cas Systems and Cas Protein Families

Kira S. Makarova and Eugene V. Koonin

Abstract The CRISPR-Cas modules are adaptive antivirus immunity systems that are present in most archaea and many bacteria. These systems function by incorporating fragments of alien genomes into specific genomic loci, transcribing the inserts and using the transcripts as guide RNAs to destroy the genome of the cognate virus or plasmid. This RNA interference-like immune response is mediated by numerous, highly diverse Cas (CRISPR-associated) proteins, several of which form the Cascade complex involved in the processing of CRISPR loci transcripts and cleavage of the target DNA. Comparative analysis of the CRISPR-Cas modules led to the classification of the CRISPR-Cas systems into three types (I, II and III) that are characterized by distinct sets of *cas* genes. Classification of Cas proteins into families and superfamilies is a non-trivial task because of the fast evolution of many *cas* genes. Exhaustive sequence comparison aided by analysis of the available crystal structures led to the delineation of approximately 30 protein families that can be further classified into several superfamilies. By far the most common domain in Cas proteins is the RNA Recognition Motif (RRM). The RRM domains show remarkable diversity within the CRISPR-Cas systems and in particular comprise the scaffold of the Cascade complex. In addition to the numerous RRM domains, including a distinct polymerase-cyclase domain, the Cas proteins contain a distinct Superfamily II helicase domain, and several diverse nuclease domains. Detailed comparative analysis of the sequences and structures of Cas proteins structures shed light on the deep relationships between Type I and Type III systems and allowed us to propose a simple evolutionary scenario for the

K. S. Makarova (✉) · E. V. Koonin
National Institutes of Health, National Center for Biotechnology Information,
National Library of Medicine, 8600 Rockville Pike, Bethesda, MD 20894, USA
e-mail: makarova@ncbi.nlm.nih.gov

E. V. Koonin
e-mail: koonin@ncbi.nlm.nih.gov

R. Barrangou and J. van der Oost (eds.), *CRISPR-Cas Systems*,
DOI: 10.1007/978-3-642-34657-6_3, © Springer-Verlag Berlin Heidelberg 2013

origin of CRISPR-Cas system. Moreover, combination of experimental structural studies and comparative analysis provides for detailed models of the structures of the Cascade complexes from different CRISPR-Cas types revealing remarkable architectural uniformity.

Contents

3.1 Introduction

The CRISPR-Cas systems mediating adaptive immunity against viruses and other forms of foreign DNA (notably plasmids) in archaea and bacteria are encoded by large, complex genomic loci that consist of cassettes of CRISPR repeats which are associated with remarkably diverse clusters of CRISPR-associated (*cas*) genes. At least 45 distinct protein families have been identified among the products of the cas genes (Haft et al. 2005). An analysis involving more sensitive methods of sequence comparison and additional evidence from genomic context has revealed distant homologous relationships between some of these families, paring down the number of distinct protein groups to approximately 25 and suggesting that additional Cas protein families might be linked subsequently thanks to the growth of genomic and structural data sets and further advances in computational analysis (Makarova et al. 2006).

The CRISPR-Cas loci combine the presence of highly conserved genes and gene blocks with extreme variability of both gene composition and operon architecture. This striking fluidity of the CRISPR-Cas system poses both fundamental and more practical challenges. Explaining the evolution of any complex biological system is a fundamental and traditionally difficult problem in evolutionary biology, starting with Darwin's scenario for the evolution of the eye. In the case of CRISPR-Cas, the difficulty is exacerbated by the unusually polymorphic, apparently loose arrangement of the system components. In the more practical plane, the diversity and fast evolution of the CRISPR-Cas systems complicate the task of classification of these systems that is important for rational design and interpretation of experimental results. In this chapter, we first discuss the recently developed classification of CRISPR-Cas systems, then describe the protein families encoded by *cas* genes, and conclude with an evolutionary scenario for the origin and diversification of adaptive immunity in archaea and bacteria.

3.2 Classification of the CRISPR-Cas Systems

The recently developed classification of CRISPR-Cas systems divides them into three distinct types (I, II and III) (Bhaya et al. 2011; Makarova et al. 2011b; Wiedenheft et al. 2012). All these systems include two universal genes: *cas1* encoding a metal-dependent DNase with no apparent sequence specificity that could be involved in the integration of the alien DNA (spacer) into CRISPR cassettes (Marraffini and Sontheimer 2009; Wiedenheft et al. 2009), and *cas2* encoding a metal-dependent endoribonuclease that also appears to be involved in the spacer acquisition stage (Beloglazova et al. 2008; Yosef et al. 2012). Otherwise, however, the three types of CRISPR-Cas systems substantially differ in their sets of constituent genes, and each is characterized by a unique signature gene. The signature genes for the three types are, respectively, *cas3* (a superfamily 2 helicase containing an N-terminal HD superfamily nuclease domain) (Sinkunas et al. 2011), *cas9* (a large protein containing a predicted RuvC-like and HNH nuclease domains) and *cas10* (a protein containing a domain homologous to the palm domain of nucleic acid polymerases and nucleotide cyclases) (Makarova et al. 2011b). Within these three types, CRISPR-Cas systems have been further classified into subtypes on the basis of several considerations that include distinct signature genes, along with the phylogeny of the universal *cas1* gene (Makarova et al. 2011b). The Cas proteins known as RAMPs (Repeat-Associated Mysterious Proteins) are present in several copies in both type I and III systems. Some of the RAMPs have been shown to possess sequence- or structure-specific RNAse activity that is involved in the processing of pre-crRNA transcripts (Brouns et al. 2008; Carte et al. 2008; Haurwitz et al. 2010). The crystal structures of several RAMPs have been solved and indicate that they contain one or two domains which display distinct versions of the RNA recognition motif (RRM) also known as ferredoxin fold (Makarova et al. 2006; Sakamoto et al. 2009; Haurwitz et al. 2010; Lintner et al. 2011; Wang et al. 2011).

3.3 Cas Protein Families

3.3.1 Cas1 and Cas2: Signature Cas Proteins Implicated in Spacer Acquisition

Two Cas proteins, Cas1 and Cas2, are represented in all CRISPR/Cas systems that are predicted to be functionally active. These proteins are thought to function as the 'information processing' module of CRISPR-Cas that is involved in spacer integration (the adaptation stage). The predicted roles of Cas1 and Cas2 in spacer acquisition are in agreement with the observations that these proteins are not involved in the antiviral defense stage of the mechanism when a spacer is already present in the CRISPR array (Brouns et al. 2008; Hale et al. 2009). The *cas1* and *cas2* genes comprise the cores of the three distinct types of CRISPR-Cas systems (Makarova et al. 2011b). The putative nuclease/integrase Cas1 is the most conserved among all Cas proteins. This protein is widely used as a marker for detection of CRISPR-Cas systems in bacterial and archaeal genomes and for construction of phylogenetic trees that provide a framework for reconstruction of CRISPR-Cas system evolution. Based on the evolutionary conservation of several acidic residues and a histidine, Cas1 has been predicted to possess nuclease activity (Makarova et al. 2006). So far two Cas1 proteins have been experimentally characterized and their respective structures have been solved (Wiedenheft et al. 2009; Babu et al. 2011). The Cas1 protein from *Pseudomonas aeruginosa* is a metal-dependent nuclease that cleaves ssDNA or dsDNA, generating approximately 80 bp DNA fragments. The conserved amino acid residues of Cas1 line up a metal-binding pocket in the α-helical domain of a novel fold (Fig. 3.1). The catalytic domain is connected to the N-terminal, mostly beta-stranded domain by a flexible linker (Fig. 3.1); Cas1 protein forms homodimers (Wiedenheft et al. 2009). Mutation of metal ion-binding amino acid residues of Cas1 inhibits Cas1-catalyzed DNA degradation. The function of the N-terminal domain is not clear. Similar properties have been reported for the Cas1 protein (YgbT) from *E. coli* (Babu et al. 2011). Additionally, nuclease activity of *E. coli* Cas1 against branched

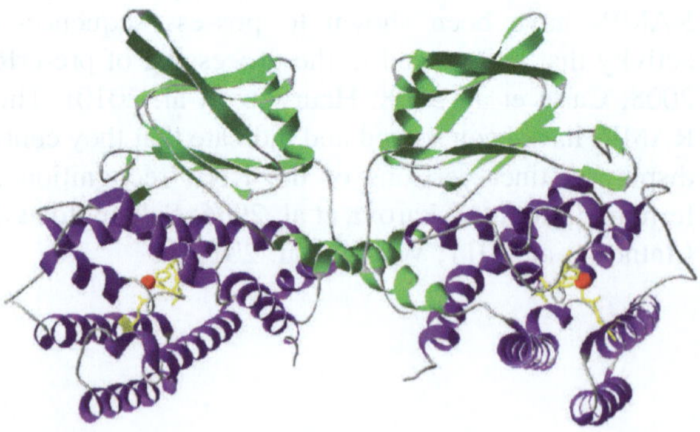

Fig. 3.1 Cas1 structure and domain fusions. The cartoon shows the dimeric structure of Cas1. The catalytic alpha-helical domain is shown in *dark violet* and the N-terminal domain is shown in *green*. Catalytic residues are *yellow* and the metal ion is *red*

DNAs including Holliday junctions, replication forks and 5′-flaps has been demonstrated (Babu et al. 2011). Furthermore, genome-wide screens have shown that YgbT physically and genetically interacts with key components of DNA repair systems such as *recB*, *recC* and *ruvB*, suggesting a dual role for Cas1 protein in bacterial antivirus immunity and DNA repair (Babu et al. 2011).

Several conserved fusions of Cas1 with other protein domains have been detected; all the genes encoding Cas1 fusion proteins belong to *cas* operons. The most common is the fusion of Cas1 with the Cas4 protein, a RecB-like nuclease (PD-(D/E)XK nuclease superfamily) containing a C-terminal three-cysteine cluster. This fusion might indicate a role for Cas4 in spacer acquisition (van der Oost et al. 2009). Several fusions of Cas1 with reverse transcriptase (RT) similarly might be indicative of involvement of RT in the function of some CRISPR-Cas systems (Makarova et al. 2006; Kojima and Kanehisa 2008). Furthermore, some RTs appear to be involved in a distinct abortive infection mechanism of antivirus defense (Kojima and Kanehisa 2008), suggesting the possibility that CRISPR-Cas systems and the abortive infection mechanism could be functionally linked.

The *cas2* gene is typically located immediately downstream of the *cas1* gene and encodes a small protein of approximately 100 amino acids; in subtype I–F CRISPR-Cas systems, *cas2* is fused to the *cas3* gene. Based on the conservation of aspartate or asparagine located after the N-terminal β-strand, the Cas2 protein has been predicted to possess nuclease activity (Makarova et al. 2006). There is structure and sequence similarity between Cas2 and the VapD toxin subunit of one of the experimentally characterized toxin-antitoxin (TA) systems (Daines et al. 2004; Makarova et al. 2006; Kwon et al. 2012). This relationship suggests a functional link between CRISPR-Cas and TA systems, with the further implication that Cas2 is likely to be an endoribonuclease with an activity similar to that of interferases, the toxin components of numerous TA systems that cleave ribosome-associated mRNAs (Yamaguchi and Inouye 2009). Several Cas2 proteins have been crystallized and studied biochemically (Beloglazova et al. 2008; Samai et al. 2010). These proteins adopt a RRM (ferredoxin) fold and form homodimers (Fig. 3.2a). For the Cas2 protein from *Sulfolobus solfataricus* (Sso1404), the ribonuclease activity has been experimentally demonstrated. It has been shown that in vitro this protein cleaves the phosphodiester linkage on the 3′-side and generates 5′-phosphate- and 3′-hydroxyl-terminated oligonucleotides, with a preference for U-rich sequences. Alanine scanning revealed a number of residues that affect the ribonuclease activity including the predicted N-terminal catalytic aspartate (Beloglazova et al. 2008). However, for Cas2 from *Desulfovibrio vulgaris* neither nuclease activity nor ssRNA or ssDNA binding have been demonstrated despite the conservation of the N-terminal aspartate (Samai et al. 2010). Currently it remains unclear whether Cas2 proteins from different organisms are indeed functionally distinct, or the differences in biochemical properties of Cas2 proteins are caused by unrecognized differences in isolation procedures and assay conditions.

Several conserved fusions of Cas2 have been detected including the fusion to Cas3 in Type I–F systems and a fusion with a DEDDh family exonuclease in

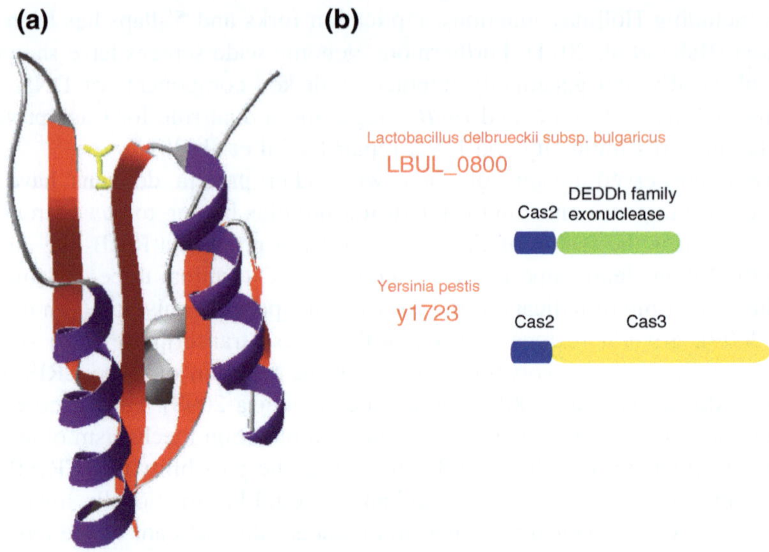

Fig. 3.2 Cas2 structure and domain fusions. **a** The cartoon shows the RRM domain of the Cas2 protein. In the RRM fold, the beta-strands are colored *red* and alpha helices are colored *violet*. Structural elements that do not belong to RRM fold are shown in *gray*. The catalytic aspartate is shown in *yellow*. **b** Two most frequent domain fusions of Cas2

several genomes with a distinct Type I-E system version, mostly in Firmicutes (Makarova et al. 2006, 2011b, Fig. 3.2b).

3.3.2 The HD Domain: A Single Strand-Specific DNAse Required for Interference

The CRISPR-associated HD nuclease is a component of all Type I and Type III systems. In most Type I systems, the HD domain forms an N-terminal fusion with the Cas3 helicase but in some Type I-A systems it appears as a stand-alone gene (*cas3″*). A few Type I–C systems (e.g. GSU0051) contain the HD domain as a C-terminal fusion with Cas3. In a limited number of Type I-E systems the Cas3 protein (HD and helicase domains) are fused to a Cascade subunit (Cse1) (Westra et al. 2012). In several Type III CRISPR-Cas systems the HD domain is fused to the Cas10 protein. In some of these Cas10-HD fusions (e.g. TM1794), the HD domain shows a circular permutation so that the N-terminal metal-binding histidine is displaced to the extreme C-terminus (Makarova et al. 2002). However, the HD domain of Cas10d (Subtype I-D) does not show the circular permutation that makes it similar to HD domain present in Cas3.

Several HD domains from different CRISPR-Cas systems have been studied experimentally, and two crystal structures have been resolved (pdb: 3S4L from *Methanocaldococcus jannaschii* and pdb: 3SKD from *Thermus thermophilus* HB8 (Beloglazova et al. 2011; Mulepati and Bailey 2011). In particular, it has been demonstrated that, in addition to the ATP-dependent helicase activity, Cas3 also shows ATP-independent nuclease activity mapped to the HD domain (Sinkunas et al. 2011); in addition it was demonstrated that the HD domain itself possesses metal-dependent single-stranded DNA endonuclease activity (Mulepati and Bailey 2011; Sinkunas et al. 2011). These analyses have recently been confirmed using the *E. coli* Cas3 (Westra et al. 2012). Earlier, however, it has been reported that HD protein SSO2001 from *Sulfolobus solfataricus* P2 CRISPR-Cas system I-A cleaves double-stranded oligonucleotides in vitro (Han and Krauss 2009). In general, the reported properties of the HD domain are compatible with the hypothesis that Cas3 functions by cleaving the DNA region exposed by Cascade upon crRNA-guided target DNA binding.

3.3.3 Cascade-Associated Proteins

Expression and transcript processing are the best characterized stage of CRISPR-Cas-mediated immunity. It has been shown that the long primary transcript of a CRISPR locus (pre-crRNA) is processed into short crRNAs. Processing of pre-crRNA is catalyzed by endoribonucleases encoded by *cas* genes that function either as subunits of a Cascade (CRISPR-associated complex for antiviral defense (Brouns et al. 2008) complex consisting of several Cas proteins, or as stand-alone enzymes, e.g. Cas6 of the archaeon *Pyrococcus furiosus* (Carte et al. 2008; Hale et al. 2009). In the latter case, the formation of a multisubunit complex (denoted Cmr complex of Type III-B system) also has been observed (Hale et al. 2009; Zhang et al. 2012). Recently, two additional Cas protein complexes have been characterized. The first one is the Csy complex associated with Type I–F system from *P. aeruginosa* (Wiedenheft et al. 2011b), which also includes the CRISPR transcript processing endoribonuclease, Cas6f (Csy4), a homolog of Cas6 (Haurwitz et al. 2010; Gesner et al. 2011; Makarova et al. 2011b). The second complex is a(rchaeal)Cascade from *S. solfataricus* which corresponds to CRISPR-Cas system Type I-A. Preliminary models of the architectures of these complexes are shown in Fig. 3.3. The general features of the Cascade complexes in Type I CRISPR-Cas systems are: (1) multiple subunits of Cas7, apparently involved in binding crRNA; (2) strong association between Cas7 and Cas5 proteins; (3) loose association of Cas6 with Cascade; Cas6 missing in some organisms; (4) loose association between the large (Cse1/CasA) and small subunits (Cse2/CasB) if present. The details of the interactions and arrangement of the *E. coli* Cascade subunits have been recently elucidated using cryo-electron microscopy (Wiedenheft et al. 2011a). It has been also shown that after finding a partial or

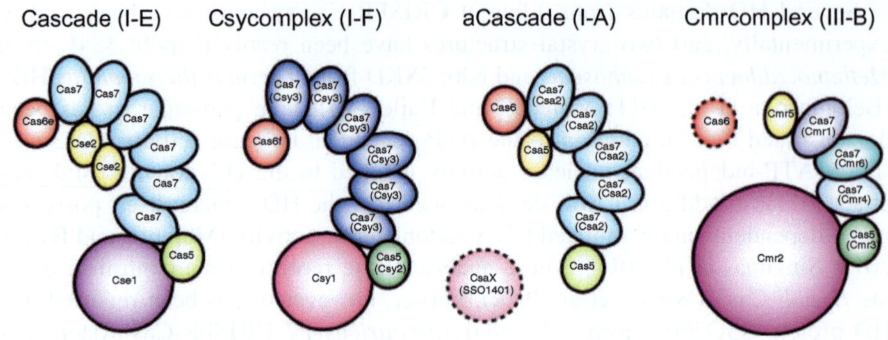

Fig. 3.3 Cascade complexes models. The models for four characterized Cascade complexes include Cascade from *E. coli* (Brouns et al. 2008; Jore et al. 2011b), Csy complex for the system Type I–F from *P. aeruginosa* (Wiedenheft et al. 2011b), aCASCADE from *S. solfataricus* (Lintner et al. 2011; Zhang et al. 2012), and Cmr complex from *P. furiosus* (Hale et al. 2009). For the first three complexes, the observed or inferred stoichiometry of subunits is reflected in the cartoons. The stably associated subunits are shown by *solid circles* and weakly associated subunits are shown by *dashed circles*. Three groups of RAMPs (Cas5, Cas6, Cas7) are indicated along with the corresponding gene names. Large subunits are shown by *magenta shades* and the small subunits by *yellow shades*

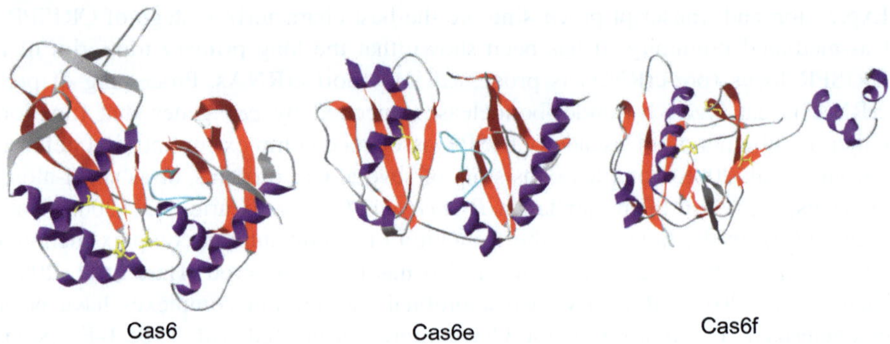

Fig. 3.4 Cas6 structures. The cartoon shows the two RRM domain-containing Cas6 family proteins. The RRM fold elements are colored as in Fig. 3.2. The glycine-rich loop is colored light *blue* and the catalytic residues, including the N-terminal histidine, are colored *yellow*

perfect match in the target DNA, the Cascade moves along the DNA molecule, occasionally selecting fragments to be incorporated into the CRISPR locus (Datsenko et al. 2012).

The crystal structures of several Cas6 homologs have been solved and the structures of the cleavage products produced by the Cascade complex have been characterized. All Cas6 homologs adopt a double RRM fold (although in the case of the available Cas6f structures the second RRM fold is heavily distorted) and feature a conserved histidine located after the first beta-strand of the N-terminal RRM domain which is crucial for nuclease activity (Fig. 3.4). All other amino acids involved in catalysis appear to differ among the Cas6 families. The cleavage of the pre-crRNA occurs within a CRISPR repeat at the 5′ side of the phosphodiester bond, generating a 5′ end hydroxyl group and either a 3′ phosphate (Cas6f) or a 2′, 3′ end cyclic phosphate group (Cas6e), and yields crRNA of approximately 60 nt size (Jore et al. 2011a; Wiedenheft et al. 2011b). In addition to its role in the processing of pre-crRNA, the Cascade complex bound to a mature crRNA appears to be involved in the interference stage by promoting R-loop formation to match a spacer within crRNA to the target ssDNA (Jore et al. 2011b).

In addition to Cas6, Cascade complexes of Type I systems typically include products of *cas7* and *cas5* genes, a large (typically, approximately 500 aa) protein and a small, mostly alpha-helical protein (Fig. 3.3). Along with Cas6, Cas5 and Cas7 proteins belong to the RAMP superfamily of RRM-containing proteins; at least four subunits of the cmr complex (Cmr1, Cmr3, Cmr4, Cmr6) also belong to the same superfamily (Makarova et al. 2006).

3.3.4 The Three Major Families of RAMPs

The first systematic sequence comparison of Cas proteins led to the identification of an extensive (super)family of diverse proteins that showed limited similarity to each other, centering on a glycine-rich loop. These proteins were denoted RAMPs (Repair-Associated Mysterious Proteins) given that Cas proteins were initially thought to represent a distinct repair system (Makarova et al. 2002). Subsequently, when the association of Cas proteins with CRISPR was realized, this superfamily has been renamed Repeat-Associated Mysterious Proteins, with the acronym RAMP surviving. Comparison of the several available crystal structures of RAMPs led to the realization that they all contained distinct forms of the RNA Recognition Motif (RRM) domain (also often described as a ferredoxin-like fold).

The report on the crystal structure of Cas7 (Csa2) from the Crenarchaeon *Sulfolobus solfataricus* (system Type I-A) (Lintner et al. 2011) was an important breakthrough that provided for a comprehensive classification of the RAMPs. This structure encompasses a single RRM domain that is structurally similar to the N-terminal RRM domain of Cas6 proteins. In addition, Cas7 contains four inserts within the RRM core and a C-terminal extension (Lintner et al. 2011). Independent sequence analysis (Makarova et al. 2011a) showed clear similarity between the Cas7 family and other RAMP families including those from Type III CRISPR-Cas systems. Several sequence blocks that are conserved in all RAMP families include the core elements of the RRM domain and an insert containing a conserved glycine that is located immediately before the second beta strand of the RRM fold (Makarova et al. 2011a).

The demonstration that the Cas7 family belongs to the RAMP superfamily prompted detailed analysis of the relationships between the RAMPs. By combining the results of comparison of all available RAMP structures, secondary

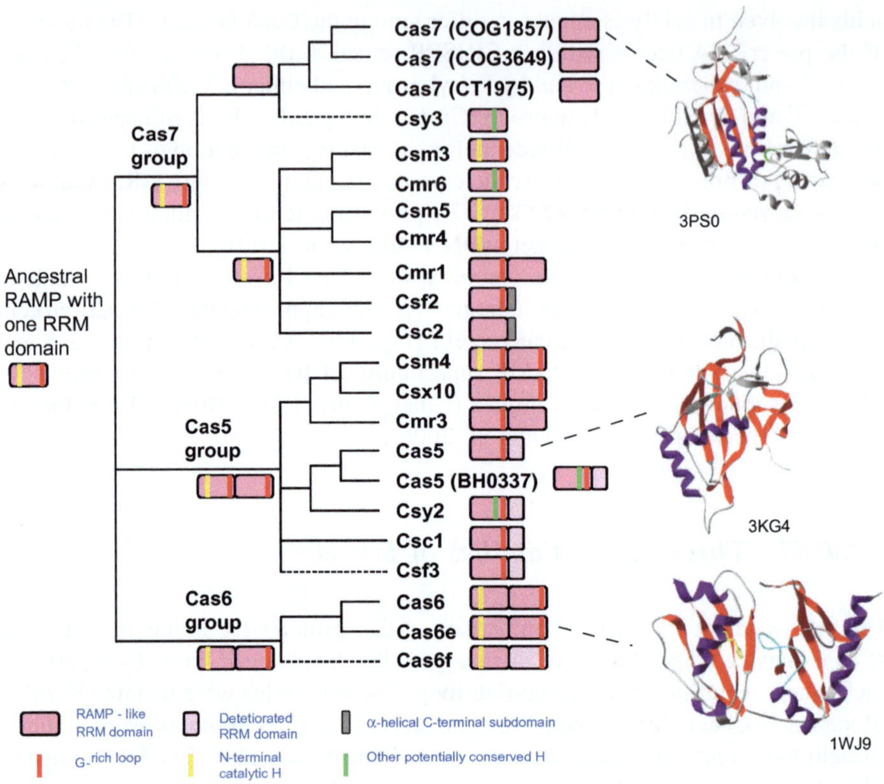

Fig. 3.5 Classification of the RAMPs. The tree-like scheme of RAMP relationships is based on the sequence similarity, structural features and neighborhood analysis described in the text, and should not be construed as a phylogenetic tree. Unresolved relationships are shown as multifurcations and tentative assignments are shown by broken lines. The catalytic activity of some of the RAMP proteins of the Cas5 and Cas7 groups involving the partially conserved histidines shown in the figure should be considered a tentative prediction. The structures for the RAMPs of the Cas5, Cas6 and Cas7 groups are shown. The RRM fold domains are depicted as in Fig. 3.4

structure prediction and sequence profile searches, the RAMP superfamily could be classified into three major families: Cas5, Cas6 and Cas7 (Fig. 3.5).

The Cas5 family RAMPs (Cas5/COG1688, Cmr3/COG1769, Csm4/COG1567, Csy2, Csc1) were unified on the basis of sequence similarity that in most cases was identifiable by profile search and the presence of a C-terminal domain downstream of the G-rich loop (Fig. 3.5). The Cas5 family consists of two distinct subfamilies one of which contains two RRM domains and the other one contains a single RRM domain (Fig. 3.5).

The Cas6 family includes Cas6 proteins proper (COG1853/COG5551) as well as highly diverged homologs from the I-E (Cas6e) and I–F (Cas6f) CRISPR-Cas subtypes. This group is supported by the available structures and is compatible with the reported functions for the representatives of each family. Most of the Cas6 proteins encompass two well-defined RRM domains that are connected by a "flange" in the extended conformation and contain a glycine-rich loop upstream of the last strand of the second RRM domain. Thus, the ancestor of the Cas6 family can be confidently inferred to have possessed two RRM domains. The Cas7 family includes Cas7 proteins present in the majority of Type I systems (COG1857) and a variety of RAMPs that mostly are associated with Type III CRISPR-Cas systems. All these proteins contain a single RRM domain with additional elaborations as demonstrated by examination of the recently reported Cas7 structure (Fig. 3.5), sequence comparison and secondary structure prediction. The diversity and weak conservation of the sequences and structures of the RAMPs hamper the elucidation of the evolutionary relationships between the three major groups. Considering only the relationship between the domain architectures of the RAMPs, the most parsimonious evolutionary scenario would involve an ancestral RAMP with a single enzymatically active RRM domain, resembling Cas7, and a single duplication in the putative common ancestor of the Cas5 and Cas6 groups, with subsequent deterioration or displacement of the C-terminal RRM domains in several Cas5 and Cas6 lineages (Fig. 3.5).

3.3.5 The Characteristic Arrangement of RAMP Genes in CRISPR-Cas Operons

Mapping the classification of RAMPs described in the preceding section onto the operons of the Type I and Type III CRISPR-Cas systems reveals a common architectural pattern. Most subtypes Type I CRISPR-Cas systems encode one RAMP of the Cas5, Cas6 and Cas7 families each. Operons of type III CRISPR-Cas system are organized similarly except that they typically encode multiple Cas7 family RAMPs. Notably, Cas5 and a Cas7 usually are encoded by adjacent genes. The Cas5 and Cas7 orthologs in two distinct CRISPR-Cas systems belong to the stable core of the Cascade complex in extremely diverse organisms, *E. coli* (Type I-E) (van der Oost et al. 2009; Jore et al. 2011b) and *S. solfataricus* (Type I-A) (Lintner et al. 2011). Taken together, these in silico findings strongly suggest physical and functional interaction between Cas5 and Cas7 as a key feature of CRISPR-Cas systems, as indeed confirmed by the Cascade structural arrangement (Wiedenheft et al. 2011a). Unclassified (U-type) CRISPR-Cas systems form operons that lack *cas5* but in which a *cas7* (*csf2*) gene is located adjacent to the csf3 gene suggesting that Csf3 could be a truncated derivative of Cas5 performing an analogous function.

3.3.6 Putative Homology Between the Large and Small Subunits of Diverse Type I and Type III CRISPR-Cas Systems

Multiple lines of evidence suggest that large subunits contained in most of the CRISPR-Cas systems could be homologous to Cas10 proteins which contain a polymerase-like Palm domain and are predicted to be enzymatically active in Type III CRISPR-Cas systems but inactivated in Type I systems (Fig. 3.6a) (Makarova et al. 2011a).

Among the large subunits of Type I CRISPR-Cas systems, significant sequence conservation has been demonstrated (Makarova et al. 2006) for several subfamilies of the Cas8 family (Cas8a1/Csa6, a subfamily of subtype I-A; Cas8b/Csh1/Cst1, a subfamily of subtype I–B; and Cas8c/Csd1, a subfamily of subtype I–C). Extensive sequence profiles comparison led to unification of other subfamilies of Cas proteins, in particular, Cmx1/Csx13/LA3191 associated with some diverged variants of Subtype I–C and Cas8a2 (Csa4/Csx9 subfamily), with Cas8 family (Makarova et al. 2011a). The large Cascade subunit of Subtype I-D shows similarity to the Zn-finger regions of the Cas8b/Cst1of I–B system and additionally is fused to an HD domain analogously to the Type III Cas10 proteins. The large subunits of subtypes I-E (Cse1) and I–F (Csy1) still appear to be unique, without any detectable sequence similarity to each other or to any Cas8 family proteins.

Type III CRISPR-Cas systems contain several subfamilies of Cas10 (Csm1, Cmr2 and Csx11 according to (Haft et al. 2005) that have been denoted CRISPR polymerases because of the presence of a readily identifiable Palm/Cyclase domain (Pei and Grishin 2001; Makarova et al. 2002; Anantharaman et al. 2010). The CRISPR polymerase consists of several domains, namely the HD domain (ssDNase), a distinct domain so far unique to this protein family, a Zn-finger domain, and the Palm domain, the signature domain of various polymerases and cyclases that adopts a distinct RRM fold (Makarova et al. 2002). The Palm domain of CRISPR polymerases is more similar to the Palm domain of cyclases than to those of 3′–5′ DNA and RNA polymerases, and it contains all typical secondary structure elements including four beta-strands of the core RRM fold (Anantharaman et al. 2010). Many structures of Palm domain-containing polymerases from all domains of cellular life and numerous viruses have been solved and compared (Steitz and Yin 2004). Most of these polymerases show a common arrangement of the core domains and the same modes of nucleic acid binding; the polymerases additionally contain a variety of editing nuclease domains and regulatory domains. The core domains (usually arranged in the same order from the N-terminus to the C-terminus) are the following: the "Fingers" domain that binds a nucleotide, the catalytic "Palm" domain that binds single-stranded nucleic acid, and the "Thumb" domain that binds double-stranded nucleic acid (Steitz and Yin 2004). Despite this structural and mechanistic uniformity, only the Palm domains of these numerous polymerase families are clearly homologous (Steitz and Yin 2004; Iyer et al. 2008). The most conserved feature of the Palm domains is the beta-hairpin formed by strands 2 and 3 of the RRM fold (Aravind et al. 2002;

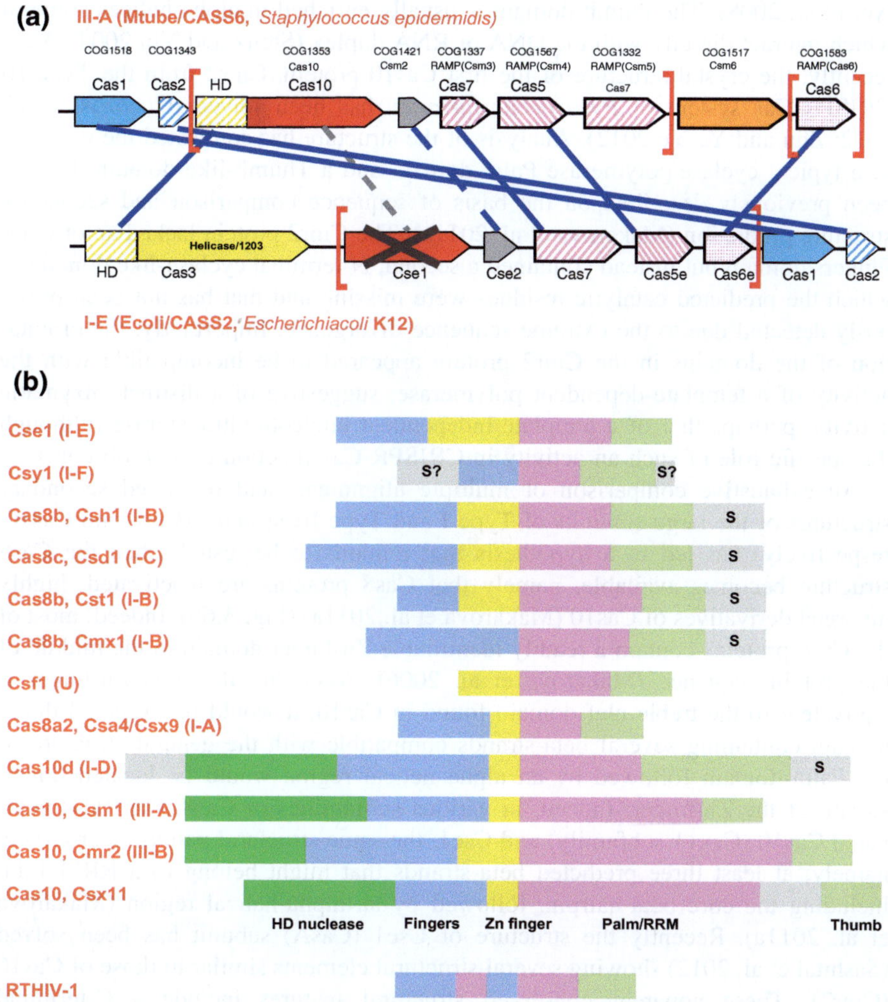

Fig. 3.6 Gene content similarity between type I-E and type III-A systems and structural organizations of large subunits of different CRISPR-Cas systems of type I and III. **a** Genes in the operons for I-E and III-A subtypes are shown by *arrows* with size roughly proportion to the size of the corresponding gene. Homologous genes are shown by the *arrows* of same color or hashing. RAMPs are shown by *pink* or *pink hashing*. *Solid lines* connect genes for which homology can be confidently demonstrated, and *dashed lines* connect genes for which homology is inferred tentatively. The Cascade complex subunits are shown by *square brackets*. Two previously published domain annotations are included for comparison. **b** Domain organization of large subunits of different type I and III CRISPR-Cas systems. Domain size is roughly proportional to correspondent sequence length. The letter "S" marks the regions that could be homologous to small subunits of Cascade complex encoded as separated genes in Type III systems, I-E subtype and some systems of I-A subtype

Iyer et al. 2008). The thumb domain is usually enriched in alpha helices some of which interact directly with the DNA or RNA duplex (Steitz and Yin 2004). Very recently, the crystal structure of the first Cas10 protein, Cmr2 from the Type III CRISPR-Cas system of *Pyrococcus furiosus*, has been solved (Cocozaki et al. 2012; Zhu and Ye Ye 2012). Analysis of the structure has confirmed the presence of a typical cyclase/polymerase Palm domain and a Thumb-like domain that has been previously described on the basis of sequence comparison and secondary structure prediction (Makarova et al. 2011a). The Cmr2 protein lacked an obvious Fingers domain but instead contained a second, N-terminal cyclase-like domain in which the predicted catalytic residues were missing and that has not been previously detected due to the extreme sequence divergence. Importantly, the orientation of the domains in the Cmr2 protein appeared to be incompatible with the activity of a template-dependent polymerase, suggestive of a distinct enzymatic activity, perhaps that of a template-independent nucleotidyltransferase (although the specific role of such an activity in CRISPR-Cas function remain obscure).

An exhaustive comparison of multiple alignments and predicted secondary structures of the large subunits of Type I and Type III systems (Cas8 and Cas10, respectively) has led to a hypothesis that remains to be tested when the Cas8 structure becomes available, namely that Cas8 proteins are inactivated, highly diverged derivatives of Cas10 (Makarova et al. 2011a) (Fig. 3.6b). Indeed, most of the Cas8 proteins contain a readily identifiable Zn-finger domain in the middle of the protein sequence (Makarova et al. 2006). Assuming that this Zn-finger is equivalent to the treble-clef domain found in Cas10, it would be expected that a domain containing several beta-strands compatible with the general structure of the Palm-domain followed by an alpha helical region would be located downstream of the Zn-finger. Indeed, in various subfamilies of Cas8, Cas10d, inactivated Cas10 (Csx11 subfamily) and Cse1, the same structural pattern is observed, namely, at least three predicted beta-strands that might belong to a RRM fold, including the core beta-hairpin, followed by an alpha-helical region (Makarova et al. 2011a). Recently the structure of Cse1 (CasA) subunit has been solved (Sashital et al. 2012) showing several structural elements similar to those of Cas10 (Cmr2). These apparent conserved structural features include a C-terminal 4-helical barrel-like domain, a beta-hairpin matching the beta-hairpin formed by strands 2 and 3 of the RRM fold in Cas10 and a loop that appears to correspond to the treble-clef domain of Cas10 although Cse1 does not contain the conserved cysteines that are typical of treble-clef. Because two other subfamilies (Csy1 and Cmx1) do not contain Zn-fingers, it is difficult to map the beginning of the putative Palm-domain within these sequences. However, sequence similarity between Cmx1 and Cas8 could be identified (Makarova et al. 2011a), and given that Cmx1 proteins possess an alpha-helical C-terminal region, it seems likely that Cmx1 is a diverged homolog of Cas8. The Csy1 protein might be homologous to Cse1 (the large subunit of the subtype I-E system) given the overall similarity in the operon organization between the I-E and I–F systems and the clustering of these systems in the Cas1 phylogeny (Makarova et al. 2011b). Like Cse1, Csy1 also contains an alpha-helical C-terminal domain and an N-terminal region with mixed alpha-helices and beta-strands. Although the pattern of the predicted secondary structure elements of Csy1 cannot be confidently aligned with either Cse1 or Cas8, the possibility that it contains a derived RRM-like fold cannot be ruled out. Most of the large subunits of type I CRISPR-Cas systems containing Zn-fingers also possess an N-terminal region with mixed beta-strands and alpha helices which is compatible with the general organization of the region following the HD domain and preceding the Zn-finger in Cas10 subfamilies. Taken together, analysis of the general secondary structure features, the presence of the Zn-finger domain in many large subunits, the similar operon organization and the experimentally demonstrated functional link to RAMPs and the Cascade complex (Brouns et al. 2008; Hale et al. 2009; Jore et al. 2011b) raise the possibility that all large subunits of CRISPR-Cas systems might be inactivated derivatives of the CRISPR polymerase (Fig. 3.6a). However, there is currently not enough evidence to rule out non-homologous displacement of some large subunits or their individual domains.

The pattern of predicted secondary structure elements in the putative Fingers domain of Cas10 and several large subunits, in particular Csx11, Cas8a2/Csa4, and Csc3, resembles the structures of the RRM domain of the RAMPs. Like the RRM core domain, many of the Fingers-like domains contain four predicted beta-strands. Furthermore, the Fingers-like domains start with a beta strand-alpha helix element and ends with an alpha helix-beta-strand element, which are the two most conserved structural patterns in the RAMPs. Thus, the Fingers domain of the large subunits might adopt an RRM fold. This prediction has been subsequently confirmed by the demonstration that the Fingers domain in the Cmr2 structure adopt a cyclase-like fold similar to the Palm domain (Cocozaki et al. 2012; Zhu and Ye 2012).

In several families of large subunits (Cas8a1, Cas8b, Cas8c, Cmx1 and Cas10d) of the I-A, B, C and D system subtypes, the C-terminal region (the predicted Thumb domain) is longer than it is in Cas10 proteins (8 alpha helices compared to 4 in Cas10). In the respective subtypes of CRISPR-Cas, the small Cascade subunit is missing. Typically, the small subunit is an alpha-helical protein containing 6 alpha helices (the structure is available for cmr5, namely 2OEB for the *Archaeoglobus fulgidus* protein AF1862, and 2ZOP for the *Thermus thermophilus* protein TTHB164; see Fig. 3.7). The size and predicted structure of the small subunit appear to be compatible with the size and structure of the extra alpha helical region at the C-termini of the longest large subunits (Fig. 3.6b). The Csy1 protein, the subtype I–F specific large subunit, contains eight predicted alpha helices at the C-terminus and four helices at the extreme N-terminus. Because none of the predicted RAMP proteins from this system contain extended alpha-helical regions compatible with the size of the small subunit, the structural and functional counterpart of the small subunit might be "hidden" within Csy1.

The demonstration that at least some of the large subunits of Type I CRISPR-Cas systems are homologous to the CRISPR polymerase suggests that all these large proteins function and interact with DNA or RNA in a mode analogous to that of other Palm domain polymerases (Table 3.1). In particular, the Palm domain probably interacts with ssDNA whereas the analog of the Thumb interacts with dsDNA. Notably, evolutionarily conserved inactivated derivatives of Palm domain

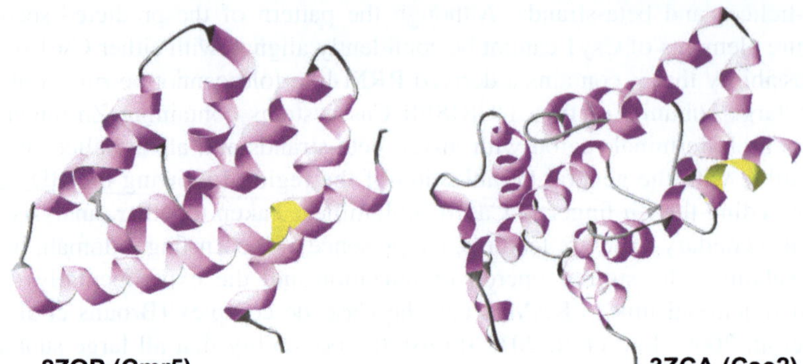

Fig. 3.7 Structures and motifs of the small subunits of CRISPR-cas systems. Two available structures for small subunits are shown. Conserved tryptophanes within the C-terminal alpha helix are shown by *yellow*

polymerases have been detected in archaea and eukaryotes although their functions remain uncharacterized (Rogozin et al. 2008; Tahirov et al. 2009).

The conservation of the complete set of catalytic residues typical of Palm domain polymerases and cyclases implies that the Palm domain of Cas10 is enzymatically active but the nature of this activity remains unknown. There is no indication that a processive polymerase is involved in any stage of the CRISPR-Cas system functioning, and as pointed out above, the structure of Cas10 is poorly compatible with such an activity (Zhu and Ye 2012). The possibility remains that Cas10 is a nucleotidyltranferase or even a nucleotide cyclase, perhaps involved in crRNA modification. This type of activity is compatible with the activity of the tRNA(His) guanylyltransferase THG1 (Jackman and Phizicky 2006) which belongs to the same branch of Palm domain proteins with Cas10 and the GGDEF diguanylate cyclases (Anantharaman et al. 2010) (see above). Another possibility is that Cas10 has a secondary role as a helicase in one or more stages of CRISPR/Cas functioning. A helicase activity dependent on the cleavage of the α-β bond in NTP during polymerization has been demonstrated for the bacteriophage T7 RNA polymerase (Steitz 2004; Steitz and Yin 2004; Yin and Steitz 2004), which is a derivative of the Palm domain DNA polymerases (Iyer et al. 2008). Remarkably, all Type I CRISPR-Cas systems in which the large subunits are inactivated Cas10 homologs (i.e. those Ca10 homologs in which some or all of the predicted catalytic residues in the Palm domain are replaced) also include the Cas3 helicase, and conversely, all Type III systems that contain Cas10 proteins predicted to be active lack Cas3 (Makarova et al. 2011b). Thus, Cas3 might compensate for the loss of the original enzymatic function of Cas10 in Type I CRISPR-Cas system whereas the inactivated derivative of Cas10 performs an accessory structural role in these systems. Notably, some Type U CRISPR-Cas systems that contain degraded versions of Cas10 and lack Cas3 include a DinG-like helicase (see below), in

Table 3.1 Structures, domain architectures and functions of the core components of CRISPR-Cas systems

Family	Biochemical/in silico evidence	Structural features	Prediction
Cas1	Metal-dependent deoxyribonuclease; (Han et al. 2009; Wiedenheft et al. 2009); deletion of Cas1 in *E. coli* results in increased sensitivity to DNA damage and impaired chromosomal segregation (Babu et al. 2011)	PDB: 3GOD, 3LFX, 2YZS Unique fold with two domains: N-terminal β stranded domain and catalytic C-terminal α-helical domain Fusions: Cas4, RecB-like nuclease and Reverse transcriptase	Involved in integration of spacer DNA into CRISPR repeats
Cas2	RNAse specific to U-rich regions (Beloglazova et al. 2008)	PDB: 2IVY, 2I8E and 3EXC RRM (ferredoxin) fold Fusions: Cas3 and DEDDh family exonuclease	Facilitates spacer selection and/or integration. Could be involved in further crRNA cleavage
Cas3 (helicase and HD domain)	Single-stranded DNA nuclease (HD domain) and ATP-dependent helicase (Sinkunas et al. 2011); required for interference (Brouns et al. 2008)		Cuts DNA during interference; promotes strand separation
Stand alone HD nuclease	Metal-dependent deoxyribonuclease specific for double-stranded oligonucleotides (Han and Krauss 2009)	PDB: 3S4L and 3SKD	Cuts DNA during interference
Cas4		RecB-like nuclease homolog with three-cysteine C-terminal cluster (Makarova et al. 2006)	Might be involved in spacer acquisition
Cas5	Subunit of Cascade complex (Brouns et al. 2008; Jore et al. 2011b)	PDB: 3KG4 RRM (ferredoxin) fold, RAMP superfamily	Might substitute for Cas6 if catalytically active. Otherwise might be involved in both interference and adaptation stages.

(continued)

Table 3.1 (continued)

Family	Biochemical/in silico evidence	Structural features	Prediction
Cas6	Metal-independent endoribonuclease that generates crRNAs, subunit of Cascade complex (Brouns et al. 2008; Carte et al. 2008; Hale et al. 2009; Haurwitz et al. 2010; Jore et al. 2011b)	PDB: 2XLJ 1WJ9 and 3I4H Double RRM (ferredoxin) fold, RAMP superfamily	
Cas7	Subunit of Cascade complex (Brouns et al. 2008); present Cascade complex of I-E systems in 6 copies (Jore et al. 2011b) and in several copies in I-A systems (Lintner et al. 2011)	PDB: 3PS0 RRM (ferredoxin) fold with subdomains, RAMP superfamily	Implicated in interference; binds crRNA; if enzymatically active, might be involved in RNA-guided RNA cleavage.
Cas8abcef,(large subunit)	Subunit of Cascade complex, involved in PAM recognition (Brouns et al. 2008; Sashital et al. 2012)	PDB: 4AN8	Inactivated Cas10 polymerase-like protein, binds DNA, interacts with HD domain and a RAMP carrying crRNA; could be involved in both interference and spacer selection stages
Cas10 (large subunit, CRISPR polymerase)	Subunit of Cascade (Cmr) complex (Hale et al. 2009)	PDB: 3UNG, 4DOZ Two domains homologous to Palm domain polymerases and cyclases (Cocozaki et al. 2012); Fusion: HD nuclease domain	Same as Cas8, but fused to HD and thus cuts ssDNA; might be involved in strand separation
Small subunit	Small, mostly alpha helical protein, subunit of Cascade complex (Brouns et al. 2008; Hale et al. 2009); present in Cascade complex of I-E systems in two copies (Jore et al. 2011b)	PDB: 2ZCA (Cse2) and 2ZOP, 2OEB (Cmr5); Both families have a unique fold with alpha helical structure	DNA binding

(continued)

Table 3.1 (continued)

Family	Biochemical/in silico evidence	Structural features	Prediction
Cas9	In Type II CRISPR-Cas systems, Cas9 is sufficient both to generate crRNA and to cleave the target DNA (Barrangou et al. 2007; Garneau et al. 2010); Both the RuvC and HNH nuclease domains of Cas9 are involved in the cleavage of the target DNA (Sapranauskas et al. 2011; Jinek et al. 2012)	Contains a predicted RuvC-like (RNAse H fold) and HNH nuclease domains	May be considered a functional analog of CASCADE and HD nuclease

further support of the possibility that a helicase activity required for the CRISPR-Cas function can be supplied by different, in some cases, unrelated proteins.

3.3.7 Type II CRISPR-Cas Systems and Homologs of Cas9

The signature protein of the type II CRISPR-Cas systems II, Cas9, does not show any detectable similarity to any proteins in Type I and Type III systems. It appears that Cas9 is sufficient both to generate crRNA (together with housekeeping RNase III) and to cleave the target DNA in Type II systems (Garneau et al. 2010; Makarova et al. 2011b; Jinek et al. 2012; Barrangou 2007, p. 26) (see Chap. 5). The large Cas9 protein ($\sim$1,000 amino acids) contains two predicted nuclease domains, namely a HNH (McrA-like) nuclease domain that is located in the middle of the protein and a -RuvC-like nuclease domain (RNAse H fold) that contains all the characteristic catalytic motifs (Aravind et al. 2000; Makarova et al. 2006, 2011a) and hence is predicted to be enzymatically active, but contains a long (approximately 450 amino acids) insert including the HNH nuclease domain (Fig. 3.8).

The roles of the two predicted nuclease domains of Cas9 in the function of the Type II CRISPR remain unclear. However, the insertion of the HNH nuclease domain into the RNAse H fold domain suggests that the two nuclease activities are closely coupled. The HNH nuclease domain, which is common in restriction enzymes and possesses DNA-endonuclease activity (Kleanthous et al. 1999; Jakubauskas et al. 2007). Recently it has been demonstrated that HNH domain is

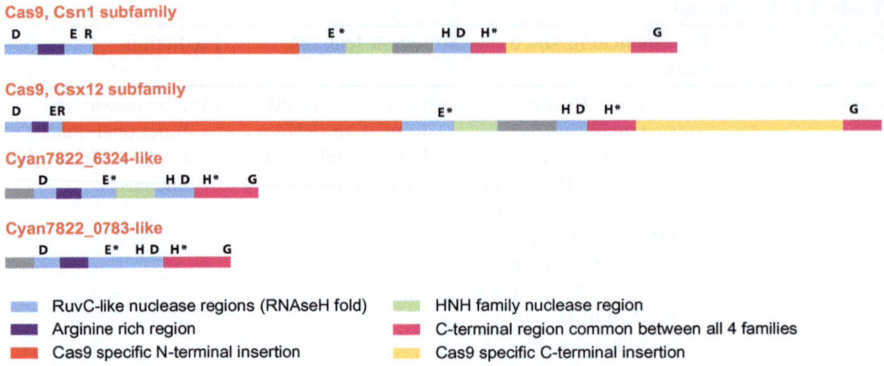

Fig. 3.8 Structural organization of Cas9 proteins and their homologs. Homologous regions are shown by the same color. Distinct sequence motifs are denoted by the corresponding conserved amino acid residues above the respective domains (when the same conserved amino acid occurs in different motifs, one is marked by an *asterisk* to avoid confusion)

responsible for nicking of one strand of the target dsDNA and the RuvC-like RNAse H fold domain is involved in cleavage of the other strand of the dsDNA target (Jinek et al. 2012). Together, these two domains each nick a strand of the target dsDNA within the proto-spacer in the immediate vicinity of the PAM, which results in blunt cleavage of the invasive DNA (Jinek et al. 2012). Mutation of the predicted catalytic amino acids of both RuvC and HNH domains of Cas9 of *Streptococcus thermophilus* abolish phage interference (Sapranauskas et al. 2011).

The Cas9 sequences show weak but statistically significant sequence similarity to a large family of prokaryotic proteins that also contain both RuvC-like and HNH-nuclease domain. This family can be divided into at least two subfamilies by domain architecture (Fig. 3.8). Analysis of the genomic context of the genes encoding these Cas9 homologs did not reveal any stable associations, and there are no CRISPR repeats in the vicinity of any of these genes. Hence, the function of these proteins remains obscure and might be distinct from the function of CRISPR-Cas. An intriguing possibility is that these Cas9 homologs represent a novel system of RNA-guided DNA interference involved in antivirus defense that in some respects could be analogous to the prokaryotic Argonaute proteins (Makarova et al. 2009b). Some of the genes encoding these proteins form large lineage-specific paralogous families (e. g. 49 genes in *Ktedonobacter racemifer* or 17 genes in *Microcoleus chthonoplastes*) suggesting at least this subset of the family could represent novel mobile elements. The *cas9* gene might have been co-opted by the CRISPR/Cas system from such mobile elements with the concomitant loss of typical CRISPR/Cas components, such as RAMPs and CRISPR polymerases resulting in the emergence of the distinctive Type II gene neighborhoods.

3.3.8 A Hypothetical Scenario for the Origin and Evolution of CRISPR-Cas Systems

Taken together, the results of comparative analysis described above allow us to propose a simple scenario for the origin and the major stages in the evolution of the CRISPR-Cas system (Fig. 3.9). The primary observations that contribute to this reconstruction of CRISPR-Cas evolution are:

1. The demonstration that Cas7 proteins represent a distinct group of RAMPs
2. Classification of the RAMP superfamily into three major families: Cas5, Cas6 and Cas7

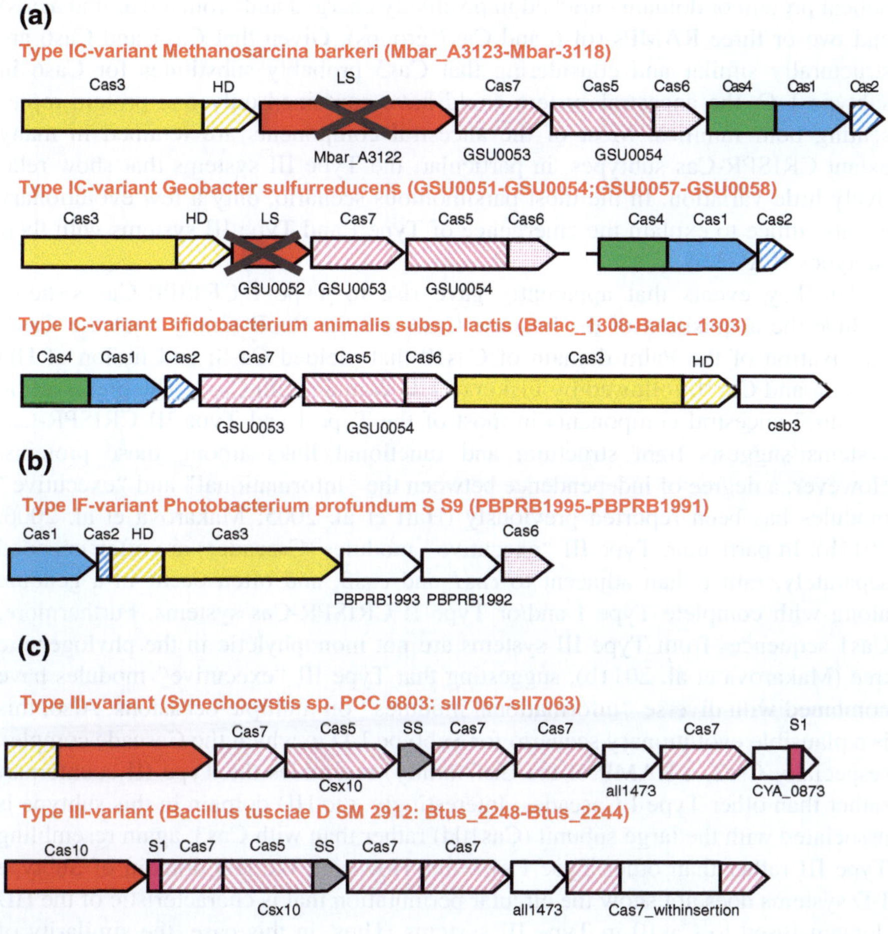

Fig. 3.9 Unusual CRISPR-Cas systems. **a** Type I–C-variants with GSU0054 (or GSU0053) signature gene. **b** Type I–F-variant. **c** Type III-variant

3. The unification, even if more tentative, of Cas8 and Cas10 in a single family of CRISPR-Cas large subunits
4. The tentative unification of small, Csm2-like subunits

Taking into account these unifying connections between the Cas proteins, comparison of the gene compositions and operon organizations of the three major types and 12 subtypes of CRISPR-Cas systems allows us to reconstruct the ancestral forms with some confidence.

The ancestral functional CRISPR-Cas system probably resembled Subtype III-A and consisted of 6 or 7 genes including the two universal *cas* genes, *cas1* and *cas2*, ("information processing" subsystem involved in the adaptation phase) along with 4 or 5 additional genes that comprised the "executive" subsystem (Cascade complex) involved in crRNA processing and interference. The "executive" module included the large subunit (Cas10/Cas8), the small subunit (an alpha-helical protein or domain enriched in positively charged and aromatic amino acids) and two or three RAMPs (of 6 and Cas7 groups). Given that Cas5 and Cas6 are structurally similar and considering that Cas5 probably substitutes for Cas6 in subtype I–C, the ancestral system could have contained only one protein representing both families. Most of the ancestral components are retained in many extant CRISPR-Cas subtypes, in particular, the Type III systems that show relatively little variation. In the most parsimonious scenario, only a few evolutionary events suffice to explain the emergence of Type I and Type III systems with their subtypes (Fig. 3.9).

The key events that apparently gave rise to Type I CRISPR-Cas systems include the acquisition of the helicase Cas3 and the RecB family nuclease Cas4; inactivation of the Palm domain of Cas10 that yielded Cas8; and fission of HD domain and Cas10 followed by fusion of HD domain with Cas3. The preservation of 6 to 7 ancestral components in most of the Type I and Type III CRISPR-Cas systems suggests tight structural and functional links among these proteins. However, a degree of independence between the "informational" and "executive" modules has been reported previously (Haft et al. 2005; Makarova et al. 2006, 2011b). In particular, Type III "executive" modules (Cascades) are often encoded separately, rather than adjacent to *cas1* and *cas2*, and often occur in a genome along with complete Type I and/or Type II CRISPR-Cas systems. Furthermore, Cas1 sequences from Type III systems are not monophyletic in the phylogenetic tree (Makarova et al. 2011b), suggesting that Type III "executive" modules have combined with diverse "informational modules" on multiple occasions. Also, this is a plausible evolutionary scenario for Subtype I-D in which the Cascade complex (especially Csc2, a RAMP of the Cas7family) resembles the Type III counterpart rather than other Type I Cascades. Interestingly, the HD domain in this subtype is associated with the large subunit (Cas10d) rather than with Cas3, again resembling Type III rather than other Type I systems. However, the HD domain of Subtype I-D systems does not show the circular permutation that is characteristic of the HD domain fused to Cas10 in Type III systems. Thus, in this case, the similarity of domain architectures seems to be convergent, i.e. the HD domain in Subtype I-D systems probably was translocated from Cas3 to inactivated Cas10 (or fused with the latter if the ancestral form was a stand-alone HD domain).

There are currently no examples of archaeal or bacterial genomes that do possess the "information processing" module without the "executive" module of the CRISPR-Cas system. Although involvement of Cas1 in various repair processes has been suggested by recent experiments (Babu et al. 2011), the tight linkage between the two CRISPR-Cas modules indicates that the primary functions of Cas1 and Cas2 depend on the Cascade complex (the "executive module"). In contrast, "Cascade only" systems (Type-U) that are not associated with CRISPR arrays have been identified, suggesting the intriguing possibility that some variants of Cascade might function as an independent (defense) system, without relying on Cas1, Cas2 or CRISPR arrays for the acquisition of spacers. Although the source of RNA guides for such a system is unclear, an interesting possibility is that this version of Cascade might recognize alien DNA molecules and process nascent alien mRNA to generate RNA guides; such mechanism would be directly analogous to the siRNA branch of the eukaryotic RNA interference systems (Csorba et al. 2009). From the evolutionary perspective, a stand-alone Cascade could be one of the antecedents of CRISPR-Cas systems.

The ancestor of the CRISPR polymerase (Cas10) could have evolved from an ancient Palm domain polymerase, such as reverse transcriptase. On the basis of a number of derived shared characters, the CRISPR polymerase has been classified as a member of a distinct group of Palm domain proteins that also includes Thg1-type 3′–5′ nucleic acid polymerases and adenylate and diguanylate cyclases (Anantharaman et al. 2010). The association with the HD domain probably goes deep into the evolutionary past given that HD family hydrolases are also commonly associated with GGDEF family diguanylate cyclases (Galperin 2006; Anantharaman et al. 2010). The ancestral function of the CRISPR polymerase that was probably associated with the HD hydrolase domain could potentially involve a distinct form of signal transduction, a role in repair and/or in antivirus defense. The latter possibility seems attractive given the tight association of this protein with the CRISPR-Cas systems.

Genomic islands, in which viral defense, mobile elements and stress response genes, such as toxin-antitoxin systems, are often present together, are likely to be "melting pots" for the emergence of new functional systems through recombination, duplication and HGT (Kawano et al. 2007; Makarova et al. 2009b, 2011c). It appears likely that the CRISPR-Cas systems evolved in such genomic environments, via combination of distinct mobile elements. The origin of RAMPs remains an enigma: these highly diverged RRM-domain proteins share derived characters that are strongly suggestive of their monophyly (such as a glycine-rich loop and a conserved histidine implicated in catalysis in numerous RAMPs) but show no significant similarity to any other proteins. An intriguing possibility is that there is a direct evolutionary connection between the CRISPR polymerase and the RAMPs given that the cores of all these proteins consist of RRM domains. The first RAMP proteins could have emerged by duplication of an inactivated polymerase followed by rapid evolution that involved the emergence of the endoribonuclease catalytic center. The ancestral RAMP might have resembled Cas7 proteins that contain a single RRM domain with structural embellishments along with a Zn-finger domain (in a subset of the Cas7 proteins), and so resemble polymerases in their domain architecture. Furthermore, several CRISPR-Cas systems apparently remain functional despite having a highly degraded form of the large subunit (type U system) or lacking the large subunit altogether (some variants of Subtype I–C and Subtype I–F) (Fig. 3.9a and b), suggesting that RAMPs might be able to substitute for the function of large subunits. The Cas6 and Cas5 families of RAMPs could have subsequently evolved from the Cas7-like RAMPs. This scenario seems plausible considering that RAMP duplications, including tandem duplications and fusions, are often present in CRISPR-Cas loci, especially among the Type III systems in which Cas7 family RAMPs are particularly prone to duplication. Interestingly, in both Type I Cascade complexes that have been characterized in detail, those from *E. coli*, *P. aeruginosa* and *S. solfataricus* (Jore et al. 2011b; Lintner et al. 2011; Wiedenheft et al. 2011a, 2011b), the Cas7 subunit is present in multiple copies. In Type III Cascades, these homo-oligomers appear to be replaced by hetero-oligomers made of paralogous Cas7 proteins. Furthermore, lineage-specific (and hence relatively recent) inactivation of the CRISPR polymerase (Cas10) was detected in some Type III systems such as MTH326-like (Fig. 3.10). All these observations attest to the dynamic character of the evolution of CRISPR-Cas systems and seem to add plausibility to the route of evolution from the CRISPR polymerase to the RAMP-based Cascade complexes (Fig. 3.10). Nevertheless, this scenario remains speculative given the absence of specific similarity between the RAMPs and CRISPR polymerases. An alternative involving recruitment of a distinct RRM-domain protein as the ancestral RAMP cannot be ruled out; identification of this potential ancestor of the RAMPs is an extremely difficult task given the high sequence divergence of these proteins that implies rapid evolution.

The CRISPR polymerase and the entire ancestral, Subtype III-A-like CRISPR-Cas system most likely evolved in hyperthermophilic archaea. Indeed, this system and in particular the *cas10* gene is present in a substantial majority of archaea and is confidently reconstructed as a set of genes present in Last Archaeal Common Ancestor (LACA) (Makarova et al. 2007). By contrast, Type III CRISPR-Cas systems are much less common in bacteria and often contain variants of Cas10 that are predicted to be inactivated (Makarova et al. 2011b). Like most antiviral defense systems, CRISPR-Cas is prone to HGT and could have rapidly spread among bacteria. Notably, many thermophilic bacteria possess Type III systems that probably have been acquired from archaea and could have started the dissemination of CRISPR-Cas among bacteria. The active Cas10 could be particularly beneficial in thermal environments, in agreement with the previous observations that identified Cas10 as a prominent genomic determinant of the thermophilic life style (Makarova et al. 2002, 2003).

The close association between Cas1 and Cas2 is more difficult to explain in terms of function or evolution. Given that Cas1 is a DNase with a Holliday junction resolvase-like activity (Wiedenheft et al. 2009; Babu et al. 2011), the

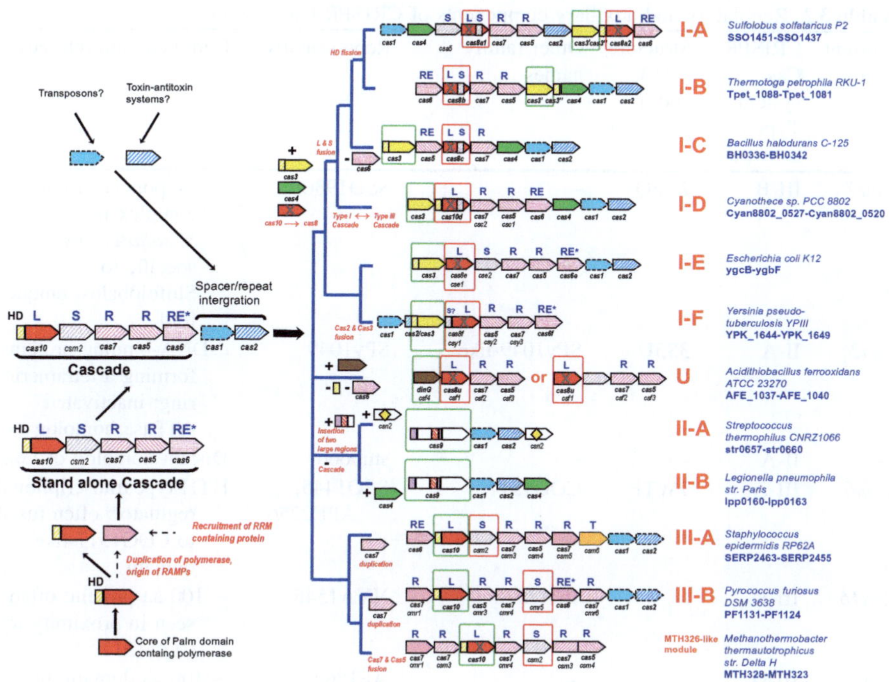

Fig. 3.10 Evolutionary scenario for the origin of CRISPR-Cas systems. Homologous genes are color-coded and identified by a family name (names follow the classification from (Makarova et al. 2011b). Names in bold are proposed systematic names including those propose in this work; "legacy names" are in regular font. The signature genes for CRISPR-Cas types are shown within *green boxes*, and for subtypes within *red boxes*. For each CRISPR-Cas subtype, a representative genome and the respective gene locus names are indicated. The bold letters above the genes show major categories of Cas proteins: *L* large CASCADE subunit, *S* small CASCADE subunit, *R* RAMP CASCADE subunit, *RE* RAMP family RNase involved in crRNA processing (experimentally characterized nucleases shown be asterisks), *T* transcriptional regulator. Genes coding for inactivated (putative) polymerases are indicated by crosses. Major evolutionary events are shown in the corresponding branches. *Broken lines* denote alternative evolutionary scenarios for the origin of RAMPs

prediction is that this protein functions as a recombinase and integrase at the spacer acquisition stage. These activities are typical of transposable elements, so the origin of Cas1 from a transposon appears likely. The endoribonuclease Cas2 might have evolved from another class of equally widespread mobile elements, namely toxin-antitoxin systems. Cas2 is yet another RRM-domain-containing component of CRISPR-Cas systems that is homologous to VapDHi, the toxin in the two-component toxin-antitoxin system vapDHi/VapX (Daines et al. 2004; Makarova et al. 2006, 2011a; Kwon et al. 2012). It remains unclear whether Cas1 and Cas2 ever formed a distinct two gene unit or have independently joined the evolving CRISPR-Cas system.

Table 3.2 Regulatory and auxiliary components of CRISPR-Cas systems

Current name	CRISPR-Cas system Type or subtype	Structure (PDB code)	Other family names	Representatives	Comment and references
cmr7	III-B	2X5Q	–	SSO1986	A component of cmr complex in *S. solfataricus*; specific to Sulfolobales; unique fold
csn2	II-A	3S5U	SPy1049-like	SPy1049	dsDNA-binding protein forming a tetrameric ring; inactivated ATPase homolog
–	II-A	–	–	stu0660	Distant homolog of *csn2*
csm6	III-A	2WTE	COG1517	SSO1445, APE2256	HTH-type transcriptional regulator; often fused to COG1517-like domain
csx16	III-U	–	*VVA1548*	VVA1548	~100 aa protein; often seen in proximity to COG1517
csx3	III-U	–	–	AF1864	~100 aa domain, in some cases fused to COG1517 family domains
csx1	III-U	1XMX, 2I71	COG1517, COG4006; *csa3*, *csx2*, *NE0113*, DxTHG motif *TIGR02710*	PF1127, MJ1666, TM1812, NE0113	All these proteins have a domain with Rossmann-like fold; many are fused to HTH domain; some have a fusion with an additional domain (e.g. RecB-family nuclease domain in 1XMX)
csx15	III-U	–	TTE2665	TTE2665	~130 aa protein, no prediction; some are fused to AAA ATPase domain
–	–	–	COG2378	sll7009	HTH-type transcriptional regulator, containing an additional C-terminal ligand binding domain

Type II systems are the only class of CRISPR-Cas for which the origin of Cascade complex components could not be confidently inferred. Nevertheless, experimental data suggests that Type II systems in general function similarly to the Cascade complexes of Type I and Type III systems (Garneau et al. 2010). Of the three types of CRISPR-Cas systems, the Type II systems have changed the most compared to the inferred ancestral form. This transformation involved replacement of the genes encoding the subunits of the ancestral Cascade complex as well as the large (polymerase) and small subunits by a single large, multidomain protein, Cas9, which contains two unrelated nuclease domains (Fig. 3.8). Both nuclease domains of Cas9 have been shown to contribute to interference by cleaving different DNA strands (Jinek et al. 2012).

3.4 Conclusions

The CRISPR-Cas systems are extremely variable in their gene composition, and most of the *cas* genes evolve fast compared to other genes in prokaryotes (Takeuchi et al. 2012). Due to this rapid evolution, comparative analysis of the Cas protein sequences and structures is a challenging task. Increasingly subtle relationship leading to unification of protein families previously thought to be unrelated have been detected thanks to the growth of the genome collection and refinement of computational methods (Makarova et al. 2002, 2006, 2011b; Haft et al. 2005), and more such findings can be anticipated. At this stage, it has become clear that Cas proteins can be classified into no more than a dozen major superfamilies including the Cas1–Cas10 proteins, another group of small subunits (perhaps to be denoted Cas11) and additionally a few regulatory protein families such as Csm6 and auxiliary protein such as Csm2 (Table 3.2).

The central structural unit of the CRISPR-Cas systems is the RRM domain that is present in numerous Cas proteins in a striking variety of structurally and functionally distinct forms. The RRM domains reach extreme diversity of enzymatically active and inactivated versions in the RAMP superfamily and in addition are also present in Cas10 and Cas2. Given the extensive diversification of RRM domains within the CRISPR-Cas systems, it appears likely that additional, barely recognizable RRM domains exist in poorly characterized Cas proteins such as the large subunits of Type I systems that might be inactivated derivatives of Cas10.

All the diversity of the CRISPR-Cas systems notwithstanding, comparative analysis of Cas protein sequences and structures and genomic organizations of the CRISPR-*cas* loci has conferred considerable order onto this apparent morass. Three major types and 10 subtypes of CRISPR-Cas systems have been delineated, each with its own signature gene(s). Moreover, these comparative studies imply a simple scenario for the origin and evolution of the CRISPR-Cas machinery in thermophilic archaea. In this scenario, the CRISPR-Cas systems originated as a large protein that combined the polymerase and HD hydrolase domain and might have functioned as a stand-alone antivirus defense system. The next step of evolution might have involved duplication of the RRM portion of the polymerase, followed either by inactivation that yielded the ancestral, Cas7-like RAMP, or by recruitment of a distinct RRM-domain protein that became the ancestral RAMP. Regardless of the exact origin of the ancestral RAMP, it has undergone a series of additional duplications and rapid diversification that yielded the stand-alone Cascade complex. The formation of the ancestral CRISPR-Cas system was then completed through the integration of Cascade with Cas1 and Cas2. The central theme of this scenario is the origin of the components of the CRISPR-Cas system from different classes of mobile elements. Other prokaryotic defense systems such as restriction-modification (Kobayashi 2001; Fukuda et al. 2008) and toxin-antitoxin systems (Makarova et al. 2009a; Van Melderen and Saavedra De Bast 2009) also comprise of such elements, indicating a major trend in the relationships between prokaryotes, viruses that infect them, other classes of selfish elements and defense mechanisms.

References

Anantharaman V, Iyer LM, Aravind L (2010) Presence of a classical RRM-fold palm domain in Thg1-type 3′- 5′nucleic acid polymerases and the origin of the GGDEF and CRISPR polymerase domains. Biol Direct 5:43

Aravind L, Makarova KS, Koonin EV (2000) Survey and summary: holliday junction resolvases and related nucleases: identification of new families, phyletic distribution and evolutionary trajectories. Nucleic Acids Res 28:3417–3432

Aravind L, Mazumder R, Vasudevan S, Koonin EV (2002) Trends in protein evolution inferred from sequence and structure analysis. Curr Opin Struct Biol 12:392–399

Babu M, Beloglazova N, Flick R, Graham C, Skarina T, Nocek B, Gagarinova A, Pogoutse O, Brown G, Binkowski A, Phanse S, Joachimiak A, Koonin EV, Savchenko A, Emili A, Greenblatt J, Edwards AM, Yakunin AF (2011) A dual function of the CRISPR-Cas system in bacterial antivirus immunity and DNA repair. Mol Microbiol 79:484–502

Barrangou R, Fremaux C, Deveau H, Richards M, Boyaval P, Moineau S, Romero DA, Horvath P (2007) CRISPR provides acquired resistance against viruses in prokaryotes. Science 315:1709–1712

Beloglazova N, Brown G, Zimmerman MD, Proudfoot M, Makarova KS, Kudritska M, Kochinyan S, Wang S, Chruszcz M, Minor W, Koonin EV, Edwards AM, Savchenko A, Yakunin AF (2008) A novel family of sequence-specific endoribonucleases associated with the clustered regularly interspaced short palindromic repeats. J Biol Chem 283:20361–20371

Beloglazova N, Petit P, Flick R, Brown G, Savchenko A, Yakunin AF (2011) Structure and activity of the Cas3 HD nuclease MJ0384, an effector enzyme of the CRISPR interference. EMBO J 30:4616–4627

Bhaya D, Davison M, Barrangou R (2011) CRISPR-Cas systems in bacteria and archaea: versatile small RNAs for adaptive defense and regulation. Annu Rev Genet 45:273–297

Brouns SJ, Jore MM, Lundgren M, Westra ER, Slijkhuis RJ, Snijders AP, Dickman MJ, Makarova KS, Koonin EV, van der Oost J (2008) Small CRISPR RNAs guide antiviral defense in prokaryotes. Science 321:960–964

Carte J, Wang R, Li H, Terns RM, Terns MP (2008) Cas6 is an endoribonuclease that generates guide RNAs for invader defense in prokaryotes. Genes Dev 22:3489–3496

Cocozaki AI, Ramia NF, Shao Y, Hale CR, Terns RM, Terns MP, Li H (2012) Structure of the Cmr2 subunit of the CRISPR-Cas RNA silencing complex. Structure 20:545–553

Csorba T, Pantaleo V, Burgyan J (2009) RNA silencing: an antiviral mechanism. Adv Virus Res 75:35–71

Daines DA, Jarisch J, Smith AL (2004) Identification and characterization of a nontypeable Haemophilus influenzae putative toxin–antitoxin locus. BMC Microbiol 4:30

Datsenko KA, Pougach K, Tikhonov A, Wanner BL, Severinov K, Semenova E (2012) Molecular memory of prior infections activates the CRISPR/Cas adaptive bacterial immunity system. Nat Commun 3:945

Fukuda E, Kaminska KH, Bujnicki JM, Kobayashi I (2008) Cell death upon epigenetic genome methylation: a novel function of methyl-specific deoxyribonucleases. Genome Biol 9:R163

Galperin MY (2006) Structural classification of bacterial response regulators: diversity of output domains and domain combinations. J Bacteriol 188:4169–4182

Garneau JE, Dupuis ME, Villion M, Romero DA, Barrangou R, Boyaval P, Fremaux C, Horvath P, Magadan AH, Moineau S (2010) The CRISPR/Cas bacterial immune system cleaves bacteriophage and plasmid DNA. Nature 468:67–71

Gesner EM, Schellenberg MJ, Garside EL, George MM, Macmillan AM (2011) Recognition and maturation of effector RNAs in a CRISPR interference pathway. Nat Struct Mol Biol 18:688–692

Haft DH, Selengut J, Mongodin EF, Nelson KE (2005) A guild of 45 CRISPR-associated (Cas) protein families and multiple CRISPR/Cas subtypes exist in prokaryotic genomes. PLoS Comput Biol 1:e60

Hale CR, Zhao P, Olson S, Duff MO, Graveley BR, Wells L, Terns RM, Terns MP (2009) RNA-guided RNA cleavage by a CRISPR RNA-Cas protein complex. Cell 139:945–956

Han D, Krauss G (2009) Characterization of the endonuclease SSO2001 from Sulfolobus solfataricus P2. FEBS Lett 583:771–776

Han D, Lehmann K, Krauss G (2009) SSO1450–a CAS1 protein from Sulfolobus solfataricus P2 with high affinity for RNA and DNA. FEBS Lett 583:1928–1932

Haurwitz RE, Jinek M, Wiedenheft B, Zhou K, Doudna JA (2010) Sequence- and structure-specific RNA processing by a CRISPR endonuclease. Science 329:1355–1358

Iyer LM, Abhiman S, Aravind L (2008) A new family of polymerases related to superfamily A DNA polymerases and T7-like DNA-dependent RNA polymerases. Biol Direct 3:39

Jackman JE, Phizicky EM (2006) tRNAHis guanylyltransferase adds G-1 to the 5′ end of tRNAHis by recognition of the anticodon, one of several features unexpectedly shared with tRNA synthetases. RNA 12:1007–1014

Jakubauskas A, Giedriene J, Bujnicki JM, Janulaitis A (2007) Identification of a single HNH active site in type IIS restriction endonuclease Eco31I. J Mol Biol 370:157–169

Jinek M, Chylinski K, Fonfara I, Hauer M, Doudna JA, Charpentier E (2012) A programmable dual-RNA-guided DNA endonuclease in adaptive bacterial immunity. Science 337(6096):816–821

Jore MM, Brouns SJ, van der Oost J (2011a) RNA in defense: CRISPRs protect prokaryotes against mobile genetic elements. Cold Spring Harb Perspect Biol 4(6)

Jore MM, Lundgren M, van Duijn E, Bultema JB, Westra ER, Waghmare SP, Wiedenheft B, Pul U, Wurm R, Wagner R, Beijer MR, Barendregt A, Zhou K, Snijders AP, Dickman MJ, Doudna JA, Boekema EJ, Heck AJ, van der Oost J, Brouns SJ (2011b) Structural basis for CRISPR RNA-guided DNA recognition by Cascade. Nat Struct Mol Biol 18:529–536

Kawano M, Aravind L, Storz G (2007) An antisense RNA controls synthesis of an SOS-induced toxin evolved from an antitoxin. Mol Microbiol 64:738–754

Kleanthous C, Kuhlmann UC, Pommer AJ, Ferguson N, Radford SE, Moore GR, James R, Hemmings AM (1999) Structural and mechanistic basis of immunity toward endonuclease colicins. Nat Struct Biol 6:243–252

Kobayashi I (2001) Behavior of restriction-modification systems as selfish mobile elements and their impact on genome evolution. Nucleic Acids Res 29:3742–3756

Kojima KK, Kanehisa M (2008) Systematic survey for novel types of prokaryotic retroelements based on gene neighborhood and protein architecture. Mol Biol Evol 25:1395–1404

Kwon AR, Kim JH, Park SJ, Lee KY, Min YH, Im H, Lee I, Lee BJ (2012) Structural and biochemical characterization of HP0315 from Helicobacter pylori as a VapD protein with an endoribonuclease activity. Nucleic Acids Res 40:4216–4228

Lintner NG, Kerou M, Brumfield SK, Graham S, Liu H, Naismith JH, Sdano M, Peng N, She Q, Copie V, Young MJ, White MF, Lawrence CM (2011) Structural and functional characterization of an archaeal CASCADE complex for CRISPR-mediated viral defense. J Biol Chem 286:21643–21656

Makarova KS, Aravind L, Grishin NV, Rogozin IB, Koonin EV (2002) A DNA repair system specific for thermophilic Archaea and bacteria predicted by genomic context analysis. Nucleic Acids Res 30:482–496

Makarova KS, Aravind L, Wolf YI, Koonin EV (2011a) Unification of Cas protein families and a simple scenario for the origin and evolution of CRISPR-Cas systems. Biol Direct 6:38

Makarova KS, Grishin NV, Shabalina SA, Wolf YI, Koonin EV (2006) A putative RNA-interference-based immune system in prokaryotes: computational analysis of the predicted enzymatic machinery, functional analogies with eukaryotic RNAi, and hypothetical mechanisms of action. Biol Direct 1:7

Makarova KS, Haft DH, Barrangou R, Brouns SJJ, Charpentier E, Horvath P, Moineau S, Mojica FJM, Wolf YI, Yakunin AF, van der Oost J, Koonin EV (2011b) Evolution and classification of the CRISPR/Cas systems. Nat Rev Microbiol 9:467–477

Makarova KS, Sorokin AV, Novichkov PS, Wolf YI, Koonin EV (2007) Clusters of orthologous genes for 41 archaeal genomes and implications for evolutionary genomics of archaea. Biol Direct 2:33

Makarova KS, Wolf YI, Koonin EV (2003) Potential genomic determinants of hyperthermophily. Trends Genet 19:172–176

Makarova KS, Wolf YI, Koonin EV (2009a) Comprehensive comparative-genomic analysis of type 2 toxin-antitoxin systems and related mobile stress response systems in prokaryotes. Biol Direct 4:19

Makarova KS, Wolf YI, Snir S, Koonin EV (2011c) Defense islands in bacterial and archaeal genomes and prediction of novel defense systems. J Bacteriol 193:6039–6056

Makarova KS, Wolf YI, van der Oost J, Koonin EV (2009b) Prokaryotic homologs of Argonaute proteins are predicted to function as key components of a novel system of defense against mobile genetic elements. Biol Direct 4:29

Marraffini LA, Sontheimer EJ (2009) Invasive DNA, chopped and in the CRISPR. Structure 17:786–788

Mulepati S, Bailey S (2011) Structural and biochemical analysis of the nuclease domain of the clustered regularly interspaced short palindromic repeat (CRISPR) associated protein 3(CAS3). J Biol Chem 286(36):31896–31903

Pei J, Grishin NV (2001) GGDEF domain is homologous to adenylyl cyclase. Proteins 42:210–216

Rogozin IB, Makarova KS, Pavlov YI, Koonin EV (2008) A highly conserved family of inactivated archaeal B family DNA polymerases. Biol Direct 3:32

Sakamoto K, Agari Y, Agari K, Yokoyama S, Kuramitsu S, Shinkai A (2009) X-ray crystal structure of a CRISPR-associated RAMP module [corrected] Cmr5 protein [corrected] from Thermus thermophilus HB8. Proteins 75:528–532

Samai P, Smith P, Shuman S (2010) Structure of a CRISPR-associated protein Cas2 from Desulfovibrio vulgaris. Acta Crystallogr Sect F: Struct Biol Cryst Commun 66:1552–1556

Sapranauskas R, Gasiunas G, Fremaux C, Barrangou R, Horvath P, Siksnys V (2011) The Streptococcus thermophilus CRISPR/Cas system provides immunity in Escherichia coli. Nucleic Acids Res 39(21):9275–9282

Sashital DG, Wiedenheft B, Doudna JA (2012) Mechanism of foreign DNA selection in a bacterial adaptive immune system. Mol Cell 46:606–615

Sinkunas T, Gasiunas G, Fremaux C, Barrangou R, Horvath P, Siksnys V (2011) Cas3 is a single-stranded DNA nuclease and ATP-dependent helicase in the CRISPR/Cas immune system. EMBO J 30(7):1335–1342

Steitz TA (2004) The structural basis of the transition from initiation to elongation phases of transcription, as well as translocation and strand separation, by T7 RNA polymerase. Curr Opin Struct Biol 14:4–9

Steitz TA, Yin YW (2004) Accuracy, lesion bypass, strand displacement and translocation by DNA polymerases. Philos Trans R Soc Lond B Biol Sci 359:17–23

Tahirov TH, Makarova KS, Rogozin IB, Pavlov YI, Koonin EV (2009) Evolution of DNA polymerases: an inactivated polymerase-exonuclease module in Pol epsilon and a chimeric origin of eukaryotic polymerases from two classes of archaeal ancestors. Biol Direct 4:11

Takeuchi N, Wolf YI, Makarova KS, Koonin EV (2012) Nature and intensity of selection pressure on CRISPR-associated genes. J Bacteriol 194:1216–1225

van der Oost J, Jore MM, Westra ER, Lundgren M, Brouns SJ (2009) CRISPR-based adaptive and heritable immunity in prokaryotes. Trends Biochem Sci 34:401–407

Van Melderen L, Saavedra De Bast M (2009) Bacterial toxin-antitoxin systems: more than selfish entities? PLoS Genet 5:e1000437

Wang R, Preamplume G, Terns MP, Terns RM, Li H (2011) Interaction of the Cas6 riboendonuclease with CRISPR RNAs: recognition and cleavage. Structure 19:257–264

Westra ER, van Erp PB, Kunne T, Wong SP, Staals RH, Seegers CL, Bollen S, Jore MM, Semenova E, Severinov K, de Vos WM, Dame RT, de Vries R, Brouns SJ, van der Oost J (2012) CRISPR immunity relies on the consecutive binding and degradation of negatively supercoiled invader DNA by cascade and Cas3. Mol Cell 46:595–605

Wiedenheft B, Lander GC, Zhou K, Jore MM, Brouns SJ, van der Oost J, Doudna JA, Nogales E (2011a) Structures of the RNA-guided surveillance complex from a bacterial immune system. Nature 477:486–489

Wiedenheft B, Sternberg SH, Doudna JA (2012) RNA-guided genetic silencing systems in bacteria and archaea. Nature 482:331–338

Wiedenheft B, van Duijn E, Bultema J, Waghmare S, Zhou K, Barendregt A, Westphal W, Heck A, Boekema E, Dickman M, Doudna JA (2011b) RNA-guided complex from a bacterial immune system enhances target recognition through seed sequence interactions. Proc Natl Acad Sci USA 108:10092–10097

Wiedenheft B, Zhou K, Jinek M, Coyle SM, Ma W, Doudna JA (2009) Structural basis for DNase activity of a conserved protein implicated in CRISPR-mediated genome defense. Structure 17:904–912

Yamaguchi Y, Inouye M (2009) mRNA interferases, sequence-specific endoribonucleases from the toxin-antitoxin systems. Prog Mol Biol Transl Sci 85:467–500

Yin YW, Steitz TA (2004) The structural mechanism of translocation and helicase activity in T7 RNA polymerase. Cell 116:393–404

Yosef I, Goren MG, Qimron U (2012) Proteins and DNA elements essential for the CRISPR adaptation process in Escherichia coli. Nucleic Acids Res 40:5569–5576

Zhang J, Rouillon C, Kerou M, Reeks J, Brugger K, Graham S, Reimann J, Cannone G, Liu H, Albers SV, Naismith JH, Spagnolo L, White MF (2012) Structure and mechanism of the CMR complex for CRISPR-mediated antiviral immunity. Mol Cell 45:303–313

Zhu X, Ye K (2012) Crystal structure of Cmr2 reveals a nucleotide cyclase-related enzyme in type III CRISPR-Cas systems. FEBS Lett 586(6):939–945

Chapter 4
Regulation of CRISPR-Based Immune Responses

Zihni Arslan, Edze R. Westra, Rolf Wagner and Ümit Pul

Abstract Nucleic acid cleaving CRISPR effector complexes, consisting of Cas protein(s) and crRNAs, provide protection against invading genetic elements, such as phage and (conjugative) plasmids. However, under some conditions, cells may experience a selective advantage if they avoid energy investment in CRISPR defense, for example, if they contain additional defense systems (e.g., R-M systems, phage exclusion systems) that provide sufficient protection. The formation of CRISPR effector complexes is a multistep process that requires (1) expression of the *cas* genes, (2) assembly of the Cas proteins into a multiprotein complex, (3) transcription of a CRISPR array into a pre-crRNA molecule, and (4) the subsequent sequence-specific processing of the pre-crRNA by a dedicated endoribonuclease, yielding crRNAs that are then loaded on the Cas protein complex. The resulting ribonucleoprotein complex may have intrinsic cleavage activity on complementary nucleic acids (e.g., the RAMP module complex of *Pyrococcus furiosus*) or may to this end require recruitment of an additional component upon target binding (e.g., Cas3 recruitment by Cascade in *Escherichia coli*). The different steps toward the formation of the final effector complexes offer several potential targets for regulation of the CRISPR system. Although studies dealing

---

Z. Arslan · R. Wagner · Ü. Pul (✉)
Heinrich-Heine-University Düsseldorf, Molecular Biology of Prokaryotes,
Universitätsstr. 1 D-40225 Düsseldorf, Germany
e-mail: pul@hhu.de

Z. Arslan
e-mail: zihni.arslan@hhu.de

R. Wagner
e-mail: r.wagner@hhu.de

E. R. Westra
Laboratory of Microbiology, Department of Agrotechnology and Food Sciences,
Wageningen University, Dreijenplein 10 6703 HB Wageningen, The Netherlands
e-mail: edze.westra@wur.nl

R. Barrangou and J. van der Oost (eds.), *CRISPR-Cas Systems*,
DOI: 10.1007/978-3-642-34657-6_4, © Springer-Verlag Berlin Heidelberg 2013

with this regulation are limited and thus far restricted to a few organisms, the number of host factors involved in CRISPR regulation increases rapidly. CRISPR defense can be regulated at the level of (*cas* gene and/or CRISPR) transcription by DNA-binding global regulators such as H-NS, LeuO, cAMP-CRP, or at the post-transcriptional level by the chaperon HtpG, which has been shown to be essential for Cas3 activity in *E. coli*. The presence of σ^{32}-dependent promoters within the *cas* operon and the involvement of the BaeSR two-component system suggest a coupling of CRISPR activity to membrane or heat stress in *E. coli*. In this chapter, we will summarize the recent findings on the regulation of the CRISPR system, mainly in *E. coli*, for which several regulatory components have been identified. We will also discuss the role of other potential regulatory mechanisms, such as translational regulation of *cas* gene expression through overlapping open reading frames on a polycistronic mRNA and the regulation of pre-crRNA stability or processing (Fig. 4.1).

Contents

4.1 Regulation of the CRISPR System in *E. coli*

The CRISPR defense system seems to be constitutively active or rapidly activated in several organisms, among others *Streptococcus thermophilus* and *Staphylococcus epidermidis*, since it prevents phage infection or plasmid conjugation when appropriate spacer sequences are provided (Barrangou et al. 2007; Marraffini and Sontheimer 2008). Moreover, in *Streptococcus pyogenes,* crRNA levels were among the highest of noncoding small RNAs (Deltcheva et al. 2011).

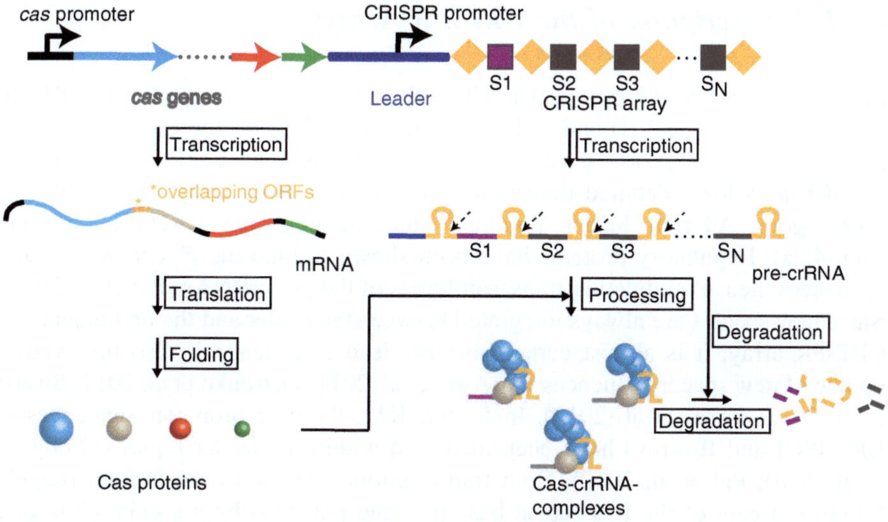

Fig. 4.1 Pathway of Cas-crRNA effector complex formation. The formation of Cas-crRNA effector complexes requires the expression of the Cas proteins, and the transcription and processing of the pre-crRNA. The individual steps toward the formation of the final effector complexes represent potential targets for regulation of the CRISPR activity. In addition to the well-documented transcriptional regulation of *cas* genes and CRISPR, post-transcriptional regulation mechanisms are also likely targets for CRISPR regulation, such as translational control of Cas proteins, dependence of Cas protein folding/stability on chaperones, or the involvement of nonCas proteins on pre-crRNA processing and stability

However, a recent shotgun proteome analysis of *S. thermophilus* DGCC7710 revealed a significant induction of Cas proteins following phage infection (Young et al. 2012). The up-regulation of type II-specific Cas9 and type I-E specific Cas7 proteins in phage-infected *S. thermophilus* cells indicates that even in constitutively active CRISPR-Cas systems, the expression of individual Cas proteins are regulated and inducible in response to phage infection (Young et al. 2012). In *E. coli,* the CRISPR expression is more tightly regulated and nearly completely shut-off, resulting in a lack of CRISPR-based immunity under normal laboratory conditions. The CRISPR system in *E. coli* K12 is silenced by the heat-stable nucleoid protein H-NS and activated by its antagonist LeuO. Although CRISPR defense is constitutively active in H-NS deletion strains (Pul et al. 2010; Westra et al. 2010) and (conditionally) active in wild-type strains under natural growth conditions (Diez-Villasenor et al. 2010), a phage infection itself does not serve as a trigger of CRISPR defense activation in wild-type *E. coli*, suggesting more complex regulation by additional inhibitors, activators, or growth conditions not yet identified. Recent studies suggest that signaling pathways related to envelope stress (e.g., BaeSR system) can activate CRISPR defense (Baranova and Nikaido 2002; Perez-Rodriguez et al. 2011).

4.1.1 Transcription of the CRISPR Arrays

E. coli K12 contains the type I-E CRISPR system, consisting of two CRISPR arrays (CRISPR I and II) and eight *cas* genes, encoding for the Cascade forming proteins Cse1, Cse2, Cas7, Cas5, and Cas6e and the proteins Cas1, Cas2, and Cas3 (see Chap. 6 for a detailed description of Cascade). Both arrays are flanked by homologous AT-rich leader regions, which contain the CRISPR promoter (Fig. 4.2a). Regulatory proteins have been shown to bind the *E. coli* K12 leader sequences, hence modulating transcript levels of the pre-crRNAs (Pul et al. 2010). Since new spacers are always integrated between the leader and the first repeat of a CRISPR array, it is almost certain that the leader sequence is also involved in uptake of new spacer sequences (Al-Attar et al. 2011; Datsenko et al. 2012; Swarts et al. 2012; Yosef et al. 2012). In *E. coli* K12, the core promoter sequences of CRISPR I and II arrays have been identified within the leader regions (Pougach et al. 2010; Pul et al. 2010). Both transcriptional start sites are located roughly 50 bp upstream of the first repeat base pair and preceded by a special class of a σ^{70}-dependent RNA polymerase promoter with an extended -10 characteristic (TGxTATAAT) (Mitchell et al. 2003). The CRISPR promoters exhibit some constitutive activity although the pre-crRNAs levels are very low under normal laboratory growth conditions due to direct repressing activity of H-NS. Low abundance might further be the result of a high turnover of CRISPR transcripts by unknown RNases. Hence, in an engineered *E. coli* K12 strain, containing spacers targeting λ phage, only low basal levels of pre-crRNA and Cascade complex could be detected resulting in a marginal CRISPR response (Pougach et al. 2010; Westra et al. 2010).

Generally, pre-crRNAs appear to be short-lived and the processing to the stable crRNAs occurs rapidly (Fig. 4.1). Efficient expression of the Cascade genes is essential for this processing step, which results in the accumulation of stable crRNAs with
$2'$–$3'$-cyclic phosphate ends (Jore et al. 2011). The strongly reduced level of mature crRNAs in *E. coli* under normal growth conditions is therefore due to a double regulation, namely transcriptional repression of the CRISPR promoter and the tight transcriptional silencing of Cascade expression, necessary for crRNA processing (Pougach et al. 2010; Westra et al. 2010).

4.1.2 Transcriptional Regulation of Cascade Complex by H-NS

In *E. coli,* a single σ^{70}-dependent promoter, termed P*cas*, directs the transcription of a polycistronic mRNA comprising the Cascade-*cas1*-*cas2* genes (Fig. 4.2a) (Pul et al. 2010). The promoter P*cas* is located in the upstream region of the *cse1* gene [referred to as IGLB, for intergenic region between *ygcL* (*cse1*) and *ygcB* (*cas3*)]. Although the P*cas* promoter is functional and purified σ^{70}-RNA

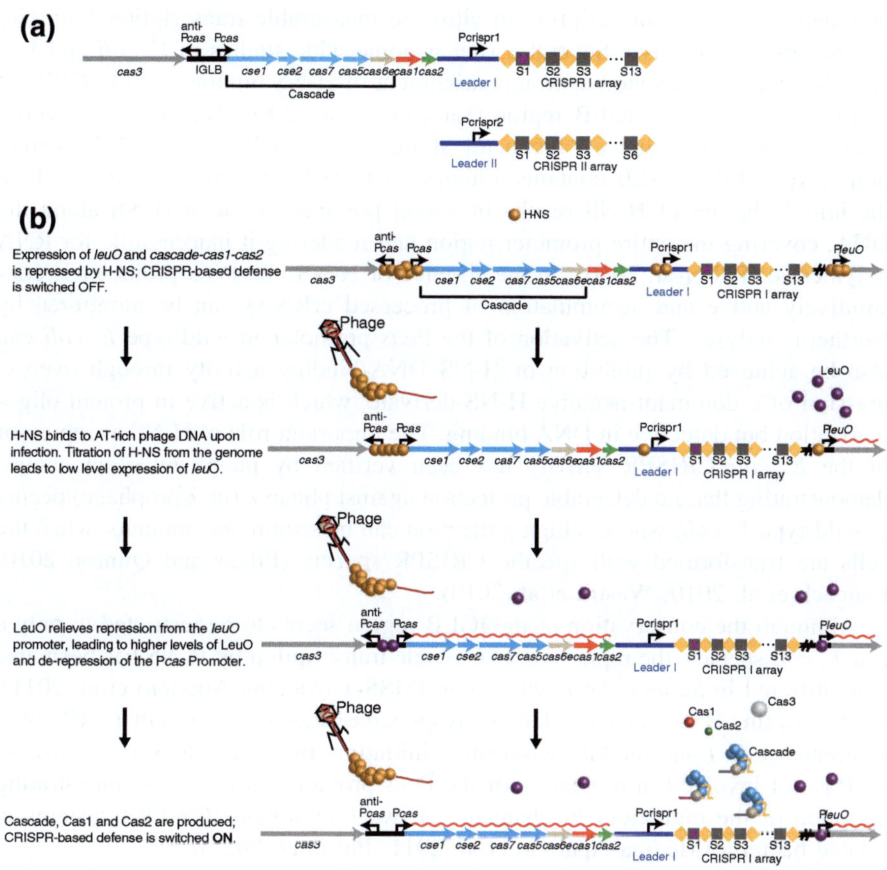

Fig. 4.2 On/Off switch of the CRISPR-Cas system in *E. coli* K12. **a** The two CRISPR arrays of *E. coli* K12 are shown schematically. The CRISPR I array consists of 14 repeat sequences (*diamonds*, colored in *orange*), 13 spacer sequences (*boxes*, colored in *dark gray*), and is preceded by Leader I, which contains the Pcrispr1 promoter, directing the transcription of the CRISPR I locus. The *cas* genes, associated with the CRISPR I array, are indicated by *gray, blue, red and green arrows*. The non-coding region between *cas3* and *cse1*, denoted as IGLB, contains the two divergent promoters P*cas* and antiP*cas*. The P*cas* promoter directs the transcription unit covering the seven downstream *cas* genes. The role of antiP*cas* and the respective anti*cas* transcript is not known. The second CRISPR array is physically unlinked to CRISPR I array and contains six spacers. The Leader II array is homologous to Leader I array and contains the promoter Pcrispr2. **b** Model for activation of CRISPR-based defense in the presence of AT-rich replicating invader DNA. During standard growth conditions, H-NS (*brown*) binds the P*cas* and Panti-*cas* promoter, the Pcrispr1 promoter, and the P*leuO* promoter, leading to repression of transcription from these promoters. Hence, CRISPR-based defense is switched off. When phage DNA (or other source of AT-rich DNA such as plasmid DNA) enters the cell, H-NS can be redistributed from the genome to the invader DNA. This may relieve repression of H-NS controlled genes, such as *leuO*. Since IGLB was reported to have very high affinity for H-NS, P*cas* may remain silenced at this stage. Since LeuO (*purple*) positively regulates its own expression, LeuO levels increase rapidly and mediate derepression of P*cas*, while the Panti-*cas* is repressed by LeuO as well. Subsequent transcription from the Pcas promoter leads to the transcription of the Cas1, Cas2, and Cascade-encoding genes. As expression of the Cas3-encoding gene is less tightly regulated, the cell contains at this stage all components for a CRISPR-based immune response. Hence, CRISPR-based defense is switched on

polymerase initiates transcription in vitro, no measurable transcriptional activity can be observed in vivo. Several earlier genome-wide studies in *E. coli* and *Salmonella enterica* indicated a strong enrichment of H-NS binding sites in CRISPR regions, including the IGLB region (Lucchini et al. 2006; Navarre et al. 2006; Oshima et al. 2006). Indeed, biochemical analyses of H-NS-IGLB DNA interactions revealed that IGLB contains a high-affinity H-NS binding site, from which the initial binding of H-NS results in lateral polymerization of H-NS along the DNA, covering the entire promoter region and rendering it inaccessible for RNA polymerase (Pul et al. 2010). In the absence of H-NS, the P*cas* promoter is constitutively active and accumulation of processed crRNAs can be monitored by Northern analyses. The activation of the P*cas* promoter in wild-type *E. coli* can also be achieved by inhibition of H-NS DNA-binding activity through overexpression of a dominant-negative H-NS-derivate, which is active in protein oligomerization but defective in DNA binding. The important role of H-NS as repressor of the *E. coli* CRISPR activity has been verified by phage infection assays, demonstrating that no detectable protection against phage λ (or λ prophage) occurs in wild-type *E. coli,* whereas high protection can be seen in *hns* mutants, when the cells are transformed with specific CRISPR spacers (Edgar and Qimron 2010; Pougach et al. 2010; Westra et al. 2010).

Although the conservation of the IGLB region seems to be restricted to only a few *E. coli* species, the repression of Cascade transcription by H-NS has been also demonstrated in *Salmonella typhimurium* IMSS-1 (Medina-Aparicio et al. 2011). In this strain, H-NS and the leucine-responsive regulatory protein (LRP) bind upstream of *cse1* and inhibit transcription initiation. In contrast to *S. typhimurium*, LRP is not involved in regulation of the P*cas* promoter in *E. coli*, demonstrating variation of the regulatory mechanisms of two homologous CRISPR systems in related bacteria (Medina-Aparicio et al. 2011; Pul et al. 2010).

In *E. coli* K12, a second divergently orientated promoter, termed antiP*cas*, has been mapped in the IGLB region which is located roughly 80 bp upstream of the P*cas* promoter (Fig. 4.2a). Like P*cas,* the antiP*cas* promoter is tightly repressed by H-NS and activated in *hns* background. Transcription from antiP*cas* leads to an RNA of approximately 200 nucleotides, complementary to the 3′-end of the *cas3* gene (Pul et al. 2010). Although the physiological significance of this transcript is not known yet, a secondary structure prediction indicates a stable structure (Pul et al. 2010), characteristic for many regulatory RNAs (Waters and Storz 2009). An antisense transcript initiated within a *cas* gene has also been found in the type III-A CRISPR-Cas system of *Mycobacterium tuberculosis* (Arnvig et al. 2011). Deep-sequencing analysis of the *M. tuberculosis* transcriptome revealed the presence of this antisense transcript at high abundance, which is initiated at the 3′-end of the *cas2* gene and extends into the neighboring *cas1* gene. The high abundance of this *cas* antisense transcript suggests a regulatory function, likely affecting *cas* gene expression or mRNA stability (Arnvig et al. 2011). It remains to be shown whether such cis-encoded antisense RNAs are generally involved in regulation of *cas* gene expression.

4.1.3 Activation of the *E. coli* CRISPR System by the Transcription Factor LeuO

H-NS is a pleiotropic regulator, which belongs to the family of nucleoid-associated proteins (NAPs). As is known for many NAP members, the DNA-binding activity of H-NS can be modulated through changes in the environment, such as temperature or osmolarity, or by other proteins influencing the DNA-binding capacity, either by protein–protein interaction or by competition with H-NS for DNA-binding sites (Dorman 2004; Stoebel et al. 2008). One of the most prominent antagonists of H-NS is the LysR-type transcription factor LeuO, a global regulator, which acts as an activator of many H-NS-repressed operons (De la Cruz et al. 2007; Shimada et al. 2011; Stoebel et al. 2008; Stratmann et al. 2008).

The H-NS-mediated repression of CRISPR-based immunity in *E. coli* K12 and the homologous type I-E CRISPR system of *S. typhimurium* can be relieved by the transcription activator LeuO (Dillon et al. 2012; Hernandez-Lucas et al. 2008; Medina-Aparicio et al. 2011; Shimada et al. 2009). It could be shown that elevated levels of plasmid-encoded LeuO cause de-repression of the Cascade-*cas1-cas2* operon, resulting in protection against phage λ infection, when the cells are transformed with CRISPR spacers complementary to the phage DNA (Westra et al. 2010). In vitro binding analyses revealed that LeuO interacts with two DNA sites flanking the H-NS nucleation site on IGLB and inhibits cooperative spreading of H-NS along the P*cas* promoter region. The de-repression of Cascade transcription by LeuO causes increased crRNA abundance due to processing of the short-lived pre-crRNA to more stable crRNAs-Cascade complexes. Although the high abundance of plasmid-encoded LeuO is able to relieve H-NS-mediated repression of CRISPR immunity, it should be noted that under laboratory growth conditions, the expression level of genomic *leuO* seems to be insufficient to overcome P*cas* inhibition by H-NS. While the regulation of LeuO expression itself is complex (e.g., *leuO* expression is negatively regulated by H-NS and positively by LeuO (Chen et al. 2005; Hommais et al. 2001), the finding may indicate that still other regulators or conditions might be involved for the activation of *leuO*, in order to trigger the induction of CRISPR immunity. A recent study has demonstrated that *leuO* transcription is activated by BglJ-RcsB heterodimers (Stratmann et al. 2012). Interestingly, RcsB is the response regulator of the Rcs two-component system, involved in membrane signaling pathway (Majdalani and Gottesman 2005). Although *leuO* expression has been shown to be induced by the alarmone guanosine tetraphosphate (ppGpp) (Fang et al. 2000; Majumder et al. 2001), activation of P*cas* transcription or formation of crRNAs was not affected by induction of the stringent response in *E. coli* (Westra et al. 2010).

Other NAPs known to act often in concert with H-NS, such as FIS, StpA, or LRP have also been analyzed for binding to the regulatory region of the P*cas* promoter. While the H-NS paralogue StpA (58 % amino acid identity) displayed IGLB binding very similar to that of H-NS, the NAPs, FIS, and LRP, did not show specific interaction in a physiological concentration range. The latter NAPs were thus considered not to participate in synergistic or antagonistic regulation of the expression of the *cas* genes in *E. coli* (Pul et al. 2010). However, as indicated above, the CRISPR-Cas system of *S. typhimurium* is repressed by LRP (Medina-Aparicio et al. 2011).

4.1.4 Xenogeneic Silencing by H-NS and CRISPR Defense

Initially identified as a major component of the bacterial nucleoid, involved in structuring and compaction of the genomic DNA, H-NS has been shown to be a pleiotropic regulator in Enterobacteriacea, involved in regulation of nearly 5 % of *E. coli* genes (Hommais et al. 2001). It plays an important role in adaptation of bacteria to altered environmental conditions but also as a specific transcriptional regulator of many genes (Dorman 2004). Genome-wide analyses have revealed an important role of H-NS in down-regulating transcription of AT-rich foreign DNA in *E. coli* and *S. typhimurium* (Lucchini et al. 2006; Navarre et al. 2006; Oshima et al. 2006), a function which has been referred to as "xenogeneic silencing" (Navarre et al. 2007). Binding of H-NS to foreign genetic elements, which are often characterized by higher AT-content relative to the host genome, leads to inhibition of foreign gene transcription by silencing potential promoters (Dillon et al. 2010; Dorman 2007; Doyle et al. 2007; Navarre et al. 2007; Stoebel et al. 2008). The precise mechanisms leading to the specific activation of CRISPR-Cas are not known thus far, although the LeuO protein has been identified to act as antisilencer in *E. coli* and *S. typhimurium.* An attractive proposition how CRISPR systems could be activated takes into account that the cellular level of H-NS becomes sequestered when foreign DNA enters the cell. This in turn could cause a re-distribution of H-NS bound to genomic DNA in favor of AT-rich invading foreign DNA, resulting in the release of H-NS from the *leuO* gene and the CRISPR operons, leading to a release of repression (Fig. 4.2b) (Mojica and Diez-Villasenor 2010; Pul et al. 2010; Westra et al. 2010). The average AT-content of M13 (59.3 %) or T4 (64.7 %) phage genomes, for example, are significantly higher than the *E. coli* K12 genome (49.2 %). Although even the average AT-content of phage lambda genome (50.1 %) is comparable to that of *E. coli*, it contains regions with high AT-content of 54 or 63 % (Skalka 1969). It is tempting to speculate, therefore, that the invasion and amplification of these AT-rich elements in *E. coli* could cause a redistribution of H-NS, and thus to cause partial derepression of *leuO* and/or P*cas* promoters (Fig. 4.2b). Although this regulatory strategy may have its limitations when dealing with virulent phages due to the lack of pre-existing effector complexes, less aggressive forms of AT-rich invading DNA, such as plasmids, may become subject to CRISPR interference. A reduction and redistribution of H-NS bound to genomic DNA has been demonstrated to occur in response to the acquisition of a 180 kb AT-rich plasmid in *S. typhimurium* (Dillon et al. 2010).

Some H-NS antisilencing proteins are encoded by invading foreign DNA to counteract the repression by H-NS (Stoebel et al. 2008). One of these proteins is the T7 phage-encoded protein gp5.5, which interacts directly with H-NS and abolishes H-NS-mediated transcriptional silencing, including host genes (Ali et al. 2011; Liu and Richardson 1993). Therefore, it is possible that T7 infection contributes to de-repression of the CRISPR system in *E. coli*.

In contrast to H-NS antisilencer, foreign DNA encoded H-NS paralogues could be involved to keep the host CRISPR system silenced. The recently identified phage-encoded H-NS protein in the bacteriophage EPV1 could therefore act in a similar way to hijack the CRISPR-Cas system of the host (Skennerton et al. 2011).

Perhaps, the *E. coli* CRISPR-Cas system may be regarded as an example of an H-NS-mediated acquisition of a new system through horizontal gene transfer by the host. Tight regulation of the CRISPR-Cas system could contribute to bacterial fitness in several ways, e.g., by minimizing consumption of host cell resources, by prevention of uncontrolled spacer uptake [and the accompanying risk of self-targeting (Stern et al. 2010)], or by reducing the levels of Cas proteins that are potentially toxic due to their various DNase and RNase activities (Makarova et al. 2006). The role of H-NS in CRISPR regulation is obviously not restricted to the interference stage. The uptake of new spacer sequences in *E. coli* has been observed in *hns* mutant strains (Swarts et al. 2012), suggesting that H-NS-mediated inhibition affects also the adaptation stage and could contribute to minimize the risk of self-targeting spacer uptake.

4.1.5 Role of Envelope and Heat-shock Stress Responses on CRISPR Activity

An alternative pathway triggering the induction of CRISPR-based immunity against invading foreign DNA may involve the sensing of membrane stress associated with injection of DNA into the cell. In *E. coli*, envelope stress response is mediated by the two-component systems Cpx, Bae, and Rcs, the alternative sigma factor σ^E, and the phage shock protein (Psp) (Darwin 2005; Raivio 2005; Rowley et al. 2005). The first evidence for an involvement of the BaeSR two-component system in triggering the *E. coli* CRISPR response was obtained by the observation that overexpression of BaeR activates transcription of *cse1* (formerly known as *ygcL*) (Baranova and Nikaido 2002). A more detailed insight into the participation of BaeSR two-component system in CRISPR activity was provided by a recent study (Perez-Rodriguez et al. 2011). The authors demonstrated that the CRISPR system is activated in cells expressing a plasmid-encoded protein, which is substrate for export by the twin-arginine translocation (Tat) pathway. In the absence of the chaperone DnaK, the aberrantly folded Tat substrate leads to activation of the BaeSR signal transduction system by unknown mechanisms, which in turn causes an induction of the CRISPR system and subsequent inactivation of the plasmid encoding for the miss-folded protein. The activation of the CRISPR system has been shown to be mediated by binding of the phosphorylated response regulator BaeR (P-BaeR) within the coding region of the *cse1* gene. Although the mechanistic details of BaeSR activation and the induction of the CRISPR system by the response regulator BaeR are not yet understood, this example represents a clear link between envelope stress and CRISPR response (Perez-Rodriguez et al. 2011; Raivio 2011).

In addition to the σ^{70}-dependent CRISPR promoters, two potential σ^{32}-dependent promoters have been mapped within coding regions of the CRISPR-Cas operon of *E. coli*: the first one is located upstream of *cas5* within the coding region of *cas7*, while the second one maps upstream of *cas2* within the coding region of *cas1* (Nonaka et al. 2006; Wade et al. 2006). The latter has been identified by microarray analysis and verified by 5′-RACE PCR (Nonaka et al. 2006). The alternative sigma factor σ^{32} mediates the heat-shock response by redirecting the specificity of the RNA polymerase from housekeeping to heat-shock gene transcription (Straus et al. 1987; Yamamori and Yura 1980). The presence of σ^{32}-dependent CRISPR promoters suggests a link between CRISPR immunity and heat-shock response. The location of the heat-shock promoters indicates that they will not induce expression of the complete Cascade operon but could serve to boost transcription of limiting *cas* gene products under specific conditions. In line with coupling *cas* gene expression to heat shock, it was found that one of the proteins induced after phage challenge is the chaperon high temperature protein G (HtpG), which has been shown to be essential for Cas3 activity (Yosef et al. 2011). Previous work by Qimron et al., had demonstrated that transformation of λ-lysogenized *E. coli* with anti-λ spacer causes cell death after induction of the CRISPR system (Edgar and Qimron 2010). The authors took advantage of this suicidal effect of CRISPR on lysogenized cells to establish a genetic screening system for the identification of genes essential for CRISPR activity (Yosef et al. 2011). This screening resulted in the detection of *htpG*, encoding a protein (HtpG) homologous to the eukaryotic chaperone Hsp90 (Bardwell and Craig 1987). The absence of HtpG dramatically reduces the activity of the CRISPR system, while the activity could be restored by transforming the Δ*htpG* cells with a plasmid encoding the Cas3 protein. The results provide strong evidence that functional Cas3 is the limiting factor for CRISPR activity in absence of HtpG. The Cas3 protein has nuclease and helicase activities and is essential for inactivation of target DNA during the interference stage (Brouns et al. 2008; Westra et al. 2012; see Chap. 6). Co-overexpression of HtpG and Cas3 protein leads to increased Cas3 levels, supporting the notion that HtpG is involved in folding and/or stabilization of the Cas protein.

In summary, the present data indicate, at least for *E. coli*, that the expression of the CRISPR system is linked to global stress responses, like envelope or heat stress by complex mechanisms including transcriptional and post-transcriptional control (Fig. 4.3). We are only starting to understand the basics of these complex regulatory networks and future work is required to explore the details how and by which signals phage–host interaction or plasmid conjugation causes envelope

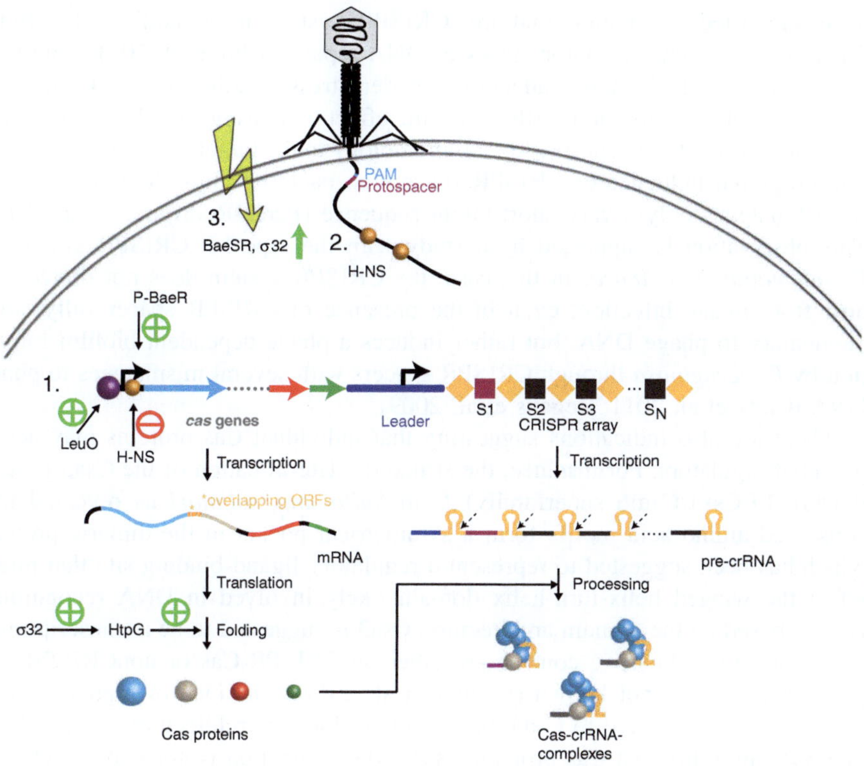

Fig. 4.3 Regulatory network of CRISPR response in *E. coli* K12. The CRISPR system of *E. coli* K12 can be activated under several conditions. (*1*) The inhibition by H-NS can be relieved by elevated levels of LeuO (indicated as magenta ball), which competes with H-NS for binding to the IGLB region. (*2*) Another possibility for the induction of the CRISPR system, though not yet proven, relies on the dissociation of IGLB-bound H-NS molecules, which are sequestered to the AT-rich invading DNA. (*3*) Sensing of membrane stress or heat-shock could activate the two-component system BaeSR and/or σ^{32} expression. The phosphorylated response regulator BaeR activates *cas* gene expression. σ^{32} activates HtpG expression, which in turn is essential for Cas3 folding or stability

stress, and how the signal transduction leading to activation of the CRISPR-Cas system functions. Unraveling the link between CRISPR expression and the bacterial global stress response will be an important goal in the future.

4.1.6 Regulatory Functions of CRISPR-Cas

The complex regulation of the CRISPR system within the frame of global bacterial stress responses could suggest that the CRISPR system has a regulatory role in addition to its function in bacterial immunity against foreign DNA. Recent studies have supported the notion that the CRISPR system bears additional cellular functions, e.g., gene regulatory tasks or DNA repair (Babu et al. 2011), and thus may contribute to bacterial adaptation under stress conditions (see Chap. 10). In fact, the observation of a crRNA-specific effect on plasmid stability with spacer sequences of relatively short (8–11 bp) homology to the target DNA could point to a more global influence of CRISPR spacers on bacterial physiology, even if the spacer matches only a very short target sequence (Perez-Rodriguez et al. 2011). This observation is supported by a study with the type I-F CRISPR system of *Pseudomonas aeruginosa*. In this case, the CRISPR system does not protect the host from phage infection, even in the presence of CRISPR spacer fully complementary to phage DNA, but rather induces a phage-dependent biofilm formation by *P. aeruginosa* through CRISPR spacers with several mismatches to phage DNA (Cady et al. 2011; Zegans et al. 2009).

There are also indications suggesting that individual Cas proteins may act as potential regulators. For instance, the structural determination of the Csa3 protein (COG1517/Csx1/Csm6 superfamily) from *Sulfolobus solfataricus* revealed that conserved amino acid motifs form a symmetrical pocket in the dimeric protein, which has been suggested to represent a regulatory ligand-binding site that might affect the winged helix-turn-helix domain likely involved in DNA recognition. Hence, based on the domain architecture, Csa3 is suggested to be a transcriptional regulator under allosteric control—whether on CRISPR-Cas or nonCRISPR-Cas gene expression is not known (Lintner et al. 2011). The idea of regulatory Cas proteins is not restricted to Csa3, but a similar domain architecture has also been identified in a different Cas protein, Csx1, whose function is not known yet.

4.2 Activation of CRISPR Transcription of *T. thermophilus* During Phage Infection and the Role of cAMP-CRP

T. thermophilus HB8 contains two CRISPR arrays on its chromosome and nine on the megaplasmid pTT27, and several *cas* genes belonging to type I-E, type III-A, and type III-B, respectively (Grissa et al. 2007). A microarray analysis of the transcriptome of *T. thermophilus* HB8 after phage infection has revealed that type I-E and type III-A operons are positively regulated by the global regulator cAMP-receptor protein (CRP) (Shinkai et al. 2007). In line with this, the expression of *cas* genes was found to be induced after infection of *T. thermophilus* with the lytic phage ΦYS40 in a cAMP-CRP-dependent manner (Agari et al. 2010). Interestingly, phage infection does not cause an induction of *crp* expression itself, suggesting that the up-regulation of the CRISPR operons after phage infection is mediated by the second messenger cAMP. It is hypothesized that phage infection increases the intracellular cAMP concentration, which in turn activates CRP and thus cAMP-CRP responsive operons. The observed activation of the CRISPR operons by CRP could also be an indirect effect, e.g., through down-regulation of repressor proteins after phage infection, which compete with cAMP-CRP for binding on the CRISPR operon. Although a phage-induced increase of cAMP synthesis remains to be shown, the supposition that the well-known second messenger cAMP could act as signaling molecule for phage infection is attractive. In vitro analysis indicated a putative CRP binding site upstream of *cas3* gene of *E. coli* K12, which lends support to this assumption but transcriptomic and genomic SELEX screening analyses did not provide evidence for a regulation of *E. coli* CRISPR system by CRP so far (Shimada et al. 2011; Zheng et al. 2004).

4.3 Modulation of CRISPR Transcription by the DNA Repeat Binding Protein Cbp1 in *Sulfolobus*

Usually, the transcription of the CRISPR array starts unidirectionally from the promoter located in the leader region, leading to formation of a long pre-crRNA covering the entire CRISPR spacers (Brouns et al. 2008; Carte et al. 2008; Pougach et al. 2010). In *S. acidocaldarius* and *S. solfataricus*, and recently in *Clostridium thermocellum*, evidence was provided that both CRISPR strands are transcribed bidirectionally and putative transcriptional start sites located within the spacer sequences result in the production of shorter primary CRISPR transcripts (Deng et al. 2011; Lillestol et al. 2009; Richter et al. 2012). Although the physiological significance of the bidirectional transcription of the CRISPR strands is not known, the potential base-pairing between the reverse strand transcripts and crRNAs could potentially be involved in regulation of CRISPR interference.

In *Sulfolobus,* a CRISPR DNA-repeat binding protein (Cbp1) has been shown to interact directly with the CRISPR array, where it probably modulates transcription of the pre-crRNA (Deng et al. 2011; Peng et al. 2003). Cbp1 homologues are present in acidothermophilic Sulfolobales and hyperthermophilic Desulfurococcales, and the proteins contain a triple repeat sequence, which shows homology to the helix-turn-helix DNA-binding motif (Deng et al. 2011). It has been demonstrated that the Cbp1 protein binds within spacer-repeat units of the CRISPR array. Moreover, the overexpression of Cbp1 results in the increased accumulation of the larger pre-crRNAs and the transcription initiation within internal promoter sites becomes inhibited by Cpb1 (Deng et al. 2011). In contrast, Cbp1 has no effect on the reverse-strand transcripts. Clearly, more work is required to understand the exact role of the reverse transcripts and the function of the interesting protein Cbp1 on CRISPR expression in *Sulfolobus*.

4.4 Overlapping *Cas* Genes: Regulation by Translational Coupling?

Regulation of gene expression in bacteria is known to occur at all steps leading to the final products. In the above sections, we have summarized examples of the regulation of CRISPR activity including regulation on the transcriptional level through small DNA-binding regulators or alternative σ-factors, which recognize specific promoter structures. Further examples for post-transcriptional regulation, involving processing of the pre-crRNAs in type II CRISPR systems by tracrRNA and RNaseIII (Deltcheva et al. 2011; see Chap. 5) or the action of the chaperone HtpG, which functions as putative regulator at the post-translational level (see Sect. 4.1.5) have been demonstrated. The latter mechanisms are executed by non-Cas proteins, which clearly indicate that CRISPR activity is closely linked to the physiology of the host (Table 4.1).

Intensive regulation of *cas* gene expression on the translational level can be expected based on the unusual stoichiometry of Cascade ($Cse1_1Cse2_2Cas7_6Cas5_1$-Cas6e) and based on the arrangement of *cas* genes within the Cascade-encoding transcript (a polycistronic mRNA of *cse1-cse2-cas7-cas5-cas6e-cas1-cas2*). Since all Cascade proteins are translated from the same single mRNA regulation on the translational level is most likely. This notion is further strengthened by the arrangement of the *cas* genes. Interestingly, their open reading frames are either overlapping (by 8 or 14 nucleotides) or separated by extremely short intergenic distances (1, 2, 12, or 15 nucleotides) (Fig. 4.4).

Overlapping genes or very short intergenic regions are considered to be important for viral or phage genomes, where the genome size is limiting but also for regulation of gene expression (Inokuchi et al. 2000; Sakharkar and Chow 2005). As shown in Fig. 4.4, overlapping open reading frames of *cas* genes are prevalent in other CRISPR-Cas subtypes and their occurrence is highly suggestive for regulation of Cas protein expression by translational coupling.

The arrangement of overlapping reading frames may have additional consequences for the proteins encoded. Apart from reflecting a general regulatory mechanism for gene expression, assuring a distinctly different stoichiometry of the Cas proteins in effector complexes, overlapping genes with the same translational open reading frame have the potential to permit the synthesis of more than one form of the encoded proteins (Yu et al. 2007). Whether this aspect might be meaningful for Cas proteins is not known.

Generally, translational coupling within polycistronic mRNAs provides a means for enhanced robustness in the expression of neighboring genes (Lovdok et al. 2009). Moreover, it may be regarded as a safeguard that synthesis of a downstream gene does not occur unless translation of the preceding gene has taken place (Berkhout et al. 1987). This could actually also account for Cas proteins and may thus point to a potential toxicity of individual Cas proteins, when expressed in nontimely or nonstoichiometric manner.

Table 4.1

Regulator	Organism	Effect	Regulated genes	Reference
H-NS	*E. coli* K12	Repression	Type I-E operon: *cse1-cse2-cas7-cas5-cas6e-cas1-cas2*	Pul et al. 2010, Pougach et al. 2010, Westra et al. 2010
		Repression	CRISPR array transcription	Pul et al. 2010
	S. typhimurium IMSS-1	Repression	Type I-E operon: *cse1-cse2-cas7-cas5-cas6e-cas1-cas2*	Medina-Aparicio et al. 2011
LeuO	*E. coli* K12	Activation	Type I-E operon: *cse1-cse2-cas7-cas5-cas6e-cas1-cas2*	Westra et al. 2010
	S. typhimurium IMSS-1	Activation	Type I-E operon: *cse1-cse2-cas7-cas5-cas6e-cas1-cas2*	Medina-Aparicio et al. 2011
LRP	*S. typhimurium* IMSS-1	Repression	Type I-E operon: *cse1-cse2-cas7-cas5-cas6e-cas1-cas2*	Medina-Aparicio et al. 2011
BaeSR	*E. coli* K12	Activation	Type I-E genes: *cse1, cas5, cas6e*	Baranova and Nikaido 2002 Perez-Rodriguez et al. 2011
σ^{32}	*E. coli* K12	Activation	Type I-E genes: *cas5, cas2*	Wade et al. 2006 Nonaka et al. 2006
		Activation	*htpG* *HtpG essential for Cas3 activity	Wade et al. 2006 *Yosef et al. 2011
cAMP-CRP	*T. thermophilus* HB8	Activation	Type III-A operon: *cas10-csm2-csm3-csm4-csm5-csx1_III-U*	Shinkai et al. 2007 Agari et al. 2010
		Activation	Type I-E operon: *cas3-cse1-cse2-cse7-cas5-cas6e-cas1-cas2*	
	E. coli K12	Activation	Type I-E gene *cas3*	Zheng et al. 2004
Cbp1	*S. islandicus, S. solfataricus*	Modulator	CRISPR array transcription	Deng et al. 2011

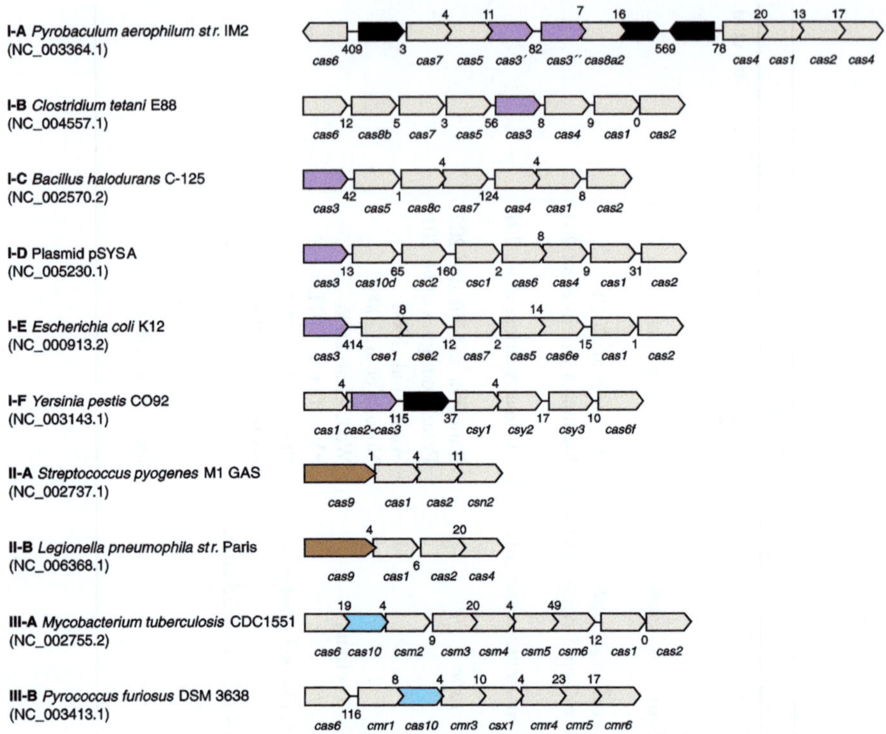

Fig. 4.4 Overlapping *cas* genes in the different CRISPR-Cas subtypes. Operon organizations of the 10 CRISPR-Cas subtypes are shown, indicating overlapping open reading frames or short intergenic regions of the *cas* genes. The numbers above the genes indicate the nucleotides shared by two genes, the numbers below give the length of the intergenic regions in nucleotides between two genes. The classification of the subtypes and the nomenclature of the *cas* genes are taken from (Makarova et al. 2011). The signature genes of the three CRISPR types are colored in *magenta* (*cas3*), *brown* (*cas9*), or *blue* (*cas10*). Genes, shown in *black*, are not annotated as CRISPR-associated

4.5 Regulated Versus Nonregulated CRISPR Systems

Although a permanently active CRISPR system could save precious time for defense, reducing the mortal thread of invaders, an unregulated CRISPR system might increase the danger of self-directed spacer uptake or nonspecific interference with endogenous nucleic acids by Cas proteins. In this chapter, we have described several examples for the coupling of CRISPR activity to different stress conditions, although many aspects of the links to environmental changes are not well understood yet. The strong repression, almost complete silencing, of the CRISPR system in *E. coli* K12 puts into question the effectiveness of CRISPR-Cas as an immune system in this organism, since the phage sensitivity of *E. coli* K12 transformed with specific spacer sequences is only marginally reduced. One plausible consideration might be that the strong down-regulation is a result of long time cultivation of *E. coli* K12 as laboratory strain in phage-free environment. Another possibility could be that CRISPR acts as a last line of defense in *E. coli,* equipped with other defense systems against foreign DNA, such as the restriction-modification systems (R-M) or widespread means of phage exclusion. On the other hand, the down-regulation of CRISPR activity may also be important to ensure the maintenance of genetic diversity by horizontal gene transfer.

At present, we do not completely understand the complex scenario for induction of the type I-E CRISPR system in *E. coli.* Recent studies add more players, likely contributing to the network of defense regulation. An interesting example for unexpected links to possible CRISPR activation stems from recent studies. It was shown for *E. coli* O157:H7 cells that the exposure to lysates of lettuce leaves leads to induction of several stress-related genes including *cas* genes (Kyle et al. 2010). In *Xanthomonas oryzae,* the expression of individual type I-C *cas* genes seems to be induced by quorum sensing (Han et al. 2011), mediated by the small secretory peptide Ax21 (Lee et al. 2009). These examples illustrate that unexpected signals may have the potential to activate a CRISPR response, indicating that much more has to be learned before we fully understand the complex networks behind the regulation of CRISPR defense.

References

Agari Y, Sakamoto K, Tamakoshi M, Oshima T, Kuramitsu S, Shinkai A (2010) Transcription profile of thermus thermophilus CRISPR systems after phage infection. J Mol Biol 395:270–281

Al-Attar S, Westra ER, van der Oost J, Brouns SJ (2011) Clustered regularly interspaced short palindromic repeats (CRISPRs): the hallmark of an ingenious antiviral defense mechanism in prokaryotes. Biol Chem 392:277–289

Ali SS, Beckett E, Bae SJ, Navarre WW (2011) The 5.5 protein of phage T7 inhibits H-NS through interactions with the central oligomerization domain. J Bacteriol 193:4881–4892

Arnvig KB, Comas I, Thomson NR, Houghton J, Boshoff HI, Croucher NJ, Rose G, Perkins TT, Parkhill J, Dougan G, Young DB (2011) Sequence-based analysis uncovers an abundance of non-coding RNA in the total transcriptome of *Mycobacterium tuberculosis*. PLoS Pathog 7:e1002342

Babu M, Beloglazova N, Flick R, Graham C, Skarina T, Nocek B, Gagarinova A, Pogoutse O, Brown G, Binkowski A, Phanse S, Joachimiak A, Koonin EV, Savchenko A, Emili A, Greenblatt J, Edwards AM, Yakunin AF (2011) A dual function of the CRISPR-Cas system in bacterial antivirus immunity and DNA repair. Mol Microbiol 79:484–502

Baranova N, Nikaido H (2002) The baeSR two-component regulatory system activates transcription of the yegMNOB (mdtABCD) transporter gene cluster in *Escherichia coli* and increases its resistance to novobiocin and deoxycholate. J Bacteriol 184:4168–4176

Bardwell JC, Craig EA (1987) Eukaryotic Mr 83,000 heat shock protein has a homologue in *Escherichia coli*. Proc Natl Acad Sci U S A 84:5177–5181

Barrangou R, Fremaux C, Deveau H, Richards M, Boyaval P, Moineau S, Romero DA, Horvath P (2007) CRISPR provides acquired resistance against viruses in prokaryotes. Science 315:1709–1712

Berkhout B, Schmidt BF, van Strien A, van Boom J, van Westrenen J, van Duin J (1987) Lysis gene of bacteriophage MS2 is activated by translation termination at the overlapping coat gene. J Mol Biol 195:517–524

Brouns SJ, Jore MM, Lundgren M, Westra ER, Slijkhuis RJ, Snijders AP, Dickman MJ, Makarova KS, Koonin EV, van der Oost J (2008) Small CRISPR RNAs guide antiviral defense in prokaryotes. Science 321:960–964

Cady KC, White AS, Hammond JH, Abendroth MD, Karthikeyan RS, Lalitha P, Zegans ME, O'Toole GA (2011) Prevalence, conservation and functional analysis of *Yersinia* and *Escherichia* CRISPR regions in clinical *Pseudomonas aeruginosa* isolates. Microbiology 157:430–437

Carte J, Wang R, Li H, Terns RM, Terns MP (2008) Cas6 is an endoribonuclease that generates guide RNAs for invader defense in prokaryotes. Genes Dev 22:3489–3496

Chen S, Iannolo M, Calvo JM (2005) Cooperative binding of the leucine-responsive regulatory protein (Lrp) to DNA. J Mol Biol 345:251–264

Darwin AJ (2005) The phage-shock-protein response. Mol Microbiol 57:621–628

Datsenko KA, Pougach K, Tikhonov A, Wanner BL, Severinov K, Semenova E (2012) Molecular memory of prior infections activates the CRISPR/Cas adaptive bacterial immunity system. Nat Commun 3:945

De la Cruz MA, Fernandez-Mora M, Guadarrama C, Flores-Valdez MA, Bustamante VH, Vazquez A, Calva E (2007) LeuO antagonizes H-NS and StpA-dependent repression in *Salmonella enterica* ompS1. Mol Microbiol 66:727–743

Deltcheva E, Chylinski K, Sharma CM, Gonzales K, Chao Y, Pirzada ZA, Eckert MR, Vogel J, Charpentier E (2011) CRISPR RNA maturation by trans-encoded small RNA and host factor RNase III. Nature 471:602–607

Deng L, Kenchappa CS, Peng X, She Q, Garrett RA (2012) Modulation of CRISPR locus transcription by the repeat-binding protein Cbp1 in *Sulfolobus*. Nucleic Acids Res 40:2470–2480

Diez-Villasenor C, Almendros C, Garcia-Martinez J, Mojica FJ (2010) Diversity of CRISPR loci in *Escherichia coli*. Microbiology 156:1351–1361

Dillon SC, Cameron AD, Hokamp K, Lucchini S, Hinton JC, Dorman CJ (2010) Genome-wide analysis of the H-NS and Sfh regulatory networks in *Salmonella Typhimurium* identifies a plasmid-encoded transcription silencing mechanism. Mol Microbiol 76:1250–1265

Dillon SC, Espinosa E, Hokamp K, Ussery DW, Casadesus J, Dorman CJ (2012) LeuO is a global regulator of gene expression in *Salmonella enterica* serovar Typhimurium. Mol Microbiol 85(6):1072–1089

Dorman CJ (2004) H-NS: a universal regulator for a dynamic genome. Nat Rev Microbiol 2: 391–400

Dorman CJ (2007) H-NS, the genome sentinel. Nat Rev Microbiol 5:157–161

Doyle M, Fookes M, Ivens A, Mangan MW, Wain J, Dorman CJ (2007) An H-NS-like stealth protein aids horizontal DNA transmission in bacteria. Science 315:251–252

Edgar R, Qimron U (2010) The *Escherichia coli* CRISPR system protects from {lambda}lysogenizatin, lysogens, and prophage induction. J Bacteriol 192(23):6291–6294

Fang M, Majumder A, Tsai KJ, Wu HY (2000) ppGpp-dependent leuO expression in bacteria under stress. Biochem Biophys Res Commun 276:64–70

Grissa I, Vergnaud G, Pourcel C (2007) The CRISPRdb database and tools to display CRISPRs and to generate dictionaries of spacers and repeats. BMC Bioinf 8:172

Han SW, Sriariyanun M, Lee SW, Sharma M, Bahar O, Bower Z, Ronald PC (2011) Small protein-mediated quorum sensing in a gram-negative bacterium. PLoS ONE 6:e29192

Hernandez-Lucas I, Gallego-Hernandez AL, Encarnacion S, Fernandez-Mora M, Martinez-Batallar AG, Salgado H, Oropeza R, Calva E (2008) The LysR-type transcriptional regulator LeuO controls expression of several genes in *Salmonella enterica* serovar Typhi. J Bacteriol 190:1658–1670

Hommais F, Krin E, Laurent-Winter C, Soutourina O, Malpertuy A, Le Caer JP, Danchin A, Bertin P (2001) Large-scale monitoring of pleiotropic regulation of gene expression by the prokaryotic nucleoid-associated protein, H-NS. Mol Microbiol 40:20–36

Inokuchi Y, Hirashima A, Sekine Y, Janosi L, Kaji A (2000) Role of ribosome recycling factor (RRF) in translational coupling. EMBO J 19:3788–3798

Jore MM, Brouns SJ, van der Oost J (2011) RNA in defense: CRISPRs protect prokaryotes against mobile genetic elements. Cold Spring Harb Perspect Biol

Kyle JL, Parker CT, Goudeau D, Brandl MT (2010) Transcriptome analysis of *Escherichia coli* O157:H7 exposed to lysates of lettuce leaves. Appl Environ Microbiol 76:1375–1387

Lee SW, Han SW, Sririyanum M, Park CJ, Seo YS, Ronald PC (2009) A type I-secreted, sulfated peptide triggers XA21-mediated innate immunity. Science 326:850–853

Lillestol RK, Shah SA, Brugger K, Redder P, Phan H, Christiansen J, Garrett RA (2009) CRISPR families of the crenarchaeal genus *Sulfolobus*: bidirectional transcription and dynamic properties. Mol Microbiol 72:259–272

Lintner NG, Frankel KA, Tsutakawa SE, Alsbury DL, Copie V, Young MJ, Tainer JA, Lawrence CM (2011) The structure of the CRISPR-associated protein Csa3 provides insight into the regulation of the CRISPR/Cas system. J Mol Biol 405:939–955

Liu Q, Richardson CC (1993) Gene 5.5 protein of bacteriophage T7 inhibits the nucleoid protein H-NS of *Escherichia coli*. Proc Natl Acad Sci U S A 90:1761–1765

Lovdok L, Bentele K, Vladimirov N, Muller A, Pop FS, Lebiedz D, Kollmann M, Sourjik V (2009) Role of translational coupling in robustness of bacterial chemotaxis pathway. PLoS Biol 7:e1000171

Lucchini S, Rowley G, Goldberg MD, Hurd D, Harrison M, Hinton JC (2006) H-NS mediates the silencing of laterally acquired genes in bacteria. PLoS Pathog 2:e81

Majdalani N, Gottesman S (2005) The Rcs phosphorelay: a complex signal transduction system. Annu Rev Microbiol 59:379–405

Majumder A, Fang M, Tsai KJ, Ueguchi C, Mizuno T, Wu HY (2001) LeuO expression in response to starvation for branched-chain amino acids. J Biol Chem 276:19046–19051

Makarova KS, Grishin NV, Shabalina SA, Wolf YI, Koonin EV (2006) A putative RNA-interference-based immune system in prokaryotes: computational analysis of the predicted enzymatic machinery, functional analogies with eukaryotic RNAi, and hypothetical mechanisms of action. Biol Direct 1:7

Makarova KS, Haft DH, Barrangou R, Brouns SJ, Charpentier E, Horvath P, Moineau S, Mojica FJ, Wolf YI, Yakunin AF, van der Oost J, Koonin EV (2011) Evolution and classification of the CRISPR-Cas systems. Nat Rev Microbiol 9:467–477

Marraffini LA, Sontheimer EJ (2008) CRISPR interference limits horizontal gene transfer in *Staphylococci* by targeting DNA. Science 322:1843–1845

Medina-Aparicio L, Rebollar-Flores JE, Gallego-Hernandez AL, Vazquez A, Olvera L, Gutierrez-Rios RM, Calva E, Hernandez-Lucas I (2011) The CRISPR/Cas immune system is an operon regulated by LeuO, H-NS and LRP in *Salmonella enterica* serovar Typhi. J Bacteriol 193:2396–2407

Mitchell JE, Zheng D, Busby SJ, Minchin SD (2003) Identification and analysis of 'extended -10' promoters in *Escherichia coli*. Nucleic Acids Res 31:4689–4695

Mojica FJ, Diez-Villasenor C (2010) The on-off switch of CRISPR immunity against phages in *Escherichia coli*. Mol Microbiol 77:1341–1345

Navarre WW, Porwollik S, Wang Y, McClelland M, Rosen H, Libby SJ, Fang FC (2006) Selective silencing of foreign DNA with low GC content by the H-NS protein in *Salmonella*. Science 313:236–238

Navarre WW, McClelland M, Libby SJ, Fang FC (2007) Silencing of xenogeneic DNA by H-NS-facilitation of lateral gene transfer in bacteria by a defense system that recognizes foreign DNA. Genes Dev 21:1456–1471

Nonaka G, Blankschien M, Herman C, Gross CA, Rhodius VA (2006) Regulon and promoter analysis of the *E. coli* heat-shock factor, sigma32, reveals a multifaceted cellular response to heat stress. Genes Dev 20:1776–1789

Oshima T, Ishikawa S, Kurokawa K, Aiba H, Ogasawara N (2006) *Escherichia coli* histone-like protein H-NS preferentially binds to horizontally acquired DNA in association with RNA polymerase. DNA Res 13:141–153

Peng X, Brugger K, Shen B, Chen L, She Q, Garrett RA (2003) Genus-specific protein binding to the large clusters of DNA repeats (short regularly spaced repeats) present in *Sulfolobus* genomes. J Bacteriol 185:2410–2417

Perez-Rodriguez R, Haitjema C, Huang Q, Nam KH, Bernardis S, Ke A, Delisa MP (2011) Envelope stress is a trigger of CRISPR RNA-mediated DNA silencing in *Escherichia coli*. Mol Microbiol 79:584–599

Pougach K, Semenova E, Bogdanova E, Datsenko KA, Djordjevic M, Wanner BL, Severinov K (2010) Transcription, processing and function of CRISPR cassettes in *Escherichia coli*. Mol Microbiol 77:1367–1379

Pul U, Wurm R, Arslan Z, Geissen R, Hofmann N, Wagner R (2010) Identification and characterization of *E. coli* CRISPR-cas promoters and their silencing by H-NS. Mol Microbiol 75:1495–1512

Raivio TL (2005) Envelope stress responses and Gram-negative bacterial pathogenesis. Mol Microbiol 56:1119–1128

Raivio T (2011) Identifying your enemies–could envelope stress trigger microbial immunity? Mol Microbiol 79:557–561

Richter H, Zoephel J, Schermuly J, Maticzka D, Backofen R, Randau L (2012) Characterization of CRISPR RNA processing in Clostridium thermocellum and Methanococcus maripaludis. Nucleic Acids Res

Rowley G, Stevenson A, Kormanec J, Roberts M (2005) Effect of inactivation of degS on *Salmonella enterica* serovar typhimurium in vitro and in vivo. Infect Immun 73:459–463

Sakharkar KR, Chow VT (2005) Strategies for genome reduction in microbial genomes. Genome Inform 16(69–75):26

Shimada T, Yamamoto K, Ishihama A (2009) Involvement of the leucine response transcription factor LeuO in regulation of the genes for sulfa drug efflux. J Bacteriol 191:4562–4571

Shimada T, Fujita N, Yamamoto K, Ishihama A (2011) Novel roles of cAMP receptor protein (CRP) in regulation of transport and metabolism of carbon sources. PLoS ONE 6:e20081

Shinkai A, Kira S, Nakagawa N, Kashihara A, Kuramitsu S, Yokoyama S (2007) Transcription activation mediated by a cyclic AMP receptor protein from Thermus thermophilus HB8. J Bacteriol 189:3891–3901

Skalka A (1969) Nucleotide distribution and functional orientation in the deoxyribonucleic acid of phage phi 80. J Virol 3:150–156

Skennerton CT, Angly FE, Breitbart M, Bragg L, He S, McMahon KD, Hugenholtz P, Tyson GW (2011) Phage encoded H-NS: a potential achilles heel in the bacterial defence system. PLoS ONE 6:e20095

Stern A, Keren L, Wurtzel O, Amitai G, Sorek R (2010) Self-targeting by CRISPR: gene regulation or autoimmunity? Trends Genet 26:335–340

Stoebel DM, Free A, Dorman CJ (2008) Anti-silencing: overcoming H-NS-mediated repression of transcription in Gram-negative enteric bacteria. Microbiology 154:2533–2545

Stratmann T, Madhusudan S, Schnetz K (2008) Regulation of the yjjQ-bglJ operon encoding LuxR-type transcription factors and the divergent yjjP gene by H-NS and LeuO. J Bacteriol 190:926–935

Stratmann T, Pul U, Wurm R, Wagner R, Schnetz K (2012) RcsB-BglJ activates the *Escherichia coli* leuO gene, encoding an H-NS antagonist and pleiotropic regulator of virulence determinants. Mol Microbiol 83:1109–1123

Straus DB, Walter WA, Gross CA (1987) The heat shock response of *E. coli* is regulated by changes in the concentration of sigma 32. Nature 329:348–351

Swarts DC, Mosterd C, van Passel MW, Brouns SJ (2012) CRISPR interference directs strand specific spacer acquisition. PLoS ONE 7:e35888

Wade JT, Roa DC, Grainger DC, Hurd D, Busby SJ, Struhl K, Nudler E (2006) Extensive functional overlap between sigma factors in *Escherichia coli*. Nat Struct Mol Biol 13:806–814

Waters LS, Storz G (2009) Regulatory RNAs in bacteria. Cell 136:615–628

Westra ER, Pul U, Heidrich N, Jore MM, Lundgren M, Stratmann T, Wurm R, Raine A, Mescher M, Van Heereveld L, Mastop M, Wagner EG, Schnetz K, Van Der Oost J, Wagner R, Brouns SJ (2010) H-NS-mediated repression of CRISPR-based immunity in *Escherichia coli* K12 can be relieved by the transcription activator LeuO. Mol Microbiol 77:1380–1393

Westra ER, van Erp PB, Kunne T, Wong SP, Staals RH, Seegers CL, Bollen S, Jore MM, Semenova E, Severinov K, de Vos WM, Dame RT, de Vries R, Brouns SJ, van der Oost J (2012) CRISPR immunity relies on the consecutive binding and degradation of negatively supercoiled invader DNA by Cascade and Cas3. Mol Cell 46:595–605

Yamamori T, Yura T (1980) Temperature-induced synthesis of specific proteins in *Escherichia coli*: evidence for transcriptional control. J Bacteriol 142:843–851

Yosef I, Goren MG, Kiro R, Edgar R, Qimron U (2011) High-temperature protein G is essential for activity of the *Escherichia coli* clustered regularly interspaced short palindromic repeats (CRISPR)/Cas system. Proc Natl Acad Sci U S A 50(108):20136–20141

Yosef I, Goren MG, Qimron U (2012) Proteins and DNA elements essential for the CRISPR adaptation process in *Escherichia coli*. Nucleic Acids Res 40:5569–5576

Young JC, Dill BD, Pan C, Hettich RL, Banfield JF, Shah M, Fremaux C, Horvath P, Barrangou R, Verberkmoes NC (2012) Phage-induced expression of CRISPR-associated proteins is revealed by shotgun proteomics in *Streptococcus thermophilus*. PLoS ONE 7:e38077

Yu JS, Kokoska RJ, Khemici V, Steege DA (2007) In-frame overlapping genes: the challenges for regulating gene expression. Mol Microbiol 63:1158–1172

Zegans ME, Wagner JC, Cady KC, Murphy DM, Hammond JH, O'Toole GA (2009) Interaction between bacteriophage DMS3 and host CRISPR region inhibits group behaviors of *Pseudomonas aeruginosa*. J Bacteriol 191:210–219

Zheng D, Constantinidou C, Hobman JL, Minchin SD (2004) Identification of the CRP regulon using in vitro and in vivo transcriptional profiling. Nucleic Acids Res 32:5874–5893

Chapter 5
crRNA Biogenesis

Emmanuelle Charpentier, John van der Oost and Malcolm F. White

Abstract Mature crRNAs are key elements in CRISPR-Cas defense against genome invaders. These short RNAs are composed of unique repeat/spacer sequences that guide the Cas protein(s) to the cognate invading nucleic acids for their destruction. The biogenesis of mature crRNAs involves highly precise processing events. Interestingly, different types of CRISPR-Cas systems have evolved distinct crRNA maturation mechanisms. The CRISPR repeat-spacer array is transcribed as a precursor CRISPR RNA molecule (pre-crRNA) that undergoes one or two maturation steps. In type I CRISPR-Cas systems, pre-crRNA is cleaved within the repeat regions by a specific Cas6-like endoribonuclease that at least in some cases is a subunit of a Cascade complex to yield the mature crRNAs. In type III systems, the standalone endoribonuclease Cas6 processes pre-crRNA by cleavage within the repeats, producing an intermediate molecule that is further trimmed to generate the mature crRNAs. Type II systems have evolved a unique crRNA biogenesis pathway,

E. Charpentier
The Laboratory for Molecular Infection Medicine Sweden (MIMS),
Umeå Centre for Microbial Research (UCMR), Department of Molecular Biology,
Umeå University, Umeå, Sweden

J. van der Oost
Laboratory of Microbiology, Wageningen University, Wageningen, The Netherlands
e-mail: john.vanderoost@wur.nl

M. F. White
Biomedical Sciences Research Complex, University of St Andrews, St Andrews,
Fife KY16 9ST, UK
e-mail: mfw2@st-andrews.ac.uk

E. Charpentier (✉)
Helmholtz Centre for Infection Research, Braunschweig, Germany
e-mail: emmanuelle.charpentier@mims.umu.se

E. Charpentier
Hannover Medical School, Hannover, Germany

R. Barrangou and J. van der Oost (eds.), *CRISPR-Cas Systems*,
DOI: 10.1007/978-3-642-34657-6_5, © Springer-Verlag Berlin Heidelberg 2013

in which a *trans*-acting small RNA (encoded by the CRISPR-Cas locus) base pairs with each repeat sequence of the pre-crRNA to form a double-stranded RNA template that is cleaved by the housekeeping endoribonuclease III in the presence of protein Cas9 (Csn1). The generated intermediates are then subjected to further maturation by a yet to be revealed mechanism. In this chapter, we present a detailed comparative analysis of pre-crRNA recognition and cleavage mechanisms involved in crRNA biogenesis in the three types of CRISPR-Cas systems.

Contents

5.1 Introduction

The core components of the CRISPR-Cas defense machinery are the short CRISPR RNAs (crRNAs) that associate with one or more Cas proteins to target and destroy invading nucleic acids. The CRISPR-Cas systems are extremely

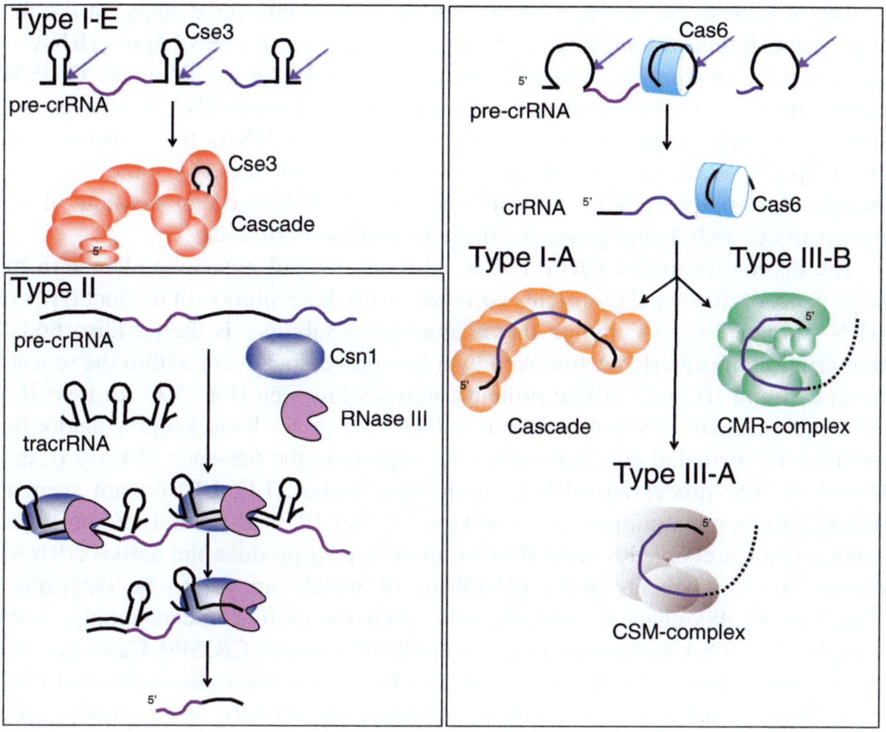

Fig. 5.1 Comparison of crRNA processing pathways in type I, II, and III systems. In the type I-E system, the palindromic repeats in pre-crRNA form hairpin structures that are recognized by the nuclease Cas6e (Cse3), which is an integral subunit of Cascade. After cleavage, the crRNA hairpin remains associated with Cas6e while other subunits bind the 5′ handle and spacer, which is used for the recognition of cognate genetic element sequences. In type II systems, pre-crRNA with unstructured repeats is bound to an RNA species known as tracrRNA that is complementary to the repeat sequence, forming an RNA duplex that is recognized and cleaved by host RNase III in the presence of Cas9 (Csn1) protein. Further processing by unknown nucleases generates mature crRNA. In type III-B systems, crRNA is generated by the Cas6 endonuclease (as mentioned for type I systems). Cas6 binds unstructured pre-crRNA, cleaving within the repeat to generate crRNA with 5′ and 3′ repeat-derived termini. These crRNAs are taken up by archaeal Cascade (homologous to a type I-A system) or alternatively loaded into the Cmr (type III-B) complex, when present. In the latter case, the 3′ repeat-derived sequence is trimmed away by unknown nucleases. The recently described Cas5d endoribonuclease of subtype I-C that also cleaves pre-crRNA within the repeats and assembles in a Cascade-like complex (Nam et al. 2012) is not represented here

variable in their *Cas* gene composition; a recent reevaluation has resulted in a classification with three main CRISPR-Cas types that are further divided into subtypes (Makarova et al. 2011a, b). Despite the *Cas* diversification, all systems share a common molecular mechanism for genome silencing in which the mature crRNAs contain a unique invader-derived partial sequence that guides the Cas protein(s) to the cognate invading nucleic acids for their eventual destruction. Critical for the activity of CRISPR-Cas is the maturation of crRNAs from the precursor transcript of the CRISPR repeat-spacer array.

The biogenesis of mature crRNAs can be divided into three steps. In the first step, transcription, a long primary transcript or precursor *crRNA* (pre-crRNA) is transcribed from a promoter located upstream of the leader preceding the CRISPR repeat-spacer array. In the second step, cleavage, the pre-crRNA is cleaved at a specific site within the repeats to yield intermediate crRNAs that consist of the entire spacer sequence flanked by partial repeat sequences. In some cases, an additional step, processing, concerns a second nucleolytic processing of the intermediate crRNA that generates the active mature crRNAs.

The diversification of CRISPR-Cas into various (sub)types together with the large panel of distinct Cas proteins correlates with the evolution of distinct types of crRNA biogenesis. A common theme among the subtypes is the (unidirectional) transcription of pre-crRNA followed by a first processing event within the repeats. In types I and III, a Cas6-like protein catalyzes this step (Fig. 5.1). In type II, a *trans*-acting small RNA directs pre-crRNA dicing by housekeeping endoribonuclease III-mediated cleavage within the repeats in the presence of Cas9 (Csn1) (Fig. 5.1). The processed crRNAs from types I (I-A, I-E, I-F) do not seem to undergo further maturation, whereas types II and III (and possibly some type I subtypes) have evolved a second maturation step to produce the active crRNAs, the distinct components and mechanisms of which are yet to be determined (Fig. 5.1). In this chapter, we review in detail the current understanding of the remarkable crRNA maturation processes in the three main CRISPR-Cas types. We discuss similarities and differences of the pathways and analogies with small RNA maturation mechanisms of interference in Eukaryotes. Finally, we provide possible perspectives toward a more complete delineation of the crRNA biogenesis mechanisms.

5.2 crRNA Biogenesis in Type I Systems

Type I systems are present in both bacteria and archaea (Makarova et al. 2011a, b). Like all CRISPR-Cas systems, types I are predicted to target mobile genetic sequences. Recently, some first experimental evidence has been provided for spacer acquisition in *Escherichia coli* (subtype I-E), and the correlating resistance against plasmid (Swarts et al. 2012; Yosef et al. 2012) and phage (Datsensko et al. 2012). In *Pseudomonas aeruginosa*, the system (subtype I-F) is required for inhibition of biofilm formation that depends on an integrated bacteriophage (Cady and O'Toole 2011) and its role in phage maintenance resistance is yet to be demonstrated. Type I systems are characterized by a Cascade (-like) ribonucleoprotein complex and a nuclease/helicase (Cas3) required for interference. Processing of the pre-crRNA transcript is catalyzed by a Cas6-like metal-independent endoribonuclease that cleaves the repeat sequence at a conserved position 8 nt upstream of the repeat-spacer boundary. The mature crRNAs end up in Cascade where they play the crucial role of guiding the complex to the complementary target DNA. In most type I systems characterized so far, the Cas6-like enzyme is a subunit of a Cascade-like complex, which is distinct from the apparent standalone version of Cas6 that may supply the intermediate or mature crRNAs to different complexes in type III systems (see below, "crRNA biogenesis in Type III"). The crRNAs of subtypes I-E and I-F have stable hairpin structures, the functions of which might be to initially expose the cleavage site to the Cas6 catalytic domain, and to subsequently assist in the stable interaction between guide crRNA and Cascade. Following Cas6-mediated cleavage within the repeats, crRNAs of subtypes I-A, I-E, and I-F are not processed any further.

5.2.1 Type I crRNAs are Expressed and Processed in vivo

Expression of type I crRNAs has been demonstrated in *Sulfolobus solfataricus* and *Thermoproteus tenax* (I-A), *Clostridium thermocellum* and *Methanococcus maripaludis* (I-B), *E. coli* and *Thermus thermophilus* (I-E), and *P. aeruginosa* (I-F), essentially by northern blot and RNA sequencing analyses (Brouns et al. 2008; Haurwitz et al. 2010; Jore et al. 2011; Juranek et al. 2012; Lintner et al. 2011; Plagens et al. 2012; Randau 2012; Richter et al. 2012). The *subtype I-A* locus is characterized by the presence of *cas6*, located in 3′ of an operon composed of *cas1*, *cas4*, *csa5*, *cas8a1*, *cas7* (*cas2*), *cas5*, *cas2*, *cas3′cas3″*, *cas8a2*. The archaeon *S. solfataricus* was shown to express subtype I-A crRNAs of 60–70 nt bound to a Cascade-like (aCascade) protein complex (Lintner et al. 2011). Expression of subtype I-A mature crRNAs processed from larger transcripts was also recently detected in the hyperthermophilic crenarcheon *T. tenax* (Plagens et al. 2012). The *subtype I-B* locus contains the gene *cas6* followed by the genes *cas8b*, *cas7*, *cas5*, *cas3′cas3″*, *cas4*, *cas1*, and *cas2*. Expression and processing of subtype I-B pre-crRNAs were detected in the bacterial species *C. thermocellum* and the archaeal species *M. maripaludis* (Richter et al. 2012). In this study, RNAs antisense to crRNAs, transcribed from spacer elements, were also detected in *C. thermocellum*, as previously described for the subtype III-B in *Sulfolobus* (Lillestol et al. 2009) and *Pyrococcus furiosus* (Hale et al. 2012) (see below). *Subtype I-E* in *E. coli* is specified by the presence of *cse1* (*casA*), *cse2* (*casB*), *cas7* (*casC*), *cas5* (*casD*), *cas6e* (*casE*), in addition to the core genes *cas1* and *cas2* and the nuclease/helicase gene *cas3*. In 2008 and 2011, Brouns and Jore identified crRNAs of 61 nt as mature species produced by the type I-E array (Brouns et al. 2008; Jore et al. 2011). Interestingly, transcription of the Cascade (see below)-encoding *cse1-cse2-cas7-cas5-cas6e* operon, a transcript antisense to *cas3* mRNA and to a certain extent of the CRISPR array is controlled by an interplay of the global transcriptional regulators H-NS (heat-stable nucleoid-structuring) and LeuO (Hommais et al. 2001; Oshima et al. 2006; Pougach et al. 2010; Pul et al. 2010; Westra et al. 2010). In addition, expression of the *E. coli* Cascade operon is positively regulated by BaeR, response regulator of the two-component system BaeSR involved in membrane stress (Baranova and Nikaido 2002; Perez-Rodriguez et al. 2011). More recently, subtype I-E crRNA expression in *T. thermophilus* was detected by differential RNA sequencing and northern blot analysis (Juranek et al. 2012). The *subtype I-F cas* operon consists of the genes *cas1*, *cas2-cas3*, *csy1*, *csy2*, *csy3*, and *cas6f* (*csy4*). In *P. aeruginosa*, crRNA fragments of this subtype were visualized by northern blot analysis of RNAs co-purified with Cas6f (Haurwitz et al. 2010).

5.2.2 *In Subtypes I-A, I-B, I-E and I-F, the Endoribonuclease Cas6/6x Cleaves Pre-crRNA Within the Repeats*

5.2.2.1 Subtype I-A in *S. Solfataricus*

An archaeal Cascade (aCascade)-like complex from subtype I-A of *S. solfataricus* has been structurally and functionally analyzed (Lintner et al. 2011). In *S. solfataricus*, Cas7 was shown to co-purify with the proteins Cas5a, Cas6, Csa5, and processed forms of crRNAs, with the dominant protein Cas7 forming a stable complex with Cas5a (Lintner et al. 2011). Transmission electron microscopy revealed helical structures of variable length, perhaps because of substoichiometric amounts of other Cascade components, similar to that observed with *E. coli* Cascade samples (Brouns, Jore and Van der Oost, unpublished). Cas7 was structurally analyzed and shown to have a crescent-shaped structure composed of a modified RNA-recognition motif (Lintner et al. 2011), in perfect agreement with the role of Cas7 in binding crRNAs (Wiedenheft et al. 2011b). In addition, *S. solfataricus* Cas6 has a metal-independent ribonuclease activity and generates crRNAs by cleavage of template pre-crRNAs at a single position within the repeat consistent with the position cleaved by *E. coli* Cas6e (subtype I-E) and also *P. furiosus* Cas6 (subtype III-B) (see below, "crRNA biogeneiss in Type III") (Lintner et al. 2011). This is also consistent with the sequencing analysis of crRNAs associated with aCascade that revealed a composition of a 5′ repeat fragment (8 nt), a complete spacer sequence, and a fragment of the 3′ end of the repeat (16–17 nt) (Lintner et al. 2011). Thus, Cas6 associated with Cas7–Cas5a in a complex generating mature crRNA products is reminiscent of *E. coli* Cascade (Lintner et al. 2011). Interestingly, the Cas7–Cas5a complex binds crRNAs and forms a ternary complex with target ssDNA, thus demonstrating an additional analogy to *E. coli* Cascade that utilizes bound crRNAs to target DNA within a ternary complex (Lintner et al. 2011).

5.2.2.2 Subtype I-B in *C. Thermocellum* and *M. Maripaludis*

Cas6 proteins from subtypes I-B of the bacterium *C. thermocellum* and the archaeon *M. maripaludis* were recently demonstrated to act as endoribonucleases cleaving at the same position within the crRNA repeats as for Cas6e from *E. coli* (see below)

(Richter et al. 2012). Cas6b requires two histidine residues for catalysis. This is in contrast to Cas6e that utilizes only one histidine residue (see below), suggesting more flexibility in the catalytic core of Cas6 I-B endoribonucleases (Richter et al. 2012).

5.2.2.3 Subtype I-E in *E. Coli*

In *E. coli* subtype I-E, Brouns et al. identified a protein complex formed by Cse1, Cse2, Cas7, Cas5, and Cas6e using affinity chromatography, which they named Cascade (Brouns et al. 2008). A subsequent combined genetic and biochemical approach was used to demonstrate that mature crRNAs were only produced when all proteins forming the Cascade complex were present (Brouns et al. 2008). Cas6e is the only Cascade subunit essential for pre-crRNA cleavage and the repeat sequence within pre-crRNA was shown to be required for the processing reaction (Brouns et al. 2008). RNA cleavage was also independent of divalent metal ions or adenosine triphosphate. An invariant histidine residue at position 20 in Cas6e is essential for the catalytic process (Brouns et al. 2008). Initially, some heterogeneity at the 3′ end of the isolated crRNAs was reported (Brouns et al. 2008), but a later study demonstrated that mature crRNAs of subtype I-E are the result of a single processing step, generally resulting in 61 nt fragments [see below; (Jore et al. 2011)]. Sequence analysis of crRNA species associated with Cascade demonstrated that the mature crRNAs are composed of (1) an 8 nt repeat fragment (5′ handle), (2) a complete spacer sequence (32 nt) and (3) a 21 nt repeat fragment consisting of a stable stem-loop of seven base pairs and a four nucleotide loop (3′ handle) (Brouns et al. 2008). Subsequent ESI-MS/MS analysis of the Cascade-bound crRNAs revealed 5′-hydroxyl and 2′-3′ cyclic phosphate termini (Jore et al. 2011). Low-resolution structure analysis of the Cascade complex further unraveled an unusual seahorse shape with conformational changes triggered upon binding of target DNA. It was demonstrated that crRNA-mediated guiding of Cascade to the target DNA relies on the specific base pairing between crRNA and its complementary DNA strand with displacement of the non-complementary strand, resulting in an R-loop (Jore et al. 2011). Cryo-electron microscopy analysis of the crRNA-Cascade complex more recently revealed display of crRNA along a backbone of six Cas7 subunits (Wiedenheft et al. 2011a). It was suggested that this arrangement would protect crRNA from degradation and position the crRNA to allow high-affinity base-pairing of invading DNA, initially with the seed sequence at the 5′ end of cognate crRNA (Semenova et al. 2011; Wiedenheft et al. 2011a).

5.2.2.4 Subtype I-E in *T. Thermophilus*

In 2006, Ebihara et al. provided the first crystal structure of the then hypothetical TTHB192 protein, now known as Cas6e of subtype I-E, from the bacterium *T. thermophilus* (Ebihara et al. 2006). Structure analysis revealed that the thermophilic protein consists of two independently folded domains exhibiting a ferredoxin-like fold and adopting an RNA recognition motif (RRM)-like domain (Ebihara et al. 2006). On this basis, the protein was predicted to function as a nucleic acid-binding protein (Ebihara et al. 2006). In 2011, the structure of Cas6e from *T. thermophilus* bound to repeat RNAs was determined (Gesner et al. 2011; Sashital et al. 2011). As for the *E. coli* counterpart, crRNAs associated to *T. thermophilus* Cas6e have a cyclic 2′, 3′-phosphodiester at the 3′ termini.

5.2.2.5 Subtype I-F in *P. Aeruginosa*

In *P. aeruginosa* subtype I-F, the Csy proteins Csy1, Csy2, Csy3, and Cas6f assemble into a ribonucleoprotein complex, the function of which is to facilitate recognition of target DNA by enhancing crRNA-DNA sequence-specific hybridization (Haurwitz et al. 2010). Similar to *E. coli* Cascade, the complex has a crescent shape (Haurwitz et al. 2010). However, unlike Cascade, the Csy-Cas6f structure does not show any large tail (Wiedenheft et al. 2011b), a distinction that might have functional relevance on the mode of target recognition (Wiedenheft et al. 2011b). The structure of Cas6f bound to crRNA revealed that Cas6f makes sequence-specific interactions in the major groove of the crRNA repeat stem-loop (Haurwitz et al. 2010). Cas6f binds tightly to pre-crRNA sequences by exclusive interactions upstream of the scissile phosphate, allowing Cas6f to sequester the crRNA for downstream targeting with DNA (Haurwitz et al. 2010; Sternberg et al. 2012). Binding of Cas6f to RNA is substrate specific and requires RNA major groove contacts that are highly sensitive to helical geometry. A strict preference for guanosine adjacent to the scissile phosphate in the active site was reported to contribute to the selectivity mechanism (Haurwitz et al. 2010; Sternberg et al. 2012). Cas6f uses the conserved serine and histidine residues to cleave the pre-crRNA at the 3′ side of a stable RNA stem-loop structure within the repeat (Haurwitz et al. 2010, 2012; Sternberg et al. 2012). Interestingly, unlike the crRNA processing by *E. coli* or *T. thermophilus* Cas6e, crRNAs produced by *P. aeruginosa* Cas6f have a non-cyclic phosphate at the 3′ end (Wiedenheft et al. 2011b).

5.2.3 *In Subtype I-C, Cas5d Acts as the Pre-crRNA Endoribonuclease*

The subtype I-C locus is characterized by the presence of *cas3*, *cas5*, *cas8c*, *cas7*, *cas4*, *cas1*, and *cas2* genes, and by the absence of a *cas6*-like gene. The molecular basis of pre-crRNA processing in subytpe I-C was recently investigated in *Bacillus halodurans* (Nam et al. 2012). Cas5d of the locus was identified as the endoribonuclease that cleaves pre-crRNA within the repeats. Cas5d recognizes both the base of the pre-crRNA stemloop and the 3′ single-stranded overhang in the pre-crRNA repeat and cleaves the substrate into unit length in a metal-independent manner (Nam et al. 2012). Thus, recognition of the 3′ overhang, which corresponds to the 5′ handle in the mature crRNA distinguishes Cas5d from the Cas6-like enzymes. Cleavage was reported to produce a 2′,3′-cyclic phosphate and a 5′ OH in the 5′ and 3′ halves of the crRNA products, respectively. The crystal structure of Cas5d revealed a ferredoxin-based architecture and a catalytic triad consisting of residues Y46, K116, and H117, indicative of a general acid-base mechanism (Nam et al. 2012). Additional biochemical and structural analysis showed that following pre-crRNA cleavage, Cas5d assembles into a 400-kDa complex together with the mature crRNA and Cas8c (Csd1) and Cas7 (Csd2), the other two Cas proteins specific to subtype I-C. Similar to Cascade, the subtype I-C crRNA-Cas complex would subsequently act in interference with DNA. Nam et al. also suggested that pre-crRNA processing by Cas5d and formation of the subtype I-C Cascade-like complex may be spatially and temporally coupled. Two copies of Cas5d appear to be present in the complex. In this setting, Cas5d may serve the equivalent functions of both type I-E Cas6e (CasE, Cse3) and Cas5e (CasD), and assemble at the opposite ends of the complex where one Cas5d would bind to the crRNA 5′ handle and the other Cas5d to the 3′ handle (Nam et al. 2012). Taken together, the structural features of Cas5d and the cleavage site on pre-crRNA show that Cas5d is distinct from the Cas6-like endoribonucleases.

5.3 crRNA Biogenesis in Type II Systems

Type II CRISPR-Cas systems are characterized by a minimal locus with only four genes (*cas9*, *cas1*, *cas2*, and either *csn2* or *cas4*) and the presence of tracrRNA in the vicinity of the *cas* operon or repeat-spacer array (Deltcheva et al. 2011; Makarova ct al. 2011b). Types II are present in bacteria but have, at this point, never been detected in archaea (Makarova et al. 2011b). The system has been studied mainly in streptococci where the first biological evidence for immunity against both cell death (mediated by lytic phages, *Streptococcus thermophilus*) (Barrangou et al. 2007) and acquisition of virulence genes (mediated by lysogenic bateriophages, *Streptococcus pyogenes*) (Deltcheva et al. 2011) was demonstrated. Type II is also active against plasmid maintenance (*S. thermophilus*; Garneau et al. 2010). In 2011, a study in the Gram-positive human pathogen *S. pyogenes* revealed a unique crRNA biogenesis pathway characteristic for type II wherein a first processing event is achieved by the coordinated action of three factors: a *trans*-acting small RNA, the host-encoded RNase III and the Cas9 protein (Deltcheva et al. 2011; Gottesman 2011). The findings are described below.

5.3.1 Type II crRNAs are Expressed and Processed in vivo

Following sequence analysis of available genomes of *S. pyogenes*, Deltcheva et al. selected one clinical isolate as model organism for type II investigation (Deltcheva et al. 2011) and applied the differential RNA sequencing (dRNA-seq) methodology (Sharma et al. 2010) allowing recovery and differentiation of genome-wide primary and processed transcripts (Deltcheva et al. 2011). Remarkably, the most abundant small RNA species collected from the libraries were the crRNA species (Deltcheva et al. 2011). Comparison of total and primary transcript-depleted libraries clearly demonstrated that type II pre-crRNAs were expressed and processed in vivo, which was also confirmed by northern blot analysis. Two RNA species corresponding to a unique intermediate form of 66 nt and distinct mature forms of 39–42 nt were detected. The 66 nt species consisted of 5′-repeat(1/3)-spacer-repeat(2/3)-3′ crRNAs. The second species of 39–42 nt corresponded to the mature crRNAs and consisted of a unique 5′ spacer-derived guide sequence of 20 nt and a 3′ repeat-derived sequence of ~19–22 nt. The remaining ~24 nt 5′-repeat-spacer-3′ fragment could not be observed by either dRNA-seq or northern blot analysis. The authors suggested that this presumably inactive species was likely unstable or rapidly degraded. dRNA-seq data also detected low abundance of 66 nt intermediates, indicating efficient, concomitant and rather fast processing events. It was concluded that crRNA biogenesis in type II occurs as a two-step process with a first cleavage within the repeats and a second cleavage within the spacers (Deltcheva et al. 2011). Given the lengths of the in vivo produced mature crRNAs, the second cleavage was predicted to occur at a specific distance from the first cleavage site (Deltcheva et al. 2011). An alternative scenario involving a trimming mechanism of the 66-nt intermediate forms to generate the mature crRNAs was also envisioned (Deltcheva et al. 2011).

5.3.2 tracrRNA Trans-*Activates Pre-crRNA Cleavage Within the Repeats*

Strikingly, the second most abundant small RNAs detected by dRNA-seq of *S. pyogenes* were RNA species encoded 210 nt upstream of the *cas* operon on the opposite strand (Deltcheva et al. 2011). It was demonstrated that trans-activating crRNA (tracrRNA) is a non-protein coding RNA expressed in four forms of approximate lengths of 171, 89, 75, and ~65 nt. Two primary species of 181 and 89 nt are transcribed from two distinct promoters to a shared transcriptional terminator. According to the total and processed transcript-enriched libraries, the third form of 75 nt corresponded to a processed fragment predicted to originate from either the 181- or 89-nt precursor RNA. Surprisingly, both 171- and 89-nt tracrRNAs share a 25-nt stretch of almost perfect (one mismatch) complementarity with each of the pre-crRNA repeats. In addition, the processing sites resulting in the production of 75-nt tracrRNA and 66-nt 5′-repeat-spacer-repeat-3′ intermediate crRNA species lay within the putative duplex region, which was a clear indication for tracrRNA and pre-crRNA co-processing upon base-pairing. Similarly, a pre-crRNA-deficient mutant failed to produce the 75-nt tracrRNA species and conversely production of processed crRNA species was abrogated in a tracrRNA-deficient mutant. Trans-complementation studies further demonstrated that activation of the co-processing occurs well *in trans* and both the 171- and 89-nt precursor forms of tracrRNA could activate pre-crRNA processing (Deltcheva et al. 2011). Moreover, the 89-nt tracrRNA was the least stable form of tracrRNA, an indication that it may be the primary species preferentially processed in vivo. Thus, the study resulted in the discovery of a novel RNA maturation pathway in bacteria wherein a non-protein-coding RNA (tracrRNA) *trans*-activates the maturation of a second non-protein-coding RNA (pre-crRNA) being itself maturated in the process (Deltcheva et al. 2011). Only one other example for such mechanism is available in the literature with the class of the so-called *trans*-acting small interfering RNAs (tasiRNAs) in Eukaryotes that require micro-RNAs (miRNAs) for their biogenesis, thus linking miRNA with siRNA pathways (Vaucheret 2005). Likewise, an unprecedented link was established between two classes of small RNAs in bacteria: *trans*-acting small RNAs (tracrRNA) and small interfering RNAs (crRNAs). Whether tracrRNA would have another biological function independent of CRISPR activation remains unknown.

5.3.3 The Housekeeping Endoribonuclease III Co-Processes tracrRNA and Pre-crRNA

When Deltcheva et al. further analyzed the dRNA-seq data, they observed that both co-processed 75 nt tracrRNA and 66 nt intermediate crRNA species carried short overhangs at the 3′ end, which is typical for cleavage by the endoribonuclease RNase III (Deltcheva et al. 2011). This observation was intriguing since no RNase-III like domains could be detected in the sequences of Cas9, Cas1, Cas2, and Csn2 proteins. This led to the hypothesis that the RNase III from the bacterial host is recruited to cleave tracrRNA and pre-crRNA upon base-pairing (Deltcheva et al. 2011). As expected, an RNase III-deficient mutant of *S. pyogenes* did not co-process tracrRNA and pre-crRNA, and *trans*-complementation of RNase III expression restored the co-processing phenotype to wild-type levels. In vitro, purified *E. coli* RNase III cleaves pre-annealed tracrRNA and pre-crRNA species to produce the typical RNase III-cleavage in either RNA. No activity of RNase III on either tracrRNA or pre-crRNA alone could be detected. Consistent with the shared complementarity of 171- and 89-nt tracrRNAs to pre-crRNA via the anti-repeat:repeat interaction, both primary transcripts promoted RNase III-cleavage of pre-crRNA within the repeats to produce the intermediate crRNA species. Similarly, in vivo, the 171- and 89-nt tracrRNAs were converted into the 75 nt form in the process (Deltcheva et al. 2011). Moreover, mutations in the complementary regions of tracrRNA or pre-crRNA abrogated the RNAse III-mediated co-processing with the respective wild-type RNA partner, yet a combination of compensatory tracrRNA and pre-crRNA mutants restored the cleavage. The authors concluded that duplex formation is a prerequisite for tracrRNA and pre-crRNA co-processing. RNase III—a general RNA processing factor in bacteria—mediates the process. These findings not only represent the first description of RNase III-mediated co-processing of two small non-coding RNAs but also are the first example of a non-Cas protein being recruited to CRISPR activity.

5.3.4 Cas9 is Required for tracrRNA and Pre-crRNA Co-Processing by RNase III

In CRISPR-Cas types I and III, the Cas proteins are the effector molecules in crRNA maturation events. Hence, a next step was to analyze whether Cas proteins are required for crRNA maturation in type II. Earlier studies with the type II system in *S. thermophilus* indicated that Cas1, Cas2 and Csn2 are not involved in the interference phase (Barrangou et al. 2007). In *S. pyogenes*, deleting either *cas1*, *cas2*, or *csn2* did not impair tracrRNA and pre-crRNA co-processing (Deltcheva et al. 2011). However, a *cas9*-deficient mutant did not produce the processed tracrRNA and crRNA species, which could be restored upon complementation with the gene *in trans* (Deltcheva et al. 2011). Some relevant details on the role of Cas9 (formerly known as Csn1 or Csx12) in the processing mechanism have recently been elucidated (see below). Cas9 is the signature protein of the type II systems and does not share any obvious similarity with the Cas proteins of type I and III systems (Makarova et al. 2011a). Cas9 is described as a large protein containing at least two predicted nuclease domains, the HNH (or McrA-like) nuclease domain and the RuvC-like (RNase H fold) resolvase domain. Interestingly, the RuvC-like domain seems to be interrupted by the insertion of a long sequence containing the HNH nuclease domain, which suggested coupling of the nuclease activities (Makarova et al. 2006, 2011a, b). This was recently confirmed by experimental demonstration that the two domains, HNH and RuvC-like, are involved in the cleavage of target DNA (Jinek et al. 2012). A model was proposed wherein Cas9 may facilitate the formation and stabilization of the tracrRNA—pre-crRNA duplex and thus enhance recognition by RNase III (Deltcheva et al. 2011). Altogether, the first processing event in type II is achieved by the coordinated action of tracrRNA, the housekeeping RNase III and Cas9. Although the mechanism responsible for the second crRNA maturation event has not been elucidated yet, it was suggested that Cas9 could act as the effector molecule in a ruler-type mechanism whereby the spacers would be cleaved at a fixed distance using the first processing site as an anchor (Deltcheva et al. 2011). Remarkably, whereas mature crRNAs in types I and III are composed of 5′-repeat-spacer-3′ derived sequences, the crRNAs of type II are produced by final trimming or cleavage that eliminates the 5′ part of the crRNA intermediate form, generating 5′-spacer-repeat-3′ derived sequences.

5.3.5 *Type II crRNAs are Active to Target Invading DNA in vivo*

In streptococcal type II systems, spacer sequences are identical to either lytic phage (*S. thermophilus*) or lysogenic phage (*S. pyogenes*) sequences (Barrangou et al. 2007; Deltcheva et al. 2011). The involvement of the type II system in immunity against lysogenic sequences in *S. pyogenes* indeed has been confirmed experimentally. Using a genetic approach, it was demonstrated that each of the four elements, pre-crRNA, tracrRNA, RNase III, or Cas9, is required for phage sequence uptake by the bacteria (Deltcheva et al. 2011). In the *S. thermophilus* Type II system, Cas9 also abolishes interference with bacteriophages (lytic in this case) (Barrangou et al. 2007; Garneau et al. 2010; Sapranauskas et al. 2011). As in *S. pyogenes* (Deltcheva et al. 2011), Cas9 is the only Cas protein required for type II-mediated immunity. In addition, genetic studies in both species provided evidence for targeting of invading DNA rather than mRNA (Deltcheva et al. 2011; Garneau et al. 2010; Sapranauskas et al. 2011). A recent in vitro study demonstrated that Cas9 guided by a dual-tracrRNA:crRNA structure functions as an endonuclease that introduces site-specifically double stranded-DNA breaks in the target DNA (Jinek et al. 2012).

5.3.6 *tracrRNA-Directed Processing of Pre-crRNA is Common Among Type II Systems*

Analysis of selected bacterial genomes with type II systems demonstrated the presence of anti-repeat sequences in the vicinity of the CRISPR-Cas loci (Deltcheva et al. 2011). Anti-repeat sequences most probably corresponding to tracrRNA homologues were found in commensal and pathogenic Gram-negative (e.g., *Campylobacter jejuni*, *Neisseria meningitidis*, *Treponema denticola*) and Gram-positive bacteria (e.g., *Listeria innocua*, *Streptococus agalactiae*, *Streptococcus mutans* and *S. thermophilus*) (Deltcheva et al. 2011). Northern blot analysis further confirmed expression and processing of both type II pre-crRNAs and tracrRNA homologues in *L. innocua*, *N. meningitidis*, *S. mutans* and *S. thermophilus* (Deltcheva et al. 2011). Although the anti-repeat and repeat sequences differ significantly in the analyzed genomes, the repeat sequences analyzed share a certain degree of similarity, especially in the terminal regions and around the putative cleavage site (Deltcheva et al. 2011). Notably, despite sequence differences, the sequence complementarity in anti-repeat:repeat base-pairing remains conserved (Deltcheva et al. 2011). A base substitution in the anti-repeat of tracrRNA homologues leads to a compensatory substitution in the cognate CRISPR repeat (Deltcheva et al. 2011). The duplexes also differ in the position and length of mismatches, which did not seem to interfere with the co-processing (Deltcheva et al. 2011). Thus, tracrRNA homologue anti-repeat and pre-crRNA repeat sequences appear to have co-evolved (Deltcheva et al. 2011). To conclude, tracrRNA:pre-crRNA base-pairing generally determines crRNA maturation in type II CRISPR-Cas systems, and based on RNA probing results, these systems seem to be constitutively activated to target and affect the maintenance of invader genomes (Deltcheva et al. 2011).

5.4 crRNA Biogenesis in Type III Systems

Type III CRISPR-Cas systems are present in both bacteria and archaea (Makarova et al. 2011a, b). This variant has mainly been studied in the archaeon *P. furiosus* (subtype III-B) by the Terns laboratory (Carte et al. 2010, 2008; Hale et al. 2008). In addition, crRNA biogenesis has recently been investigated in the Gram-positive bacterial pathogen *Staphylococcus epidermidis* (subtype III-A) by the Marraffini group (Hatoum-Aslan et al. 2011). In archaeal species, subtype III-B spacers are predicted to target viruses although no in vivo experiment has yet proven the full activity of the system in the limitation of virus propagation. However, recent evidence for targeting of a small RNA, antisense to pre-crRNA, was demonstrated in *P. furiosus* (Hale et al. 2012). In *S. epidermidis*, however, the subtype III-A was demonstrated to be critical for horizontal dissemination of antibiotic resistance by directly targeting invading conjugative plasmid DNA (Marraffini and Sontheimer 2008). The hallmark of crRNA production in type III is the protein Cas6, which is also present in type I. As mentioned above, in type I systems, Cas6-like endoribonucleases are either an integral component of the Cascade complexes (for example Cas6e and Cas6f in *E. coli* and *P. aeruginosa*, respectively (Brouns et al. 2008; Haurwitz et al. 2010)), or are weakly associated with the complex (for example Cas6 in *S. solfataricus* aCascade (Lintner et al. 2011)). In contrast, Cas6 of subtype III-B seems to function as a standalone CRISPR repeat RNA-specific endoribonuclease in *P. furiosus*, *S. solfataricus* and presumably in other systems III of many archaea and possibly bacteria. The origin, evolutionary path and specialization of Cas proteins is discussed in Chap. 3. crRNA maturation in type III occurs in two steps. A first processing event involves dicing of pre-crRNA by Cas6-mediated cleavage within the repeats to generate 1X intermediate units that undergo further maturation to produce the active mature crRNAs. Another feature of the CRISPR-Cas type III is the presence of *csm* and *cmr* genes encoding repeat-associated mysterious proteins (RAMP) proteins in subtype III-A and III-B, respectively. The functions of these Cas proteins remains to be clarified although some recent studies have indicated that they may function in crRNA biogenesis and/or targeting of invading nucleic acids (DNA in the case of subtype III-A and RNA in the case of subtype III-B).

5.4.1 Type III crRNAs are Expressed and Processed in vivo

5.4.1.1 The Bacterial Subtype III-A System

In 2008, Marraffini and Sontheimer demonstrated the production of an intermediate crRNA form of 71 nt in the bacterium *S. epidermidis*. Based on primer extension analysis, the authors suggested that pre-crRNA was cleaved at the base of a potential stem-loop structure within each repeat to produce an intermediate form that was further trimmed to smaller mature crRNA fragments (Marraffini and Sontheimer 2008). In 2010, the data were verified by northern blotting with the detection of mature crRNAs of 49 nt together with partially processed forms of pre-crRNA of 145 and 75 nt, which were observed only in conditions of pre-crRNA overexpression (Marraffini and Sontheimer 2010).

5.4.1.2 The Archaeal Subtype III-B System

Expression of subtype III-B crRNA species has been detected in four archaeal and one bacterial species by RNA cloning, northern blotting, or more recently deep-sequencing. In 2002, Tang et al. reported that clusters of short tandem repeats, then known as short regularly spaced repeats (SRSRs), were transcribed in the archaeon *Archaeoglobus fulgidus* (Tang et al. 2002). Northern blot analysis revealed ladders of RNA corresponding in length to 1, 2, 3, or more repeat-spacer units. Similar ladders of repeat-derived RNA were also detected (Tang et al. 2005) in the crenarchaeote *S. solfataricus*, an observation confirmed in 2005 in the same species and later in 2006 and 2009 in *S. acidocaldarius* by the Garrett laboratory (Chen et al. 2005; Lillestol et al. 2006, 2009). The authors proposed that SRSRs were transcribed as a precursor RNA that was further processed to generate the unit length small RNAs. Already in these publications, it was suggested that a first cleavage within the repeat sequences would produce monomers or multimers of the repeat motif that would undergo progressive trimming by exonucleases. This represented the first experimental evidence for CRISPR RNA processing, in this case by the (then unknown) archaeal Cas6 nuclease.

In 2008, pre-crRNA expression and processing was investigated in *P. furiosus* by the Terns lab (Hale et al. 2008). All seven CRISPR loci in the *P. furiosus* genome were transcribed into abundant and stable small species primarily composed of 39- and 45-nt invader-targeting sequences as well as to less abundant 1X and 2X intermediate forms. The 1X intermediate of about 65 nt in length corresponded to pre-crRNA presumably cleaved within the repeat sequences (Hale et al. 2008). More recently, northern blot and deep-sequencing analysis of the mature crRNAs associated with the RAMP PfuCmr complexes confirmed the data (Hale et al. 2012). In *S. solfataricus*, the White group deep-sequenced mature crRNAs isolated from complexes with SsoCmr RAMP proteins and showed the presence of RNA molecules with variable sizes centered on 46 nt that would originate from a first cleavage within each repeat and a likely exonucleolytic digestion of the 3′ end (Zhang et al. 2012).

In 2011, the group of Clouet-d'Orval described mature crRNAs expressed from two of four CRISPR loci of *Pyrococcus abyssi* (Phok et al. 2011). Interestingly, Northern blotting and RNA mapping experiments in *S. acidocaldarius* and *S. solfataricus* revealed expression and processing of RNA molecules from complementary strands of repeat-spacer arrays into discrete short RNAs of length distinct from that of the mature crRNAs (Lillestol et al. 2009). The authors of the study suggested that the antisense RNAs could either serve as neutralizers of crRNAs in the absence of invading elements or alternatively be required for the slicing activity of the invaders (Lillestol et al. 2009). In *S. solfataricus*, RNA isolated from a purified CMR complex corresponding to the reverse-complement of pre-crRNA were also identified, however, they represented only 0.01 % of total RNA sequences and thus, may not be functionally relevant (Zhang et al. 2012). In addition, pre-crRNA antisense transcription likely arising from functional promoter sequences within spacers and at a significant level compared to crRNA products was detected in *P. furiosus* (Hale et al. 2012). Northern blot and deep-sequencing analysis revealed distinct 45- and 39-nt antisense species that were demonstrated to function as endogenous target RNA of the system (Hale et al. 2012).

More recently, expression and processing of subtype III-A and III-B pre-crRNAs in *T. thermophilus* were also observed by differential RNA sequencing confirmed by northern blot analysis (Juranek et al. 2012).

5.4.2 The Endoribonuclease Cas6 Cleaves Pre-crRNA Within the Repeats

5.4.2.1 The Bacterial Subtype III-A System

In 2011, the Marraffini group further analyzed crRNA biogenesis in the subtype III-A system of *S. epidermidis* (Hatoum-Aslan et al. 2011). First, they confirmed some data already obtained in 2008, showing that pre-crRNA undergoes a first cleavage within each repeat at the base of a putative stem-loop structure (Hatoum-Aslan et al. 2011; Marraffini and Sontheimer 2008). Using primer extension and conjugation experiments with a series of pre-crRNA mutants, they provided evidence that both the RNA hairpin formation within the repeats and the sequence GGGACG at the base of the potential stem-loop structure are required for efficient primary processing (Hatoum-Aslan et al. 2011). The *Cas* operon of system III-A in *S. epidermidis* is composed of *cas1-cas2-cas10-csm2-csm3-csm4-csm5-csm6-cas6*. Northern blotting and primer extension analysis of in-frame gene deletion mutants in *S. epidermidis* revealed that not only Cas6 but surprisingly also Cas10 and Csm4 appear to be critical for the production of the 71 nt intermediate form in vivo (Hatoum-Aslan et al. 2011). They further suggested that possible roles for Cas10 and Csm4 would be to activate or assist Cas6 in its function, or maintain the stability of pre-crRNAs (Hatoum-Aslan et al. 2011).

5.4.2.2 The Archaeal Subtype III-B System

The Terns lab went on to demonstrate that the endoribonuclease responsible for crRNA processing in the subtype III-B of *P. furiosus* was Cas6, one of the core Cas proteins (Carte et al. 2008). The Cas6 cleavage site was mapped to a defined position located 8 nt from the 3′ end of the repeat sequence, thus generating unit length crRNAs (1X intermediates) with a central spacer flanked by 8 nt of repeat-derived sequence at the 5′ end and a longer repeat sequence ($\sim$22 nt) at the 3′ end (Carte et al. 2008). By testing the RNA binding and cleavage activity of Cas6 with a range of RNA sequences, it was shown that Cas6 binds specifically to a 12 nt sequence motif at the 5′ end of the repeat and cleaves the RNA near the 3′ end, leaving an 8 nt repeat-derived 5′ "handle" on the crRNA (Carte et al. 2008). Structure determination of Cas6 bound to crRNA later indicated that the RNA was rather unstructured, wrapping around Cas6 (Wang et al. 2011). The length of intervening sequence between the 5′ binding site and 3′ cleavage site was also important for maximal activity (Carte et al. 2008). In 2010, additional RNA footprinting mapping refined the Cas6-crRNA interaction to nt 2–8 located near the 5′ end of the repeat and 14 nt upstream of the cleavage site (Carte et al. 2010).

The White group recently demonstrated that the mature crRNAs of *S. solfataricus* also contain the 5′ 8-nt tag derived from the CRISPR repeat with spacer-derived sequence at the 3′ end (Zhang et al. 2012). The 3′ termini of the sequenced crRNAs showed some variability, with some spacer-derived sequences displaying short 3′ handle and others containing little repeat-derived sequences (Zhang et al. 2012). This is in contrast to mature crRNAs isolated from *S. solfataricus* Cascade complexes (subtype I-E) that include the 3′ repeat-derived sequence and do not seem to undergo a second processing event (Lintner et al. 2011).

5.4.3 Insights into the Structure of the Endoribonuclease Cas6

5.4.3.1 The Archaeal Subtype III-B System

The crystal structure of *P. furiosus* Cas6 revealed a duplicated RNA recognition motif, ferredoxin-like (RRM) fold (Fig. 5.2), with the two halves of the protein separated by a cleft that was presumed to play a role in crRNA binding (Carte et al. 2010), which is also discussed in (Fig. 3.4). Cas6 belongs to the RAMP family of proteins but is distinguishable from the other members by a predicted G-rich loop motif (consensus GhGxxxxxGhG, where h is hydrophobic and xxxxx has at least one lysine or arginine) at the C-terminus (Haft et al. 2005; Makarova et al. 2002). The predicted active site of the enzyme is similar to that of archaeal tRNA splicing

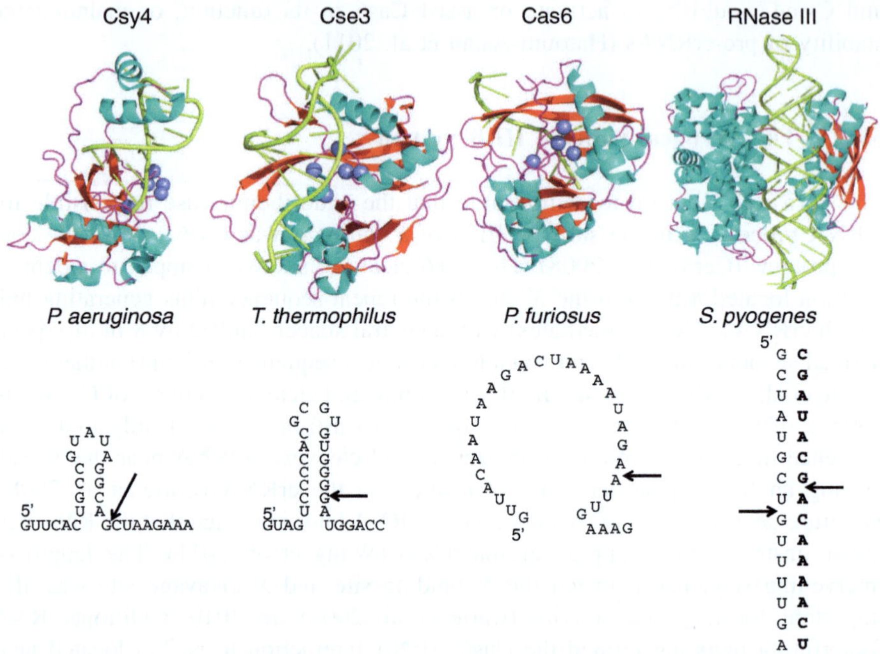

Fig. 5.2 Structures of crRNA processing enzymes **a** Structure of the type I-F crRNA endonuclease Cas6f (Csy4) from *P. aeruginosa* (PDB 2XLK) in complex with its cognate palindromic crRNA. The enzyme has a single RRM fold. **b** Crystal structure of Cas6e (Cse3) (Type I-E) endonuclease from *T. thermophilus* (PDB 2Y8 W) complexed with palindromic crRNA. **c** Structure of Cas6 from *P. furiosus* (PDB 3PKM) with a fragment of crRNA, which is non-palindromic and is thought to wind around the outside of the enzyme. Pfu Cas6 generates crRNA for both the type I-A and type III-B systems. **d** Structure of RNase III from *Aquifex aeolicus* (PDB 2EZ6) bound to a dsRNA substrate. In type II systems, the RNase III enzyme cleaves pre-crRNA in a concerted action with tracrRNA and the protein Cas9. For all structures, RNA is colored yellow and proteins are colored according to their secondary structure. The glycine rich C-terminal region that is characteristic of Cas6 superfamily crRNA endonucleases is indicated by magenta spheres. Beneath each structure is a cartoon showing the sequence, conformation, and cleavage site of each crRNA repeat

endonucleases, and accordingly, Cas6 activity is metal-independent (Carte et al. 2008). Overall, the fold is related to the Cas6e subunit of the subtype I-E Cascade complex (van der Oost et al. 2009), which performs the same function and produces unit length crRNAs with the same 8-nt repeat-derived 5′ tag (Brouns et al. 2008). Like Cas6, Cas6e also cleaves RNA in a metal-independent manner. In contrast to Cas6 having a duplicated ferredoxin fold, the RNA-bound Cas6f of the subtype I-F contains a single ferredoxin fold (Haurwitz et al. 2010). Within the cleft of Cas6, a putative catalytic triad was detected, consisting of Tyr-31, His-46 and Lys-52, which are conserved in some other Cas6 proteins (Carte et al. 2008). The triad of Cas6 is similar to that of archaeal tRNA intron splicing endonucleases, which function via an acid–base catalytic mechanism (Carte et al. 2008). Carte et al. already suggested in 2008 that the G-rich loop signature would be located immediately above the putative catalytic triad and might facilitate the placement of CRISPR repeat RNA substrates (Carte et al. 2008). Mutation of any one of the catalytic residues to alanine was shown to abolish or severely reduce the endonuclease activity of Cas6, but the mutants could still bind to the crRNA (Carte et al. 2010). An active site histidine has also been implicated in the Cas6e and Cas6f nucleases (Brouns et al. 2008; Haurwitz et al. 2010). Curiously however, there is no conserved histidine in the crenarchaeal Cas6 orthologs (Lintner et al. 2011).

A subsequent structure of *P. furiosus* Cas6 bound to crRNA revealed that the RNA is unstructured and winds around the outside of the protein, like the string on a yo-yo (Wang et al. 2011). The first 10 nt of crRNA, which was the only part visualized in the crystal structure, makes sequence-specific interactions with a conserved binding interface in Cas6 on the face opposite the catalytic site. The RNA is predicted to loop around the top of the protein, possibly without making strong interactions, before re-engaging with the protein at the glycine-rich active site, which cleaves the crRNA between nt A22 and A23. The middle, linker region of the crRNA between residues 10 and 20 can accommodate point mutations, insertions, and deletions without abrogating Cas6 activity (Wang et al. 2011).

5.4.4 Possible Mechanisms for the Second Processing Event Yielding the Mature crRNAs

5.4.4.1 The Bacterial Subtype III-A System

In *S. epidermidis*, the 71-nt intermediate crRNA species produced by a first cleavage within the CRISPR repeats were suggested to undergo additional nucleolytic cleavage of the 3′ end to generate smaller mature crRNAs of 43 and 37 nt in length (Marraffini and Sontheimer 2008, 2010). In 2011, Hatoum Aslan et al. further investigated the maturation event (Hatoum-Aslan et al. 2011). Northern blot analysis of a series of crRNA mutants combined with crRNA capture experiments demonstrated that the maturation was independent of the sequence, structure, and length of the 71-nt intermediate crRNA (Hatoum-Aslan et al. 2011). Hatoum Aslan et al. then analyzed a series of intermediate crRNA mutants containing deletions and insertions at the 5′ end while keeping the 3′ end constant. Changing the position of the 5′ primary processing site resulted in extended or diminished maturation at the 3′ end that generated mature crRNAs of constant length (Hatoum-Aslan et al. 2011). The authors suggested that maturation of the intermediate crRNA species occurs by a ruler mechanism whereby the 5′ end primary processing site serves as a reference point to measure the distance between both ends of the crRNA intermediate (Hatoum-Aslan et al. 2011). Using northern blot analysis, Hatoum-Aslan further showed that *csm2*-, *csm3*-, and *csm5*-deficient mutants did not produce crRNA mature forms while they could still express the 71-nt intermediate species, indicating that Csm2, Csm3, and Csm5 may be involved in the maturation process that generates the mature crRNAs (Hatoum-Aslan et al. 2011). Interestingly, while the *P. furiosus* subtype III-B mature crRNAs have 3′-phosphate or 2′-3′-cyclic phosphate ends (Hale et al. 2012), subtype III-A in *S. epidermidis* produce crRNAs that are likely to contain 3′-hydroxyl groups (Hatoum-Aslan et al. 2011).

5.4.4.2 The Archaeal Subtype III-B System

Because in *P. furiosus*, the mature crRNAs lack repeat sequences at their 3′ ends, Carte et al. suggested in 2008 that the Cas6 "1X intermediate" cleavage products were further processed (Carte et al. 2008). Cas6 was shown to remain bound to the repeat sequences at the 3′ end of the cleavage product, and on this basis, it was suggested that the protein might influence the subsequent 3′ end processing to generate the mature crRNAs (Carte et al. 2010, 2008). The repeat sequence at the 3′ end of the 1X intermediate RNA would be trimmed by an unidentified nuclease to produce the mature crRNA species. The existence of a second processing event was confirmed in 2009 and 2012 by sequence analysis of the mature crRNAs present in complexes together with the Cmr effector proteins that cleave target RNA (see the section "Type III mature crRNAs target both RNA and DNA" below) (Hale et al. 2009). The mature crRNAs in the complexes are produced as 39- and 45-nt species that maintain a common 5′ end (8-nt tag) generated by Cas6 while further processing at the 3′ end generates two crRNA species containing either 37 or 31 nt of guide sequence set at defined distance from the 5′ tag or from the absolute length of the crRNAs (Hale et al. 2012, 2009). Of note, native Cas6 does not seem to be stably associated with the mature crRNA species found in the Cmr complexes (Carte et al. 2010) and no co-purification of Cas6 with Cmr complexes could be detected (Hale et al. 2009; Zhang et al. 2012).

The cleavage mechanism by Cas6 was predicted to involve activation of the 2′-hydroxyl on the ribose of pre-crRNA that would lead to nucleophilic attack of the phosphodiester bond on the 3′-hydroxyl and subsequent strand scission to yield the observed 5′-hydroxyl and 2′, 3′-cyclic phosphate ends of the mature crRNAs (Carte et al. 2008). End-radiolabeling of PfuCmr2-immunopurified crRNAs recently showed that the same mature crRNAs were produced most probably following exonucleolytic digestion of the crRNA intermediates generated by Cas6 (Hale et al. 2012). Note that in type I, the 61-nt crRNAs produced by Cas6e-mediated cleavage within the pre-crRNA repeats also have 5′-hydroxyl and 2′, 3′-cyclic phosphate termini, which was experimentally confirmed by ESI-MS/MS analysis (Jore et al. 2011).

After cleaving crRNA, Cas6 remains associated with the cleaved product (Carte et al. 2010). This has parallels with Cas6e (subtype I-E) and Cas6f (subtype I-F), both of which remain bound to product crRNA (Brouns et al. 2008; Haurwitz et al. 2010). However, Cas6 does not appear to interact stably with either Cascade

(Lintner et al. 2011) or the Cmr complex (Carte et al. 2010). Presumably this allows Cas6 to generate crRNAs for both these complexes—something that would not be possible if it was an intrinsic subunit of Cascade. It is not yet clear if the handover is due to simple diffusion of the crRNA or involves a specific intermolecular interaction. What is clear is that crRNA passed to Cascade remains intact, with repeat derived sequences at both the 5′ and 3′ ends, while crRNA passed to the Cmr complex is further processed to remove the 3′ repeat sequence (Carte et al. 2008; Lintner et al. 2011). An obvious candidate for the exonucleolytic processing in subtype III-B could be the exosome, which degrades RNA with a 3′–5′ polarity, though this has not yet been demonstrated. The logical conclusion is that Cascade binds and therefore protects a longer crRNA sequence, possibly by varying the number of Cas7 subunits involved (Lintner et al. 2011), whereas the Cmr complex protects a smaller sequence, allowing 3′ trimming. This leaves open the question: why is *P. furiosus* Cmr associated with crRNAs of two defined lengths, 39 and 45 nt (Hale et al. 2009)?

5.4.5 Type III Mature crRNAs Target Either DNA (III-A) or RNA (III-B)

In vitro studies suggest that the subtype III-B crRNAs of the archaea *P. furiosus* and *S. solfataricus* can target artificially designed cognate RNAs whereas in vivo studies demonstrated that the CRIPSR-Cas subtype III-A in *S. epidermidis* recognizes DNA.

5.4.5.1 The Bacterial Subtype III-A System

In *S. epidermidis*, in vivo interference experiments demonstrated that the DNA rather than the mRNA of a nickase gene from a natural invading conjugative plasmid was directly targeted by the subtype III-A CRISPR-Cas machinery (Marraffini and Sontheimer 2008). The nature of the Cas proteins involved in the process and the exact mechanism leading to genome destruction are yet to be elucidated (refer to Chap. 8).

5.4.5.2 The Archaeal Subtype III-B System

In *P. furiosus*, the mature crRNAs of subtype III-B were shown to associate and form complexes with the RAMP module (PfuCmr) of Cas proteins in vivo (Hale et al. 2009). RNA sequencing analysis of the complexes revealed two mature crRNA species of ~45 and ~39 nt in size that share the common 8-nt repeat segment sequence at the 5′ end but have distinct 3′ ends corresponding to the spacer-derived guide sequence downstream of the repeat (Hale et al. 2009). All seven PfuCmr proteins (2x PfuCmr1, PfuCmr2, PfuCmr3, PfuCmr4, PfuCmr5, and PfuCmr6) encoded by the CRISPR-Cas locus were found associated with both crRNA species (Hale et al. 2009). This was recently confirmed by deep-sequence analysis of crRNAs isolated from crRNA-PfuCmr complexes immuno-precipitated with anti-PfuCmr2 antibodies (Hale et al. 2012). Biochemical experiments further demonstrated that the crRNA/PfuCmr protein complexes can cleave complementary target single-stranded RNA in vitro, in a process requiring divalent metal ions and generating products with 5′ hydroxyl and 3′ phosphate (or 2′-3′ cyclic phosphate) end groups (Hale et al. 2009). Using a combination of synthesized artificial target RNAs together with native crRNA-PfuCmr complexes, it was then established that cleavage of the target RNA occurred at two distinct sites located at a fixed distance from the 3′ end of the crRNAs (Hale et al. 2009). In vitro reconstitution of the crRNA-PfuCmr complexes with purified Cmr proteins and synthesized crRNAs further indicated that each of the mature crRNAs has the ability to form functional complexes with the PfuCmr proteins. RNA targeting would function like a ruler-type mechanism (Hale et al. 2009). The crRNA-guided target RNA cleavage was shown to occur at exactly 14 nt from the 3′ end of either crRNA, thus providing an explanation for the two cleavage sites that were observed while using native crRNA-PfuCmr complexes in the experiments (Hale et al. 2009).

Moreover, five (Cmr1, 3, 4, 5, and 6) of the six PfuCmr proteins seem to be required for the cleavage process by a mechanism that remains to be established (Hale et al. 2009). Interestingly, the crRNA/PfuCmr system of *P. furiosus* seems to recognize and cleave endogenous complementary RNAs in vivo (Hale et al. 2012). Northern blot and deep sequencing analysis of species co-immunopurified with the PfuCmr complex detected distinct short RNAs of 45 and 39 nt in size that corresponded to 5′ products generated from a 140 nt transcript antisense to pre-crRNA (Hale et al. 2012). The 45- and 39-nt antisense RNAs do not possess the 5′ repeat tag and did not function as guide RNAs in vitro. The Terns group concluded that these RNAs correspond to RNA fragments produced by endogenous crRNA-guided cleavage of an RNA antisense to pre-crRNA (Hale et al. 2012). The 5′ repeat tag of the mature crRNAs was shown to be critical for functional crRNA/PfuCmr complexes (Hale et al. 2012). Furthermore, the antisense RNA that is cleaved in vivo does not possess the 5′ repeat tag, however, it contains a sequence complementary to the tag of mature crRNAs (Hale et al. 2012). Thus, in contrast to the subtype III-B system of *S. epidermidis* that recognizes DNA, RNA targeting by the crRNA/PfuCmr system does not involve a self versus non-self-discrimination by a lack of complementarity between the invading target and the crRNA repeat tag (Hale et al. 2012).

In *S. solfataricus*, the SsoCmr complex composed of SsoCmr1-7 proteins also recognizes and cleaves target RNA in the presence of a crRNA containing the 5′ 8-nt repeat-derived tag and cognate spacer-derived sequence in vitro (Zhang et al. 2012). Cleavage of DNA targets could not be observed. The cleavage reaction required manganese and was stimulated in the presence of ATP. The 8-nt tag and the presence of an unpaired flap at the 3′ end of the target RNA were both essential for the cleavage activity. This was intriguing since the flap sequence was designed to contain the motif corresponding to the PAM-like sequence, already described in DNA-targeting

Cascade systems of Archaea (Zhang et al. 2012). This indicates that as in *P. furiosus*, the *S. solfataricus* subtype III-B system targeting RNA has evolved a mode of target recognition that is functionally distinct from other CRISPR-Cas types I, II, and III-A. Sequence mapping experiments further demonstrated a strong cleavage of the target RNA at a UA dinucleotide with weaker cleavage at AA dinucleotides (Zhang et al. 2012). In addition, the SsoCmr-mediated cleavage products contained 3′-hydroxyl termini that could be extended by PolyA polymerase, reminiscent of metal-dependent RNase H-type activity observed for Piwi and Argonaute proteins in Eukaryotes (Zhang et al. 2012). The SsoCmr2 containing a permuted HD nuclease domain is thought to be the active site of the complex, although implication of other RAMP subunits was also suggested (Zhang et al. 2012). The cleavage mechanism of target RNA by the crRNA-SsoCmr complex is distinct from pre-crRNA or target RNA cleavage by the endonucleases Cas6 or PfuCmr that both generate 3′-cyclic phosphate products. The reaction required a 3′ sequence overhang on the target RNA (Zhang et al. 2012). Although cleavage of both RNAs occurred, cleavage of the guide crRNA did not seem to be essential for target RNA catalysis (Zhang et al. 2012).

5.5 Conclusions and Perspectives

CRISPR-Cas-mediated immunity functions as a unique adaptive RNA-guided silencing system that uses homology-dependent interference of invading nucleic acids. Past invaders are memorized through incorporation of spacers into the CRISPR repeat-spacer array of the bacterial host, which then uses the RNA pathway as a memory device to recognize the same encounters, ultimately leading to their destruction. As for the small RNAs that guide antisense targets in Eukaryotic interference, the CRISPR-Cas pathway requires nucleolytic processing of the precursor pre-crRNA to generate the mature crRNAs. Although the defense strategy is commonly shared by all CRISPR-Cas systems, the different subtypes have evolved distinct mechanisms for crRNA biogenesis.

5.5.1 Subtype-dependent Cas Proteins for Pre-crRNA Processing

The molecular mechanisms involved in the first processing event within the pre-crRNA repeats are distinct from those responsible for subsequent maturation of crRNAs. Different Cas proteins characteristic for the subtype play distinct catalytic or assisting functions in these mechanisms. Types I and III both use Cas6-like endoribonucleases for the first cleavage within the repeats. In addition to the Cas6-like protein and the core Cas1 and Cas2, both types encode a module of several other Cas proteins, which in the case of type I form complexes with Cas6-like enzymes. For example, type I-E (*E. coli* or CASS2) encodes Cse1 (CasA), Cse2 (CasB), Cas7 (CasC), and Cas5 (CasD), which together with Cas6e (CasE) and crRNA form the Cascade complex

(Brouns et al. 2008; Ebihara et al. 2006; Gesner et al. 2011; Jore et al. 2011; Sashital et al. 2011; Wang et al. 2011; Wiedenheft et al. 2011a). The *trans*-acting nuclease Cas3 is then recruited to the complex to cleave the invading DNA (Beloglazova et al. 2011; Howard et al. 2011; Mulepati and Bailey 2011; Sinkunas et al. 2011; Wiedenheft et al. 2011a). Type I-F (Ypest or CASS3) encodes Csy1, Csy2, and Csy3, which together with Cas6f (Csy4) and crRNA form a ribonucleoprotein complex, which is likely to recruit the DNA-cleaving enzyme Cas3 as for type I-E (Haurwitz et al. 2010; Wiedenheft et al. 2011b). The type III systems encode a set of RAMP proteins, sharing Cas10 (formerly Csm1, Cmr2 and Csx11) as signature. In type III-B, Cas6 seems to function as a standalone endoribonuclease, and the associated RAMP proteins Cmr1, Cas10, Cmr3, Cmr4, Cmr5, and Cmr6 interfere downstream of the Cas6-mediated processing event in target RNA interference (Carte et al. 2008, 2010; Hale et al. 2008,2009, 2012; Wang et al. 2011). Although the enzyme(s) responsible for maturation of Cas6-generated crRNA intermediates into mature crRNAs have not been identified, it may be that the maturation step involves a trimming process catalyzed by housekeeping exoribonucleases. In type III-A, recent data have indicated that some of the associated RAMP proteins, Cas10, Csm2, Csm3, Csm4, Csm5, and Csm6, may be required for Cas6 endoribonuclease activity and/or for further maturation to produce the mature crRNAs (Hatoum-Aslan et al. 2011). Whether Cas6 of type III-A is embedded in a ribonucleoprotein complex with RAMP proteins has not yet been demonstrated. The subtype I-C does not encode a Cas6-like endoribonuclease. Instead, the protein Cas5d is the endoribonuclease, which uses a mechanism distinct from that of Cas6-like proteins to cleave pre-crRNA within the repeats (Nam et al. 2012). Like Cas6 proteins of other subtypes I, Cas5d assembles with crRNA and two other Cas proteins, Cas8c and Cas7, to form a Cascade-like interference complex (Nam et al. 2012).

In contrast, the type II system encodes a minimal number of four Cas proteins with Cas9 as signature. Remarkably, Cas9 is the only Cas protein that is required for crRNA biogenesis and for interference with invading DNA. Instead of using additional Cas proteins, the system has evolved a *trans*-acting small RNA, tracrRNA and takes advantage of the housekeeping endoribonuclease III to catalyze tracrRNA-directed cleavage within the pre-crRNA repeats (Deltcheva et al. 2011). As mentioned above, type II is exclusively present in bacteria and the absence of genes encoding endoribonuclease III-like activities in archaea may provide an explanation for the life domain restriction of type II.

5.5.2 Sub(Type)-Dependent Composition and Length of Mature crRNAs

In type I-B (*C. thermocellum* and *M. maripaludis*), type I-E (Cas6e, *E. coli*), type I-F (Cas6f, *P. aeruginosa*), type III-A (Cas6, *S. epidermidis*), and type III-B (Cas6, *P. furiosus*), mature crRNAs are composed of 8 nt of repeat sequence in 5′ directly followed by invader-targeting spacer-derived sequence (Brouns et al. 2008;

Carte et al. 2008; Haurwitz et al. 2010; Marraffini and Sontheimer 2008; Richter et al. 2012). Accordingly, *C. thermocellum* and *M. maripaludis* Cas6b, *E. coli*, *S. solfataricus* and *T. thermophilus* Cas6e, *P. aeruginosa* Cas6f and *P. furiosus* Cas6 all cleave exactly 8 nt upstream of the repeat-spacer junction within the pre-crRNA repeats (Brouns et al. 2008; Ebihara et al. 2006; Gesner et al. 2011; Haurwitz et al. 2010; Richter et al. 2012; Sashital et al. 2011). In contrast to types II and III, Cas6-like-generated type I (*E. coli*, *P. aeruginosa*, *S. solfataricus*) crRNAs do not undergo additional maturation and thus are composed of the 8-nt repeat tag at the 5′ end, complete sequence of the spacer in the middle, and the remainder of the repeat fragment, generally forming a hairpin structure, at the 5′ end (Brouns et al. 2008; Haurwitz et al. 2010). Whereas type III (*S. epidermidis*, *P. furiosus*) mature crRNAs have repeat-derived sequence at the 5′ end and spacer-derived sequence at the 3′ end (Carte et al. 2008; Marraffini and Sontheimer 2008), type II mature crRNAs are characterized by a reverse configuration with a 5′ spacer-derived sequence and a 3′ repeat-derived sequence (Deltcheva et al. 2011). In addition, type I, II, and III systems produce mature crRNAs of distinct sizes (Carte et al. 2008; Marraffini and Sontheimer 2008). Intriguingly, maturation in both types III-A and III-B generate two distinct crRNAs species. Whether both species have equivalent targeting functions has not been investigated yet. Finally, the crRNAs have different terminal configurations. Type I-C crRNAs in *B. halodurans* and type I-E crRNAs in *E. coli* have 5′-hydroxyl group and 2′-3′ cyclic phosphate (Jore et al. 2011; Nam et al. 2012) while in *P. aeruginosa*, type I-F crRNAs have 5′-hydroxyl group and 3′ phosphate (not cyclic) (Haurwitz et al. 2010). Type III-A crRNAs (*S. epidermidis*) seem to end with 3′-hydroxyl groups (Hatoum-Aslan et al. 2011) whereas type III-B crRNAs terminate with either 3′-phosphate or 2′-3′-cyclic phosphate ends (Carte et al. 2008).

5.5.3 Differential Expression Levels of Individual Mature crRNAs

Deep and differential RNA sequencing studies in types I, II, and III indicate that the most recently acquired sequences at the leader end of the CRISPR loci appear to correspond to the most abundant crRNA species (Richter et al. 2012). There is no clear explanation for this observation, however, it has been suggested that differences in pre-crRNA transcription rates, processing or stability may be involved.

5.5.4 To Fold or not to Fold

An interesting phenomenon is the property of pre-crRNA repeats to fold or not to fold. In 2007, Kunin et al. carried out a systematic analysis of the sequences and RNA folding stabilities of the rapidly expanding examples of CRISPR repeat arising from genomic sequencing (Kunin et al. 2007). They classified 12 major clusters based on conserved sequence features, and noted that CRISPR repeats in some clusters have a pronounced ability to fold into a stable hairpin structure while others lack this property. CRISPRs were therefore divided into "folded" and "unfolded" categories with the former predominating in the bacteria and the latter concentrated in the archaea. Presciently, the authors suggested that the hairpin structures might serve as a recognition motif for Cas proteins. The type I CRISPR repeats fall into the "folded" category, whereas type II and type III repeats are considered "unfolded". Type I pre-crRNA repeats do not share significant sequence similarity but contain palindromic sequences that have been predicted to form stable hairpin structures terminating upstream of the cleavage site. Structural analysis confirmed that *P. aeruginosa* Cas6f interacts specifically with the hairpin to place the cleavage site at the base of the stem-loop within the enzyme active site (Haurwitz et al. 2010). In 2010, Carte et al. indicated that based on the publication of Kunin et al. 2007, the CRISPR repeats of type III-B in *P. furiosus* were members of a group of repeat sequences predicted to be unstructured with the potential to form weak stem-loops (Carte et al. 2010). Along these lines, in 2010, Carte et al. showed that in the absence of proteins, the pre-crRNA is predominantly unstructured in solution (Carte et al. 2010). The structure of crRNA-bound Cas6 also indicated that pre-crRNA wraps around the surface of the endoribonuclease, which is consistent with the lack of folded structure (Wang et al. 2011). Although putative Cas6 orthologs share extremely low sequence identity, Cas6 recognition and cleavage of unstructured crRNA using a "wrap around" mechanism could also apply to type III-A. However, it was recently suggested that type III-A repeats of *S. epidermidis* form internal hairpins that would enhance crRNA processing at the binding and/or nucleolytic level (Hatoum-Aslan et al. 2011). In the case of type II, Kunin et al. also noticed that type II repeats lack the potential to form stem-loop structures (Kunin et al. 2007). Base-pairing of pre-crRNA to tracrRNA may compensate this deficiency by providing an intermolecular structure that directs the processing within pre-crRNA repeats.

5.5.5 Analogies to Eukaryotic RNA Interference Pathways

All genomes in the three kingdoms of life are potential targets of parasite genomes and have evolved RNA-guided defense systems to fight the intruders. CRISPR-Cas is the only RNA-mediated system of interference with invading genetic elements known to date in bacteria and archaea. Eukaryotes employ distinct RNA-guided gene silencing pathways to combat viruses or transposable elements, which in contrast to CRISPR-Cas do not require a prior step of adaptation to the targeted genome. A common theme in RNA-mediated interference pathways is the production of short-interfering RNAs through maturation of precursor RNA molecules and nucleic acid targeting by proteins that use the short RNAs as guides. Although crRNA-guided interference involves unique proteins (Cas) and sequence homology-based targeting, some CRISPR-Cas mechanisms are evocative of RNA interference in Eukaryotes. Recruitment of the host endoribonuclease III for the catalysis of type II tracrRNA-directed pre-crRNA cleavage is unique as an RNA maturation mechanism and also reminiscent of the roles of Dicer and Drosha in the biogenesis of small-interfering RNAs (siRNAs) and micro-RNAs (miRNAs) (Deltcheva et al. 2011). However, in contrast to type II CRISPR-Cas, the small RNA maturation by the eukaryotic enzymes does not necessitate a *trans*-acting small RNA. The ruler-type mechanism involved in the maturation of type III-A crRNAs was also proposed to resemble that of the eukaryotic pathways (Hatoum-Aslan et al. 2011). Analogies were suggested with the miRNA processing pathway in which interaction with DGCR8 positions primary miRNA for cleavage by Drosha, and with the function of Dicer as a molecular ruler in the biogenesis of miRNAs and siRNAs. crRNA-guided RNA cleavage by the Cmr protein complex of type III-B is another CRISPR-Cas pathway that to some extent resembles eukaryotic RNA interference in which siRNAs, miRNAs and piwiRNAs guide Argonaute-like proteins to target mRNAs (Hale et al. 2009). Cmr proteins and Argonaute 2 are not homologous. Although both use a ruler-type mechanism to select the cleavage site on the RNA target, the anchorage of proteins on the crRNAs seems to occur at different ends.

To conclude, CRISPR-Cas uses unique pre-crRNA recognition mechanisms to discriminate pre-crRNA from other cytosolic RNAs and maturation mechanisms to specifically produce the mature crRNA guides. There are numerous variations of the crRNA biogenesis pathway, each mediated by distinct components and mechanisms, which we have begun to understand only recently. Future studies are needed to decipher the details of the mechanisms by which crRNAs are produced, including the nature of proteins or complexes associated to the processing and maturation events and their interactions with pre-crRNA and crRNA. As in the case of eukaryotic RNA interference, CRISPR-Cas constitutes a tremendous source of distinct RNA-guided gene silencing pathways. On this basis, novel types of ribonucleases and novel mechanisms of RNA–protein interactions and RNA maturation are expected to emerge.

References

Baranova N, Nikaido H (2002) The BaeSR two-component regulatory system activates transcription of the *yegMNOB* (*mdtABCD*) transporter gene cluster in *Escherichia coli* and increases its resistance to novobiocin and deoxycholate. J Bacteriol 184:4168–4176

Barrangou R, Fremaux C, Deveau H, Richards M, Boyaval P, Moineau S, Romero DA, Horvath P (2007) CRISPR provides acquired resistance against viruses in prokaryotes. Science 315:1709–1712

Beloglazova N, Petit P, Flick R, Brown G, Savchenko A, Yakunin AF (2011) Structure and activity of the Cas3 HD nuclease MJ0384, an effector enzyme of the CRISPR interference. EMBO J 30:4616–4627

Brouns SJ, Jore MM, Lundgren M, Westra ER, Slijkhuis RJ, Snijders AP, Dickman MJ, Makarova KS, Koonin EV, van der Oost J (2008) Small CRISPR RNAs guide antiviral defense in prokaryotes. Science 321:960–964

Cady KC, O'Toole GA (2011) Non-identity targeting of *Yersinia*-subtype CRISPR-prophage interaction requires the Csy and Cas3 proteins. J Bacteriol 193:3433–3445

Carte J, Pfister NT, Compton MM, Terns RM, Terns MP (2010) Binding and cleavage of CRISPR RNA by Cas6. RNA 16:2181–2188

Carte J, Wang R, Li H, Terns RM, Terns MP (2008) Cas6 is an endoribonuclease that generates guide RNAs for invader defense in prokaryotes. Genes Dev 22:3489–3496

Chen L, Brugger K, Skovgaard M, Redder P, She Q, Torarinsson E, Greve B, Awayez M, Zibat A, Klenk HP et al (2005) The genome of *Sulfolobus acidocaldarius*, a model organism of the crenarchaeota. J Bacteriol 187:4992–4999

Datsenko KA, Pougach K, Tikhonov A, Wanner BL, Severinov K, Semenova E (2012) Molecular memory of prior infections activates the CRISPR/Cas adaptive bacterial immunity system. Nat Commun 3:945

Deltcheva E, Chylinski K, Sharma CM, Gonzales K, Chao Y, Pirzada ZA, Eckert MR, Vogel J, Charpentier E (2011) CRISPR RNA maturation by trans-encoded small RNA and host factor RNase III. Nature 471:602–607

Ebihara A, Yao M, Masui R, Tanaka I, Yokoyama S, Kuramitsu S (2006) Crystal structure of hypothetical protein TTHB192 from *Thermus thermophilus* HB8 reveals a new protein family with an RNA recognition motif-like domain. Protein Sci 15:1494–1499

Garneau JE, Dupuis ME, Villion M, Romero DA, Barrangou R, Boyaval P, Fremaux C, Horvath P, Magadan AH, Moineau S (2010) The CRISPR/Cas bacterial immune system cleaves bacteriophage and plasmid DNA. Nature 468:67–71

Gesner EM, Schellenberg MJ, Garside EL, George MM, Macmillan AM (2011) Recognition and maturation of effector RNAs in a CRISPR interference pathway. Nat Struct Mol Biol 18:688–692

Gottesman S (2011) Microbiology: dicing defence in bacteria. Nature 471:588–589

Haft DH, Selengut J, Mongodin EF, Nelson KE (2005) A guild of 45 CRISPR-associated (Cas) protein families and multiple CRISPR/Cas subtypes exist in prokaryotic genomes. PLoS Comput Biol 1:e60

Hale C, Kleppe K, Terns RM, Terns MP (2008) Prokaryotic silencing (psi)RNAs in *Pyrococcus furiosus*. RNA 14:2572–2579

Hale CR, Majumdar S, Elmore J, Pfister N, Compton M, Olson S, Resch AM, Glover CV 3rd, Graveley BR, Terns RM et al (2012) Essential features and rational design of CRISPR RNAs that function with the Cas RAMP module complex to cleave RNAs. Mol Cell 45:292–302

Hale CR, Zhao P, Olson S, Duff MO, Graveley BR, Wells L, Terns RM, Terns MP (2009) RNA-guided RNA cleavage by a CRISPR RNA-Cas protein complex. Cell 139:945–956

Hatoum-Aslan A, Maniv I, Marraffini LA (2011) Mature clustered, regularly interspaced, short palindromic repeats RNA (crRNA) length is measured by a ruler mechanism anchored at the precursor processing site. Proc Natl Acad Sci U S A 108:21218–21222

Haurwitz RE, Sternberg SH, Doudna JA (2012) Csy4 relies on an unusual catalytic dyad to position and cleave CRISPR RNA. EMBO J 31(12):2824–2832

Haurwitz RE, Jinek M, Wiedenheft B, Zhou K, Doudna JA (2010) Sequence- and structure-specific RNA processing by a CRISPR endonuclease. Science 329:1355–1358

Hommais F, Krin E, Laurent-Winter C, Soutourina O, Malpertuy A, Le Caer JP, Danchin A, Bertin P (2001) Large-scale monitoring of pleiotropic regulation of gene expression by the prokaryotic nucleoid-associated protein, H–NS. Mol Microbiol 40:20–36

Howard JA, Delmas S, Ivancic-Bace I, Bolt EL (2011) Helicase dissociation and annealing of RNA-DNA hybrids by *Escherichia coli* Cas3 protein. Biochem J 439:85–95

Jinek M, Chylinski K, Fonfara I, Hauer M, Doudna JA, Charpentier E (2012) A programmable dual-RNA-guided DNA endonuclease in adaptive bacterial immunity. Science Jun 28, (Epub ahead of print)

Jore MM, Lundgren M, van Duijn E, Bultema JB, Westra ER, Waghmare SP, Wiedenheft B, Pul U, Wurm R, Wagner R et al (2011) Structural basis for CRISPR RNA-guided DNA recognition by cascade. Nat Struct Mol Biol 18:529–536

Juranek S, Eban T, Altuvia Y, Brown M, Morozov P, Tuschl T, Margalit H (2012) A genome-wide view of the expression and processing patterns of *Thermus thermophilus* HB8 CRISPR RNAs. RNA 18(4):783–794

Kunin V, Sorek R, Hugenholtz P (2007) Evolutionary conservation of sequence and secondary structures in CRISPR repeats. Genome Biol 8:R61

Lillestol RK, Redder P, Garrett RA, Brugger K (2006) A putative viral defence mechanism in archaeal cells. Archaea 2:59–72

Lillestol RK, Shah SA, Brugger K, Redder P, Phan H, Christiansen J, Garrett RA (2009) CRISPR families of the crenarchaeal genus *Sulfolobus*: bidirectional transcription and dynamic properties. Mol Microbiol 72:259–272

Lintner NG, Kerou M, Brumfield SK, Graham S, Liu H, Naismith JH, Sdano M, Peng N, She Q, Copie V et al (2011) Structural and functional characterization of an archaeal clustered regularly interspaced short palindromic repeat (CRISPR)-associated complex for antiviral defense (CASCADE). J Biol Chem 286:21643–21656

Makarova KS, Aravind L, Grishin NV, Rogozin IB, Koonin EV (2002) A DNA repair system specific for thermophilic Archaea and bacteria predicted by genomic context analysis. Nucleic Acids Res 30:482–496

Makarova KS, Grishin NV, Shabalina SA, Wolf YI, Koonin EV (2006) A putative RNA-interference-based immune system in prokaryotes: computational analysis of the predicted enzymatic machinery, functional analogies with eukaryotic RNAi, and hypothetical mechanisms of action. Biol Direct 1:7

Makarova KS, Aravind L, Wolf YI, Koonin EV (2011a) Unification of Cas protein families and a simple scenario for the origin and evolution of CRISPR-Cas systems. Biol Direct 6:38

Makarova KS, Haft DH, Barrangou R, Brouns SJ, Charpentier E, Horvath P, Moineau S, Mojica FJ, Wolf YI, Yakunin A et al (2011b) Evolution and classification of the CRISPR-Cas systems. Nat Rev Microbiol 9:467–477

Marraffini LA, Sontheimer EJ (2008) CRISPR interference limits horizontal gene transfer in staphylococci by targeting DNA. Science 322:1843–1845

Marraffini LA, Sontheimer EJ (2010) Self versus non-self discrimination during CRISPR RNA-directed immunity. Nature 463:568–571

Mulepati S, Bailey S (2011) Structural and biochemical analysis of nuclease domain of clustered regularly interspaced short palindromic repeat (CRISPR)-associated protein 3 (Cas3). J Biol Chem 286:31896–31903

Nam KH, Haitjema C, Liu X, Ding F, Wang H, Delisa MP, Ke A (2012) Cas5d protein processes pre-crRNA and assembles into a cascade-like interference complex in subtype I-C/Dvulg CRISPR-Cas system. Structure Jul 24, (Epub ahead of print)

Oshima T, Ishikawa S, Kurokawa K, Aiba H, Ogasawara N (2006) *Escherichia coli* histone-like protein H-NS preferentially binds to horizontally acquired DNA in association with RNA polymerase. DNA Res 13:141–153

Perez-Rodriguez R, Haitjema C, Huang Q, Nam KH, Bernardis S, Ke A, DeLisa MP (2011) Envelope stress is a trigger of CRISPR RNA-mediated DNA silencing in *Escherichia coli*. Mol Microbiol 79:584–599

Phok K, Moisan A, Rinaldi D, Brucato N, Carpousis AJ, Gaspin C, Clouet-d'Orval B (2011) Identification of CRISPR and riboswitch related RNAs among novel noncoding RNAs of the euryarchaeon *Pyrococcus abyssi*. BMC Genomics 12:312

Plagens A, Tjaden B, Hagemann A, Randau L, Hensel R (2012) Characterization of the CRISPR/Cas subtype I-A system of the hyperthermophilic crenarchaeon *Thermoproteus tenax*. J Bacteriol 194(10):2491–2500

Pougach K, Semenova E, Bogdanova E, Datsenko KA, Djordjevic M, Wanner BL, Severinov K (2010) Transcription, processing and function of CRISPR cassettes in *Escherichia coli*. Mol Microbiol 77:1367–1379

Pul U, Wurm R, Arslan Z, Geissen R, Hofmann N, Wagner R (2010) Identification and characterization of *E. coli* CRISPR-cas promoters and their silencing by H–NS. Mol Microbiol 75:1495–1512

Randau L (2012). RNA processing in the minimal organism *Nanoarchaeum equitans*. Genome Biol 13(7):R63, (Epub ahead of print)

Richter H, Zoephel J, Schermuly J, Maticzka D, Backofen R and Randau L (2012). Characterization of CRISPR RNA processing in *Clostridium thermocellum* and *Methanococcus maripaludis*. Nucleic Acids Res Aug 8, (Epub ahead of print)

Sapranauskas R, Gasiunas G, Fremaux C, Barrangou R, Horvath P, Siksnys V (2011) The *Streptococcus thermophilus* CRISPR/Cas system provides immunity in *Escherichia coli*. Nucleic Acids Res 39:9275–9282

Sashital DG, Jinek M, Doudna JA (2011) An RNA-induced conformational change required for CRISPR RNA cleavage by the endoribonuclease Cse3. Nat Struct Mol Biol 18:680–687

Semenova E, Jore MM, Datsenko KA, Semenova A, Westra ER, Wanner B, van der Oost J, Brouns SJ, Severinov K (2011) Interference by clustered regularly interspaced short palindromic repeat (CRISPR) RNA is governed by a seed sequence. Proc Natl Acad Sci U S A 108:10098–10103

Sharma CM, Hoffmann S, Darfeuille F, Reignier J, Findeiss S, Sittka A, Chabas S, Reiche K, Hackermuller J, Reinhardt R et al (2010) The primary transcriptome of the major human pathogen *Helicobacter pylori*. Nature 464:250–255

Sinkunas T, Gasiunas G, Fremaux C, Barrangou R, Horvath P, Siksnys V (2011) Cas3 is a single-stranded DNA nuclease and ATP-dependent helicase in the CRISPR/Cas immune system. EMBO J 30:1335–1342

Sternberg SH, Haurwitz RE, Doudna JA (2012) Mechanism of substrate selection by a highly specific CRISPR endoribonuclease. RNA 18(4):661–672

Swarts DC, Mosterd C, van Passel MW, Brouns SJ (2012) CRISPR interference directs strand specific spacer acquisition. PLoS ONE 7(4):e35888

Tang TH, Bachellerie JP, Rozhdestvensky T, Bortolin ML, Huber H, Drungowski M, Elge T, Brosius J, Huttenhofer A (2002) Identification of 86 candidates for small non-messenger RNAs from the archaeon *Archaeoglobus fulgidus*. Proc Natl Acad Sci U S A 99:7536–7541

Tang TH, Polacek N, Zywicki M, Huber H, Brugger K, Garrett R, Bachellerie JP, Huttenhofer A (2005) Identification of novel non-coding RNAs as potential antisense regulators in the archaeon *Sulfolobus solfataricus*. Mol Microbiol 55:469–481

van der Oost J, Jore MM, Westra ER, Lundgren M, Brouns SJ (2009) CRISPR-based adaptive and heritable immunity in prokaryotes. Trends Biochem Sci 34:401–407

Vaucheret H (2005). MicroRNA-dependent trans-acting siRNA production. Sci STKE 2005, pe43

Wang R, Preamplume G, Terns MP, Terns RM, Li H (2011) Interaction of the Cas6 riboendonuclease with CRISPR RNAs: recognition and cleavage. Structure 19:257–264

Westra ER, Pul U, Heidrich N, Jore MM, Lundgren M, Stratmann T, Wurm R, Raine A, Mescher M, Van Heereveld L et al (2010) H-NS-mediated repression of CRISPR-based immunity in *Escherichia coli* K12 can be relieved by the transcription activator LeuO. Mol Microbiol 77:1380–1393

Wiedenheft B, Lander GC, Zhou K, Jore MM, Brouns SJ, van der Oost J, Doudna JA, Nogales E (2011a) Structures of the RNA-guided surveillance complex from a bacterial immune system. Nature 477:486–489

Wiedenheft B, van Duijn E, Bultema JB, Waghmare SP, Zhou K, Barendregt A, Westphal W, Heck AJ, Boekema EJ, Dickman MJ et al (2011b) RNA-guided complex from a bacterial immune system enhances target recognition through seed sequence interactions. Proc Natl Acad Sci U S A 108:10092–10097

Yosef I, Goren MG, Qimron U (2012) Proteins and DNA elements essential for the CRISPR adaptation process in *Escherichia coli*. Nucleic Acids Res 40(12):5569–5576

Zhang J, Rouillon C, Kerou M, Reeks J, Brugger K, Graham S, Reimann J, Cannone G, Liu H, Albers SV et al (2012) Structure and mechanism of the CMR complex for CRISPR-mediated antiviral immunity. Mol Cell 45:303–313

Chapter 6
Distribution and Mechanism of the Type I CRISPR-Cas Systems

Raymond H. J. Staals and Stan J. J. Brouns

Abstract Although the CRISPR type I system encompasses six different subtypes (I-A to I-F), only three subtypes have been studied in detail to date. This review includes an analysis of the distribution of CRISPR-Cas systems among the different bacterial and archaeal lineages, and will focus on our mechanistic understanding of the type I-E system of *Escherichia coli*. We will cover the overall organization of this system, starting with a detailed description of a typical type I-E gene cluster and its associated CRISPR array. In addition, we will describe recent insights on the three different stages in CRISPR-Cas type I-mediated defense: adaptation, expression and crRNA maturation, and interference. A comparison will be presented of the physical and functional characteristics of CRISPR effector complexes from the various subtypes.

Contents

R. H. J. Staals · S. J. J. Brouns (✉)
Laboratory of Microbiology, Department of Agrotechnology and Food Sciences,
Wageningen University, Dreijenplein 10 6703 HB Wageningen, The Netherlands
e-mail: Stan.Brouns@wur.nl

R. H. J. Staals
e-mail: Raymond.Staals@wur.nl

R. Barrangou and J. van der Oost (eds.), *CRISPR-Cas Systems*,
DOI: 10.1007/978-3-642-34657-6_6, © Springer-Verlag Berlin Heidelberg 2013

6.1 Introduction

The discovery of the first CRISPR locus dates back to 1987, when sequencing of a chromosomal fragment from *E. coli* K12 revealed the presence of identical, 29 nucleotide repeating DNA sequences, which were separated from each other by an array of nonrepetitive, 32 nucleotide DNA sequences (Ishino et al. 1987). Aided by the advances in sequencing technology, it is now evident that as much as 48 % of the bacterial and 85 % of the archaeal genomes contain at least one CRISPR array (an online database of all currently known CRISPR arrays can be found at: http://crispr.u-psud.fr/crispr/). The CRISPR name was introduced several years later when Jansen et al. discovered the co-occurrence of a set of genes (the *cas* genes) in close proximity of the repeating sequences, suggesting a functional relationship between the two (Jansen et al. 2002). The observation that some of the sequences spacing the repeats were homologous to sequences in phages and plasmids, gave rise to the hypothesis that these sequences could participate in a novel prokaryotic defense system against invading nucleic acids (Bolotin et al. 2005; Mojica et al. 2005; Pourcel et al. 2005). Indeed, strong evidence supporting this hypothesis came from a study, where "natural" phage resistance was accompanied by the expansion of the CRISPR locus with sequences homologous to the infecting phage in *Streptococcus thermophilus* (Barrangou et al. 2007).

This set the stage to elucidate the mechanism of CRISPR-mediated immunity, including how the incorporation of virus and/or phage-derived DNA sequences into the CRISPR locus could result in the observed resistance phenotype.

6.1.1 Type I CRISPR-Cas Loci

A typical arrangement of a Type I-E CRISPR-Cas locus is depicted in Fig. 6.1a. The locus consists of a set of 8 Cas genes: Cas3, Cse1, Cse2, Cas7, Cas5, Cas6e, Cas1, and Cas2 (Table 6.1), which are located upstream of the CRISPR array of repeats and spacers. Studies addressing the expression of the different components of this CRISPR-Cas locus have shown that with the exception of Cas3, all *cas* genes are under the control of the same promoter located immediately upstream of the Cse1 gene, giving rise to a long polycistronic transcript (Pougach et al. 2010; Pul et al. 2010; Westra et al. 2010). Cas3 has its own constitutive promoter, suggesting that its expression is somewhat differentially regulated when compared

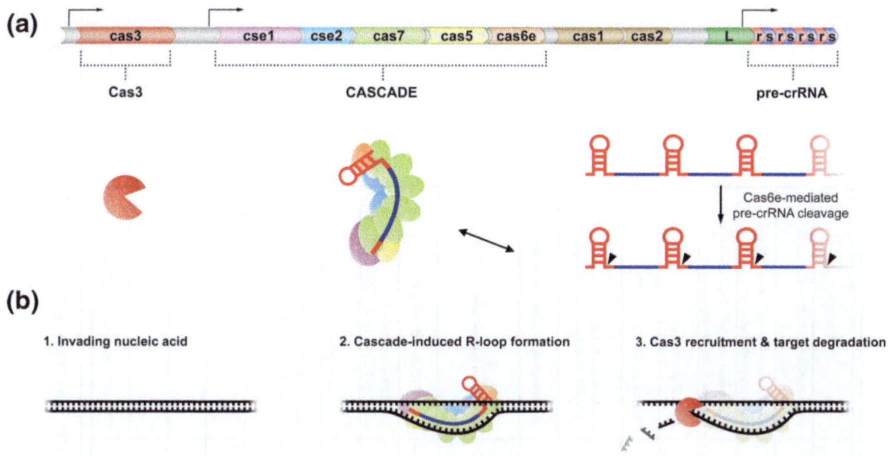

Fig. 6.1 Schematic representation of the expression and interference stages of CRISPR-Cas type I-E systems. **a** In the expression stage, the transcription gives rise to the three major components of CRISPR-based interference: Cas3, Cascade (composed of Cse1, Cse2, Cas7, Cas5, and Cas6e), and a pre-crRNA. The pre-crRNA is cleaved by Cas6e to produce mature crRNA which remain bound to the Cascade complex. **b** In the interference stage, the crRNA-loaded Cascade complex will scan and bind invading nucleic acids that are complementary to its bound crRNA. The binding of Cascade to the invading nucleic acid will result in R-loop formation, which in turn induces a conformational change of the Cascade complex that is believed to serve as a signal for Cas3 recruitment. Finally, the invading nucleic acid will be neutralized by the nuclease activities of Cas3

to the other *cas* genes present in this locus. Illustrative for this is the finding that the nucleoid-structuring protein H-NS appears to act as a key regulator of the CRISPR-Cas system, inhibiting the expression of the long polycistronic transcript, while *cas3* expression is barely affected (more details about the regulation of CRISPR-Cas systems can be found in Chap. 4). In addition to Cas3, the CRISPR array has its own promoter, which is located within an AT-rich stretch of nucleotides, called the 'leader', immediately upstream of the first repeat-spacer unit. Note that this particular promoter is also influenced by the above-mentioned H-NS protein, albeit to a lesser extent. The leader sequence might also serve as directional cues for the spacer integration, as incorporation of a novel spacer always occurs at the interface between the leader and the first spacer (see below).

6.1.2 Type I Spacers and Repeats

Although four different CRISPR arrays can be discerned within the *E. coli* K12 genome, only two of these (CRISPR loci 2.1 and 2.3) seem to be active, as judged by the capability both to integrate novel spacer sequences as well as to confer resistance toward invading nucleic acids (Diez-Villasenor et al. 2010; Swarts et al.

Table 6.1 Components of CRISPR-Cas type I subtype

Components	Subtype I-A	I-B	I-C	I-D	I-E	I-F	Synonyms	Part of complex?	RAMP?	Variants	Remarks
Cas1	[green]	[green]	[green]	[green]	[green]	[green]	-	-	-		- Metal-dependent DNA endonuclease - Putative integrase
Cas2	[green]	[green]	[green]	[green]	[green]	* [green]	-	-	-	* Cas2-Cas3 fusion # Switched HD and SF2 domains	- Metal-dependent endo-ribonuclease - Contains RRM fold
Cas3	# [green]	# [green]	[green]	[green]	[green]	* [green]	-	-	-	* Cas2-Cas3 fusion # Switched HD and SF2 domains	- HD domain and SF 2 helicase domain - Involved in target degradation
Cas4	[green]	[green]	[green]	[green]	[red]	[red]	Csa1	-	-		- RecB-like nuclease
Cas5	[green]	[green]	[green]	[red]	[green]	[red]	Cas5a, Cas5d ,Cas5e, Cas5h, Cas5p, Cas5t, Cmx5, CasD	(a)Cascade	✔		- Structural role?
Cas6	[green]	[green]	[red]	[green]	e [green]	f [green]	Cmx6 e CasE, Cse3 f Csy4	(a)Cascade / Csy	-	e Cas6e f Cas6f	- Contains RRM fold - Endoribonuclease, cleaves pre-crRNA
Cas7	[green]	[green]	[green]	[red]	[green]	[red]	Cse4, Csd2, Csh2, Cst2, Csa2, Csp1, CasC	(a)Cascade	✔		- Structural role in (a)Cascade complex
Cas8	a1/a2 [green]	b [green]	c [green]	[red]	[red]	[red]	a1 Csx8, Cmx1, Cst1, CxxC-CxxC, Csx13 a2 Csa4, Csx9 b Csh1, TM1802 c Csd1, Csp2	-	-	a1 Cas8a1 a2 Cas8a2 b Cas8b c Cas8c	- Inactivated polymerase (PALM)?
Cas10d	[red]	[red]	[red]	[green]	[red]	[red]	Csc3	-	-		- PALM-HD domain fusion
Csa5	[green]	[red]	[red]	[red]	[red]	[red]		aCascade	-		- Small a-helical protein - contains HTH domain (DNA binding)
Cse1	[red]	[red]	[red]	[red]	[green]	[red]	CasA	Cascade	-		- Involved in target binding? - Possible interaction with Cas3
Cse2	[red]	[red]	[red]	[red]	[green]	[red]	CasB	Cascade	-		- Small a-helical protein
Csc1	[red]	[red]	[red]	[green]	[red]	[red]		-	✔		
Csc2	[red]	[red]	[red]	[green]	[red]	[red]	Csc1	-	✔		
Csy1	[red]	[red]	[red]	[red]	[red]	[green]		Csy	-		
Csy2	[red]	[red]	[red]	[red]	[red]	[green]		Csy	✔		
Csy3	[red]	[red]	[red]	[red]	[red]	[green]		Csy	✔		- Structural role in Csy complex

Colored boxes indicate whether (green) or not (red) a component belongs to a particular CRISPR-Cas subtype. Signature proteins for the different subtypes are underlined

2012). Recent analysis of CRISPR spacers from 263 natural *E. coli* isolates exposed to various environments and isolated over a 20-year period from humans and animals, suggests that spacer turn over is a rare but radical event, rather than a gradual process (Touchon et al. 2011). Type I CRISPR arrays typically consist of 29 nucleotide palindromic repeats that are separated from each other by 32 or 33 nucleotide spacer sequences. The nature of the repeat sequences seems to correlate well with the type of CRISPR-Cas system being used (Kunin et al. 2007). As such, the repeats of CRISPR loci 2.1, 2.2, and 2.3 of *E. coli* K12 have been classified as repeat cluster 2, which can exclusively be found in type I-E systems. The conservation of nucleotides contributing to the palindromic nature of these repeats indicates that some (but not all) repeat clusters are capable of forming secondary structures, as it has been shown experimentally for repeat cluster 2 (Haurwitz et al. 2010; Sternberg et al. 2012). The presence of secondary structures within the repeat sequences seems to be widespread among type I systems, although they appear to be absent in type I-A (Apern) and type I-B (Tneap-Hmari) systems. The conservation of these secondary structures might be important for crRNA maturation by Cas6e and/or loading/attachment of the mature crRNA (3′ handle) on the Cascade complex (see below).

6.2 Distribution of the CRISPR-Cas Subtypes in Bacteria and Archaea

CRISPR arrays can be found in about 40 % of the bacterial and up to 80 % of the archaeal genomes. Because not all CRISPR arrays may be actively exchanging their spacer content as we have seen for *E. coli* K12 (Diez-Villasenor et al. 2010), these numbers not necessarily reflect the fraction of species with an active CRISPR-Cas system. As such, this analysis was extended by searching for the presence of signature *cas* genes, which define the type of the CRISPR-Cas subtype (Makarova et al. 2011). For instance, the type I-E subtype is characterized by the presence of the Cse1 gene, which together with Cse2, is not present in any of the other CRISPR-Cas (sub)types.

In order to determine the distribution of all CRISPR-Cas subtypes over all bacterial and archaeal lineages, we analyzed all completely sequenced genomes for the presence of subtype-specific Cas proteins, encoded by either the so-called signature genes or by using other genes that are unique for that particular subtype (Makarova et al. 2011). To this end, protein family entries of all the different signature genes of the different CRISPR-Cas subtypes were collected from the Interpro database (Hunter et al. 2012), resulting in a list of species in which members of this particular protein family can be found. After scoring the lineages to which the identified species belong, this approach allowed us to determine the distribution of the different CRISPR-Cas subtypes among the different lineages in both bacterial and archaeal superkingdoms, by expressing the number of species found as a percentage of the total number of genomes analyzed (Table 6.2).

Table 6.2 Taxonomic distribution of the different CRISPR-Cas systems in bacteria and archaea

Lineage	n=	Example species	% Type I, II or III	% Type I	% Type II	% Type IIII
Bacteria	1456	-	39.7%	31.1%	7.4%	10.3%
:: Acidobacteria	5	*A. capsulatum*	40.0%	20.0%	0.0%	20.0%
:: Actinobacteria	150	***M. tuberculosis***	38.0%	27.3%	4.7%	11.3%
:: Aquificae	9	*H. thermophilus*	66.7%	66.7%	0.0%	44.4%
:: Bacteroidetes	59	*B. helcogenes*	28.8%	11.9%	15.3%	8.5%
:: Chlamydiae	29	*C. abortus*	0.0%	0.0%	0.0%	0.0%
:: Chlorobi	11	*C. tepidum*	100.0%	81.8%	0.0%	54.5%
:: Chloroflexi	15	*C. aggregans*	73.3%	66.7%	0.0%	53.3%
:: Cyanobacteria	40	*A. variabilis*	32.5%	27.5%	0.0%	17.5%
:: Deinococcus-Thermus	14	***T. thermophilus***	78.6%	78.6%	0.0%	50.0%
:: Firmicutes	316	***S. thermophilus, S. epidermidis***	43.7%	30.1%	14.2%	15.2%
:: Fusobacteria	5	*F. nucleatum*	60.0%	20.0%	40.0%	0.0%
:: Nitrospirae	1	*T. yellowstonii*	100.0%	100.0%	0.0%	100.0%
:: Planctomycetes	5	*P. brasiliensis*	40.0%	20.0%	0.0%	20.0%
:: Proteobacteria	687	-	38.4%	34.2%	4.9%	3.8%
- Alphaproteobacteria	150	*Z. mobilis*	18.7%	15.3%	4.7%	1.3%
- Betaproteobacteria	107	*N. meningitidis*	27.1%	18.7%	11.2%	2.8%
- Deltaproteobacteria	45	***D. vulgaris***	66.7%	62.2%	0.0%	17.8%
- Epsilonproteobacteria	55	***C. jejuni***	23.6%	5.5%	20.0%	0.0%
- Gammaproteobacteria	329	***E. coli, P. aeruginosa***	49.5%	48.6%	1.2%	3.6%
:: Spirochaetes	35	*T. denticola*	37.1%	22.9%	5.7%	8.6%
:: Synergistetes	2	*A. colombiens*	0.0%	0.0%	0.0%	0.0%
:: Thermobaculum	1	*T. terrenum*	100.0%	100.0%	0.0%	100.0%
:: Thermotogae	11	***T. maritima***	90.9%	72.7%	0.0%	72.7%
:: Tenericutes	44	*M. gallisepticum*	13.6%	0.0%	13.6%	0.0%
:: Verrucomicrobia	4	*A. muciniphila*	25.0%	0.0%	25.0%	0.0%
Archaea	114	-	69.3%	35.1%	0.0%	49.1%
:: Crenarchaeota	37	-	83.8%	10.8%	0.0%	81.1%
- Thermoprotei	37	***S. solfataricus***	83.8%	10.8%	0.0%	81.1%
:: Euryarchaeota	73	-	64.4%	49.3%	0.0%	34.2%
- Archaeoglobi	4	***A. fulgidus***	75.0%	25.0%	0.0%	50.0%
- Halobacteria	14	*H. volcanii*	50.0%	50.0%	0.0%	0.0%
- Methanobacteria	8	*M. thermoautotrophicum*	62.5%	37.5%	0.0%	25.0%
- Methanococci	14	***M. jannaschii***	78.6%	57.1%	0.0%	64.3%
- Methanomicrobia	16	*M. hungatei*	62.5%	62.5%	0.0%	25.0%
- Methanopyri	1	*M. kandler*	100.0%	0.0%	0.0%	100.0%
- Thermococci	11	***P. furiosus***	54.5%	45.5%	0.0%	36.4%
- Thermoplasmata	3	*T. volcanium*	66.7%	33.3%	0.0%	66.7%
:: Korarchaeota	1	*K. cryptofilum*	100.0%	0.0%	0.0%	100.0%
:: Nanoarchaeota	1	*N. equitans*	0.0%	0.0%	0.0%	0.0%
:: Thaumarchaeota	2	*C. symbiosum*	0.0%	0.0%	0.0%	0.0%

Scoring of the different CRISPR-Cas systems among the bacterial and archaeal lineages was performed by searching the Interpro protein family database for Cas genes, which are unique for that particular subtype. The following proteins families were used to score the various subtypes, the Interpro accession codes are given within parentheses: Type I-A: CXXC–CXXC (IPR010180), Csx8 (IPR013487), NE0113/Csx13 (IPR019092); Type I-B: Csh1 (IPR013420), TM1802 (IPR013389); Type I-C: Csd1 (IPR010144), Csp2 (IPR020029); Type I-D: Csc1 (IPR017576), Csc2 (IPR017574), and Csc3 (IPR017589); Type I-E: Cse1 (IPR013381), Cse2 (IPR013382), and Cse3 (IPR010179); Type I-F: Csy1 (IPR013397), Csy2 (IPR013398), Csy3 (IPR013399), and Csy4 (IPR013396); Type II-A: Csn2 (IPR010146); Types II-A and B: CAS1-NMENI (IPR019855); Type III-A: Csm3 (IPR013412), Csm6 (IPR013489), and APE2256 (IPR013442); Type III-B: Cmr3 (IPR019117), Cmr4 (IPR013410), and Cmr5 (IPR010160). The "example species" column lists representatives that contain a CRISPR-Cas system of a particular lineage (except for the lineages for which no CRISPR-Cas system could be detected: Chlamydiae, Synergistetes, Nanoarchaeota, and Thaumarchaeota). The CRISPR-Cas systems of the species listed in bold have previously been described in detail as part of a scientific publication

This revealed that around 40 % of all bacteria and 69 % of all archaea analyzed by this method possess at least one of the main three types of CRISPR-Cas systems. Note that the percentage for the archaeal superkingdom is slightly smaller compared to the above-mentioned incidence of CRISPR arrays (80 %) as well as a previous analysis (Makarova et al. 2011) where the signature genes of the three different CRISPR-Cas types were used (cas7 and cas3 for type I, cas9 for type II, and cas10 for type III systems), rather than the signature and unique genes of the subtypes themselves. Despite this difference in methodology, the percentages for the three major CRISPR-Cas types, with the exception of type I systems in archaea (74.6 versus 35.1 % from this study), are in general agreement with each other. The relative low score for the type I systems in archaea could indicate the presence of more subtypes within the archaeal CRISPR-Cas type I systems. Although rather speculative, these new subtypes would be more abundant in the archaeal lineages, as the percentages for the bacterial type I systems (38.3 versus 31.1 % from this study) are less divergent.

As previously noted, the type I systems are more prevalent in bacteria (~31 %), while type III systems are more common in the archaeal lineages (~49 %). The type III systems seem to have a high incidence in thermophiles, which is especially evident from the low percentage found for the proteobacteria (only 3.8 % from the 687 genomes analyzed) when compared to the high percentage found in e.g., the Crenarchaeota (81.1 % of the 37 genomes analyzed).

Also, within the type I system, there appears to be quite some differences in the distribution of the five different subtypes between bacteria and archaea (Fig. 6.2). As shown in Fig. 6.2, bacteria have a bias for type I-C, I-E, and I-F subtypes, while type I-A, I-B, and I-D subtypes are more commonly found in archaea. In contrast

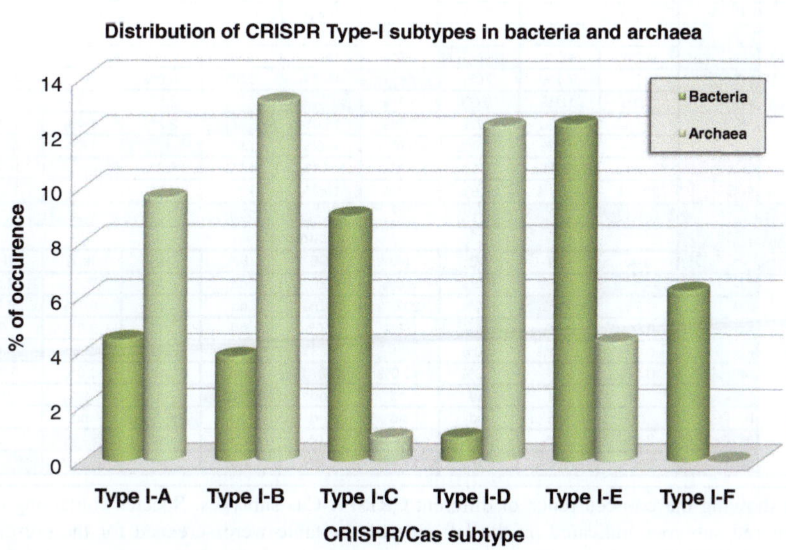

Fig. 6.2 Distribution of CRISPR type I subtypes in bacteria and archaea

to the bacterial lineages, which seem to cover every type I subtype, not a single archaeal species was found to contain the I-F subtype.

The type II system appears to be the least widespread, as only a few bacterial lineages were found to contain the corresponding genes, while the system seems to be absent from archaea altogether. In contrast, two bacterial lineages were found to exclusively contain the type II system: Tenericutes (containing the well-known *Mycoplasma* species) and Verrucomicrobia (e.g. *Verrucomicrobium spinosum* and the gut-bacterium *Akkermansia muciniphila*), although the small number of genomes analyzed here (only four species representing the Verrucomicrobia) should be taken to consideration. Another important observation was that a couple of bacterial and archaeal lineages seem to be devoid of any CRISPR-Cas system, which is especially evident for the Chlamydiae lineage considering the amount of species that were analyzed in this study. Of course, the previously mentioned note of caution when interpreting the scores of lineages with only few complete genomes should also be taken into account when lineages, which are more prone to be fully sequenced and/or annotated, are regarded. For example, strains that are commonly used in experimental settings as well as human pathogens and related species have a tendency to be overrepresented in these fully sequenced genomes, and might therefore bias the conclusions presented in this section.

Since the distribution pattern of the different CRISPR-Cas systems and their subtypes are now known, another important aspect to look at is the co-occurrence of different subtypes (Table 6.3). In both the bacterial as well as the archaeal

Table 6.3 Co-occurrence between the different CRISPR-Cas subtypes

	Co-occurence								
Species containing	I-A	I-B	I-C	I-D	I-E	I-F	II	III-A	III-B
Bacteria									
I-A		21.5%	18.5%	1.5%	9.2%	6.2%	0.0%	44.6%	66.2%
I-B	25.0%		10.7%	1.8%	3.6%	0.0%	3.6%	44.6%	62.5%
I-C	9.2%	4.6%		0.0%	14.5%	5.3%	13.7%	14.5%	22.1%
I-D	7.7%	7.7%	0.0%		15.4%	0.0%	0.0%	15.4%	46.2%
I-E	3.3%	1.1%	10.6%	1.1%		7.8%	2.8%	7.8%	13.9%
I-F	4.4%	0.0%	7.7%	0.0%	15.4%		2.2%	4.4%	7.7%
II	0.0%	1.9%	16.7%	0.0%	4.6%	1.9%		5.6%	5.6%
III-A	29.9%	25.8%	19.6%	2.1%	14.4%	4.1%	6.2%		97.9%
III-B	29.1%	23.6%	19.6%	4.1%	16.9%	4.7%	4.1%	64.2%	
Archaea									
I-A		9.1%	0.0%	0.0%	0.0%	nd	nd	27.3%	27.3%
I-B	6.7%		0.0%	20.0%	0.0%	nd	nd	6.7%	26.7%
I-C	0.0%	0.0%		0.0%	0.0%	nd	nd	0.0%	0.0%
I-D	0.0%	21.4%	0.0%		14.3%	nd	nd	42.9%	28.6%
I-E	0.0%	0.0%	0.0%	40.0%		nd	nd	0.0%	0.0%
I-F	nd	nd	nd	nd	nd		nd	nd	nd
II	nd	nd	nd	nd	nd	nd		nd	nd
III-A	7.0%	2.3%	0.0%	14.0%	0.0%	nd	nd		51.2%
III-B	8.6%	11.4%	0.0%	11.4%	0.0%	nd	nd	62.9%	

Table showing the co-occurrence of different CRISPR-Cas subtypes. Species containing one of the colored subtypes indicated on the left side of the table were screened for the presence of another CRISPR-Cas subtype. The number of co-occurrences is given as a percentage of the total number of species containing that particular subtype domains, a very high co-occurrence was found for both type III CRISPR-Cas systems. The most extreme example to illustrate this is the subtype III-A system in bacteria: nearly all (~98 %) species with this particular subtype also contained the subtype III-B. In addition, a large fraction of species containing a type III system co-occurs with subtypes I-A, I-B, and I-D, which are the most abundant subtypes of the type I CRISPR-Cas system in archaea, as mentioned earlier (Fig. 6.2). The observation that a similar co-occurrence pattern is also present in the bacterial domain, suggests that there might be functional link between these subtypes. At this stage, it might be too early to properly speculate about the biological significance of these findings. However, since type III systems are currently the only CRISPR-Cas systems found to target RNA rather than DNA (Hale et al. 2009, 2012), the idea of increasing the versatility of CRISPR-Cas-mediated defense by combining the two systems to be able to target both types of nucleic acids, is certainly appealing.

6.3 Mechanisms of CRISPR Type I-Mediated Defense

6.3.1 CRISPR Adaptation

Mechanistic insight into the adaptation stage of type I-E systems was recently obtained by three different groups using three different model systems (Datsenko et al. 2012; Goren et al. 2012; Swarts et al. 2012; Yosef et al. 2012). Using a plasmid curing system, it was described that the CRISPR-Cas-active strain *E. coli* K12 *hns*⁻, which contains de-repressed *cas* genes (see Chap. 4), is cured from a high copy number plasmid under nonselective conditions by acquiring spacers against the plasmid (Swarts et al. 2012). Integration of these spacers led to the efficient curing of the plasmid, and prohibited retransformation of the target plasmid. The loss of the high copy number plasmid in these clones resulted in a growth advantage under these nonselective conditions, causing them to become dominant in the population. Up to five antiplasmid spacers were integrated directly downstream from the leader flanking repeat, and these spacers were equally distributed over both genomic CRISPR loci. New spacers appeared to be non randomly selected to target protospacers with a CTT protospacer adjacent motif (PAM, see below) (Mojica et al. 2009) in 78 % of plasmid interfering mutants (PIMs). Interestingly, when multiple spacers were integrated in a single clone, all spacers targeted the same strand of the plasmid, implying that CRISPR interference caused by the first integrated spacer directs additional spacer acquisition events in a strand-specific manner. This positive feedback loop between active spacers in a cluster—the first acquired spacer in this experiment—and spacers acquired thereafter, enables a rapid expansion of the spacer repertoire against an actively present DNA element that is already targeted. The presence of multiple spacers against the same target amplifies the CRISPR interference effect (Barrangou et al. 2007; Brouns et al. 2008) and limits possibilities for bacteriophages and plasmids to escape the immune system by introducing point mutations at critical positions of the PAM or protospacer sequence (Semenova et al. 2011), as point mutations at multiple protospacer locations simultaneously occur at lower frequencies. The strand-specific nature of the positive feedback loop may be the result of the mode of DNA degradation by Cascade and Cas3 guided by an active spacer. This mode involves ATP-dependent unwinding of the plasmid by the Cas3 helicase domain in one direction from the protospacer, and single-stranded DNA cleavage activity of the HD nuclease domain (Beloglazova et al. 2011; Sinkunas et al. 2011; Westra et al. 2012). Most likely, one strand of the target DNA is cleaved mostly exonucleolytically, whereas the other strand of the DNA is cleaved endonucleolytically, providing the strand-specific precursors for spacer integration. Interestingly, this positive feedback loop, now known as "priming" (Datsenko et al. 2012), also occurs when target DNA sequences are mutated to avoid targeting (see below) as has been shown using a phage M13 system. Apparently, the CRISPR memory of prior infections, even if this information is not completely up to date, activates the CRISPR immune system to quickly respond to a re-emerged and slightly changed threat.

6.3.1.1 Cas Gene Requirements for Spacer Integration

The requirements of spacer acquisition have long been enigmatic. Recently, two groups were able to show that both the *cas1* and *cas2* genes are required for the process of spacer integration (Datsenko et al. 2012; Yosef et al. 2012). Contrary to most other *cas* genes, these two genes are the only two that are present in all CRISPR-Cas types (Makarova et al. 2011). Biochemical analyses of Cas1 from *Pseudomonas fluorescens* have shown that Cas1 is a metal-dependent DNA-specific endonuclease that produces double-stranded DNA (dsDNA) fragments (Wiedenheft et al. 2009), which could be precursors for new spacers. The *E. coli* Cas1 enzyme exhibits nuclease activity against single-stranded and branched DNAs including Holliday junctions, replication forks, and 5′-flaps, and interacts with protein components from DNA repair systems (Babu et al. 2011). Cas2 on the other hand was shown to be an endoribonuclease with a preference for U-rich regions (Beloglazova et al. 2008). How these DNase and RNase activities are involved in integrating new spacers in the CRISPR array, however, currently remains unknown.

6.3.1.2 Mechanism of Spacer Integration

Apart from Cas1 and Cas2, spacer integration also requires a single repeat that must be flanked upstream by what is known as the leader sequence (Yosef et al. 2012). This approximately 60 nucleotide AT-rich sequence which is directly preceding the first repeat may provide a signal to use that repeat as the template for the integration of a new spacer unit. It was also shown that only one repeat sequence is necessary to integrate new spacers and duplicating the repeat. The observation that mutations of

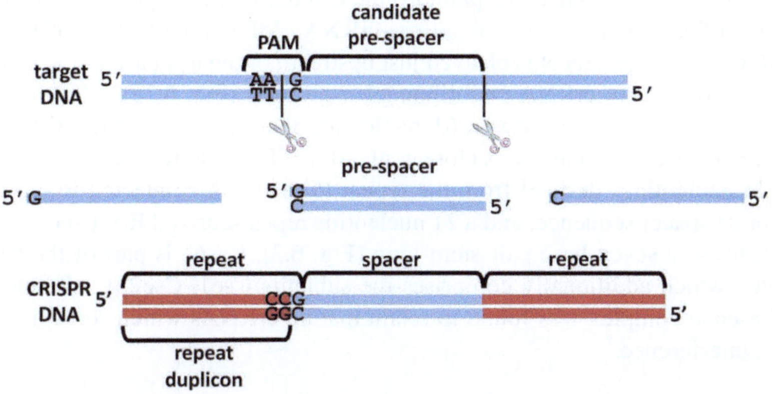

Fig. 6.3 Proposed scheme of CRISPR adaptation. A candidate pre-spacer flanked by a normal PAM is selected after which it is processed into the pre-spacer, which contains at least the last nucleotide of the PAM. The pre-spacer is then integrated into the CRISPR locus by unknown mechanisms. The nucleotide derived from the PAM forms the last nucleotide of the repeat. The pre-spacer, the processed intermediate prior to actual insertion, could be a double-stranded or single-stranded DNA molecule from either strand of the target DNA

the last nucleotide of the preceding repeat (C, A, or T instead of the common G) are never propagated in newly synthesized repeats and always correspond to the first nucleotide from the PAM, revealed a mechanism in which the last nucleotide from the repeat is derived from the PAM during integration of a new spacer (Fig. 6.3) (Datsenko et al. 2012; Goren et al. 2012; Swarts et al. 2012). The part of the repeat (28 out of 29 nucleotides) that serves as a template for the new repeat was termed the "duplicon" of the repeat (Goren et al. 2012). The fact that the PAM is still attached to the to-be-integrated spacer, i.e., the spacer precursor or pre-spacer (Fig. 6.3) (Al-Attar et al. 2011), and provides the last nucleotide of the repeat ensures that the spacer is integrated in the correct orientation in the array. Interestingly, a recent study in *S. solfataricus* and type II A in *Streptococcus agalactiae* has shown that spacers are occasionally and probably erroneously integrated in the opposite orientation (Erdmann and Garrett 2012; Lopez-Sanchez et al. 2012; Westra and Brouns 2012). This suggests that a mechanism that enables orienting the precursor spacer during integration is lacking in these systems.

6.3.2 CRISPR Expression and crRNA Maturation

Type I-E CRISPR-Cas systems, reviewed in Ivančić-Baće et al. (2012), are associated with type 2 repeats (Kunin et al. 2007). The *E. coli* K12 genome contains two type 2 CRISPR arrays (Diez-Villasenor et al. 2010) which are unidirectionally transcribed into precursor CRISPR RNA (pre-crRNA) (Brouns et al. 2008;

Pougach et al. 2010). Due to the palindromic nature of type 2 repeats, stem loops are formed in the repeat regions of the pre-crRNA. After transcription of the entire CRISPR locus, each repeat is cleaved just upstream of the last base of the stem loop by the dedicated pre-crRNA endoribonuclease Cas6e (Fig. 6.1a). Cleavage of the pre-crRNA typically generates a 61 nucleotide mature crRNA with 5′-hydroxyl- and 2′, 3′-cyclic phosphate ends (Jore et al. 2011). The mature crRNA is composed of eight nucleotides derived from the repeat (5′-handle), connected to a 32 or 33 nucleotide spacer sequence, and a 21 nucleotide repeat-derived RNA section which terminates in a seven base pair stem-loop (Fig. 6.3). Cas6e is part of the Cascade complex which additionally comprises the subunits Cse1, Cse2, Cas7, and Cas5. The Cascade complex was found to retain mature crRNAs which serve as a guide during interference.

6.3.2.1 crRNA-Guided Cas Protein Complexes

Apart from *E. coli* Cascade, Cas protein complexes from three distinct type I CRISPR-Cas subtypes have now been described: a type I-A complex from *Sulfolobus solfataricus* (Lintner et al. 2011b) and *Thermoproteus tenax* Plagens (Plagens et al. 2012), a type I-C complex from *Bacillus halodurans* (Nam et al. 2012), and a type I-F complex from *Pseudomonas aeruginosa* (Wiedenheft et al. 2011b) (Table 6.4). Generally, these complexes consist of multiple different protein subunits encoded by 3–6 *cas* genes, and carry one crRNA molecule. Furthermore, most of these Cascade-like complexes contain six copies of the Cas7 subunit, which form the backbone of these complexes and accommodate the crRNA. In addition to Cascade-like complexes from type I, two type III-B complexes have been characterized: one from *P. furiosus* (Hale et al. 2009, 2012) and one *S. solfataricus* (Zhang et al. 2012), respectively (Table 6.4). Interestingly, these so-called CMR complexes generally contain crRNA molecules of two different lengths.

While type I and type III complexes consist of multiple different Cas proteins, type II crRNA complexes are smaller and comprise only one Cas protein (Table 6.4) (Jinek et al. 2012); more details can be found in Chap. 5. In addition to the crRNA, this crRNA–protein complex requires an additional small RNA molecule called the tracrRNA (trans-encoded crRNA) to bind the target DNA. When the target DNA is bound, two distinct nuclease domains come into action and cleave the target DNA (Jinek et al. 2012).

Despite the fact that the Cas protein composition of these complexes is different, all type I and III variants appear to be around 400 kDa in size and contain a mature crRNA with a 5′ handle including 8 or 11 nucleotides of the repeat. While several complexes have been shown to target DNA (Gudbergsdottir et al. 2011; Jinek et al. 2012; Jore et al. 2011; Lintner et al. 2011b; Manica et al. 2011), only the type III-B CMR complex has been shown to cleave RNA molecules (Hale et al. 2009, 2012). By contrast, type I complexes appear to be catalytically inert with respect to target interference; they only bind their target molecule, and recruit the

Table 6.4 Comparison between "effector"-complexes from different CRISPR subtypes

	Type I-A aCASCADE	Type I-C Cascade	Type I-E CASCADE	Type I-F CSY-complex	Type II-A Cas9	Type III-B CMR complex
Organism	*S. solfataricus, T. tenax*	*B. halodurans*	*E. coli*	*P. aeruginosa*	*S. pyogenes, S.*	*thermophilus*
P. furiosus, S.	*solfataricus*					
Haft name	Csa	Csd	Cse	Csy	Csn	Cmr
crRNA length (nt)	58–69	64	61–62	60	42	39 and 45
5′-handle (nt)	8	11	8	8	Absent	8
crRNA spacer (nt)	34–44	32	32 or 33	32	20	31 and 37
3′-handle (nt)	16–17	21	21	20	22	Absent
Seed sequence			Spacer nt 1–5, 7–8	Spacer nt 1–8	Spacer nt 13–20	
Subunit	composition	Csa2 (Cas7), Cas5a, Csa5, and crRNA (+Cas3, Cas3′, and Cas8a2)[a]	Cas5d, Csd2 (Cas7), Csd1, and crRNA	Cse1, Cse2, Cas7, Cas5, Cas6e, and crRNA	Csy1, Csy2, Csy3 (Cas7), Cas6f (Csy4), and crRNA	Cas9
Cmr1-	6(+Cmr7)[b], crRNA					
Stoichiometry	Csa2 major constituent	2:6:1/1 crRNA	1:2:6:1:1/1 crRNA	1:1:6:1/1 crRNA	1/1 crRNA/1	tracrRNA
Cmr7 is most abundant						

(continued)

Table 6.4 (continued)

	Type I-A aCASCADE	Type I-C Cascade	Type I-E CASCADE	Type I-F CSY-complex	Type II-A Cas9	Type III-B CMR complex
Mass (kDa)	350–500	400	405	350	150	430
Target	DNA	DNA	DNA	DNA	DNA	RNA
Shape and	dimensions (nm)	Multiunit helical filaments of variable length, width 6 nm	Elongated arched shape	Seahorse-shaped particle of 20 × 10 nm	Crescent-shaped particle of	12 × 15 nm
	Clamp with crab-claw of 16 × 12 × 11 nm					
Cleaves			No cleavage (Cas3 responsible for cleavage)		DNA	ssRNA, crRNA[a]
PAM[c]	Protospacer-NGG		Protospacer-CTT, Protospacer-CAT, Protospacer-CCT, Protospacer-CTC	Protospacer-GG	CCN-	Protospacer
No PAM						
References	1, 2, 3, 4, 5, 6, and 7	8	9, 10, 11, 12, 13, and 14	15, 16	17, 18, and 19	20, 21, and 22

[a] In *Themoproteus tenax*

[b] In *Sulfolobus solfataricus*

[c] PAM displayed in the crRNA-complementary strand from 5′ to 3′

References used: 1 (Lillestøl et al. 2009), 2 (Gudbergsdottir et al. 2011), 3 (Manica et al. 2011), 4 (Lintner et al. 2011a), 5 (Lintner et al. 2011b), 6 (Plagens et al. 2012), 7 (Erdmann and Garrett 2012), 8 (Nam et al. 2012), 9 (Brouns et al. 2008), 10 (Jore et al. 2011), 11 (Semenova et al. 2011), 12 (Wiedenheft et al. 2011a), 13 (Mojica et al. 2009), 14 (Westra et al. 2012), 15 (Wiedenheft et al. 2011b), 16 (Cady et al. 2012), 17 (Jinek et al. 2012), 18 (Garneau et al. 2010), 19 (Sapranauskas et al. 2011), 20 (Hale et al. 2009), 21 (Hale et al. 2012), and 22 (Zhang et al. 2012) nuclease Cas3 for target DNA cleavage (Westra et al. 2012). Our distribution analysis has shown that the CMR complex co-occurs in almost all of the cases (98 %) with type III-A systems which target DNA (Marraffini and Sontheimer 2008), suggesting synergistic effects of targeting both invader DNA and RNA. Two type I systems (I-E and I-F) and the type II system have been shown to be governed by seed sequences (see below): a stretch of critical nucleotides within the crRNA spacer directly adjacent to the PAM (see below) (Jinek et al. 2012; Semenova et al. 2011; Wiedenheft et al. 2011b). These nucleotides need to perfectly pair with the target DNA sequence as mismatches of the seed sequence with the target DNA lead to loss of interference. It should be noted that in case of type II, the seed and PAM are located at the other end of the crRNA spacer sequence (Table 6.4) (Makarova et al. 2011).

6.3.2.2 Pre-crRNA Endoribonucleases

Cleavage of the pre-crRNA to generate mature crRNAs was demonstrated to be a requirement for CRISPR-mediated resistance (see below). Substitution of a highly conserved histidine in Cas6e resulted in noncleaved pre-crRNA and loss of antiphage immunity (Brouns et al. 2008). Structural analysis of Cas6e from *Thermus thermophilus* has shown that this class of pre-crRNA endonucleases binds the stem-loop structure and residues downstream of the cleavage site within the repetitive segment of pre-crRNA (Gesner et al. 2011; Sashital et al. 2011). Cas6e recognizes the major groove of the RNA A-form helix of the stem-loop by electrostatic interactions between the phosphate backbone of the 3′ strand of the RNA strand at the 3′ end of the stem-loop and a positively charged cleft along two ferredoxin-like domains. The fold of these domains has also been designated the RNA recognition motif (RRM) (Ebihara et al. 2006).

Functional homologs of Cas6e can be found in all but type II CRISPR-Cas systems (Deltcheva et al. 2011; Makarova et al. 2011), although they may display extremely low sequence identities to each other. Furthermore, pre-crRNA endonucleases appear to not always be part of crRNA-containing Cas protein complexes in systems other than type I-E. In *Pyrococcus furiosus,* Cas6 displays specific endoribonuclease activity that results in cleavage of the pre-crRNA in the repeats (Carte et al. 2008, 2010; Wang et al. 2011). Despite their poor sequence similarity, both Cas6e and Cas6 consist of double ferredoxin-like domains and exhibit similar activities. At least in the crRNA-bound state, the type I-F Cas6 functional homolog Cas6f (Csy4) from *P. aeruginosa* and *P. atrosepticum* displays a single N-terminal ferredoxin-like domain and a C-terminal domain comprising two α-helices connected to the ferredoxin-like domain by an extended linker sequence (Haurwitz et al. 2010; Przybilski et al. 2011). It was recently shown that Cas6f recognizes its pre-crRNA substrate with extremely high affinity, and that the recognition is mediated by sequence and structure specific interactions upstream of the scissile phosphate in the major groove of the RNA stem-loop (Haurwitz et al. 2010; Sternberg et al. 2012). This extremely specific mode of recognition of cognate RNA substrates ensures accurate selection of CRISPR transcripts while avoiding spurious off-target RNA binding and cleavage (Sternberg et al. 2012). In contrast to the hairpin-mode of binding by Cas6e and Cas6f, Cas6 from *P. furiosus* recognizes an unfolded repeat by clasping nucleotides 2–9 of the repeat RNA using its two ferredoxin-like domains, and tethering the distal cleavage site of the RNA between nucleotides 22 and 23 to the predicted enzyme active site on the opposite side of the ferredoxin-like domains (Wang et al. 2011).

Remarkably, not all type I systems appear to encode a *cas6* gene. It was recently shown for the type I-C system from *B. halodurans* that Cas5d can substitute for Cas6 by cleaving the pre-crRNA in each repeat (Nam et al. 2012). The structure of Cas5d revealed a single ferredoxin-like domain, but now one with an insertion of a two-stranded β-sheet in the center of the domain.

The emerging picture is that these three pre-crRNA endoribonucleases (Cas6, Cas6e, Cas6f, and Cas5d) have diverged to the extent that not only their domain architecture but also their proposed active site residues (and therefore the cleavage mechanism) are no longer conserved (Sashital et al. 2011), possibly due to the substantial differences in sequence and structure of their repeat RNA substrates, and the metal-independent mechanism of cleavage.

6.3.3 CRISPR Interference

The requirements for CRISPR-based resistance of type I-E systems were originally determined by introducing different sets of *cas* genes and CRISPR arrays from *E. coli* K12 into *E. coli* BL21 which is devoid of *cas* genes. Artificial CRISPRs were designed without taking PAMs into account and targeted four different genes of phage lambda. Although overexpression of Cascade and CRISPR RNA was sufficient to yield mature crRNAs, no resistance was observed when the CRISPRs targeting phage lambda were co-expressed with Cascade only. However, the co-expression of Cas3 did result in a dramatic increase of resistance against phage lambda infection (Brouns et al. 2008). Resistance was obtained when crRNAs were generated that would be complementary to either DNA strand of the phage genome, suggesting that interference was taking place at the level of the phage DNA. Except for type III-B CRISPR-Cas systems which target RNA in vitro and in vivo, all other systems characterized to date appear to target invader DNA (Garneau et al. 2010; Hale et al. 2009, 2012; Marraffini and Sontheimer 2008; Zhang et al. 2012). Follow-up studies with the *E. coli* type I-E system revealed that all Cascade subunits are essential for immunity (Jore et al. 2011). In addition, the high-temperature protein G (HtpG), a homolog of the eukaryotic chaperone/heat-shock protein Hsp90, is also required for immunity under endogenous *cas* gene expression conditions in *E. coli* K12. HtpG was found to be a positive modulator of the CRISPR system and essential for maintaining functional levels of Cas3 (Yosef et al. 2011).

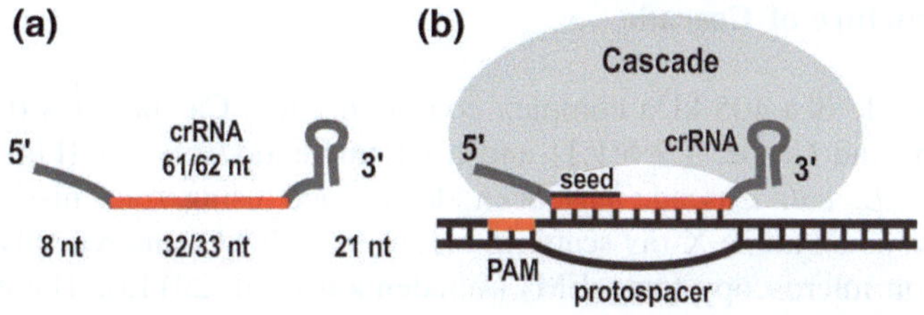

Fig. 6.4 R-loop formation by Cascade. **a** Schematic representation of the crRNA with repeat-derived crRNA in gray and spacer-derived crRNA in red. **b** Target DNA recognition by Cascade-mediated R-loop formation

6.3.3.1 R-Loop Formation

The mechanism of target nucleic acid recognition was studied in vitro using native gel-shifts (Jore et al. 2011). Cascade was shown to bind complementary nucleic acids, such as single-stranded DNA (ssDNA) and single-stranded RNA (ssRNA) molecules. Interestingly, the complexes are also able to bind sequence specifically to double-stranded target DNA molecules, which are likely the physiologically relevant substrates in the cytoplasm of prokaryotes. Double-stranded target RNA molecules on the other hand were not bound. The crRNA binds dsDNA targets by base pairing with the complementary DNA strand, while displacing the noncomplementary strand to form an R-loop (Fig. 6.4). Surprisingly, target DNA recognition by Cascade takes place without ATP consumption, suggesting that continuous invader DNA surveillance takes place without energy investments. Both R-loop formation and binding to ssDNA targets are strongly enhanced by the Cse1 subunit, which is also responsible for nonspecific DNA binding (Jore et al. 2011). It was later shown that Cse1 is in fact essential for binding dsDNA targets, and recognizes PAM sequences through a flexible loop (Mulepati et al. 2012; Sashital et al. 2012). It was proposed that target DNA recognition may initiate at the PAM and may facilitate strand invasion of the crRNA over the seed sequence to form the final target DNA-Cascade complex. R-loop formation by Cascade does not result in cleavage of the target DNA, and does not require ATP or other cofactors. The topology of target DNA was recently shown to be an important parameter for Cascade binding. Gel shift experiments using a physiologically relevant plasmid target substrate indicated that Cascade only binds negatively supercoiled (nSC) target DNA with high affinity, while protospacers in relaxed (linear or open circular) target DNA molecules of several kilobasepairs are hardly recognized (Westra et al. 2012). The preference for nSC targets can be explained by considering that the energy to melt the DNA strands over the length of a protospacer, which is required during the formation of an R-loop, is nearly 50 % reduced in nSC targets compared to a relaxed target DNA. As intracellular DNA in nonthermophilic prokaryotes is normally nSC, Cascade can scan most cellular DNA content for the presence of target sequences without investing ATP.

6.3.3.2 Structure of Cascade

E. coli Cascade is a 405 kDa complex comprising five Cas proteins (Cse1, Cse2, Cas7, Cas5, and Cas6e, 1:2:6:1:1) and a 61 nucleotide crRNA (Fig. 6.1a). The structure of *E. coli* Cascade has been determined using transmission electron microscopy, small angle X-ray scattering (Jore et al. 2011), and recently also using cryo-electron microscopy (cryo-EM) (Wiedenheft et al. 2011a). These structures reveal an unusual 10 × 20 nm, seahorse-shaped particle in which the crRNA is bound along a helical arrangement of six Cas7 subunits terminating at the 3′ end with the Cas6e subunit. The crRNA is well shielded from RNase degradation in this configuration, yet it is sufficiently exposed to allow for base pairing with complementary single-stranded nucleic acids and dsDNA.

The cryo-EM structure of Cascade bound to complementary RNA has revealed that base pairing of the crRNA spacer is achieved through a series of short helical segments (3–5 bp), which reduce the overall length of the crRNA and trigger a conformational change of the complex in which the Cse1, the Cse2 dimer, and Cas6e are repositioned (Wiedenheft et al. 2011a). This conformational change may serve as a signal to recruit a trans-acting nuclease (Cas3) for destruction of invading nucleic acid sequences. R-loop formation through the formation of short helical segments of the crRNA spacer with the target DNA is the biological solution that prevents major steric problems associated with the crRNA: DNA helix formation, and allows both ends of the crRNA to stay bound to the complex during target recognition.

6.3.3.3 Cas3

In addition to Cascade and antitarget crRNA, Cas3 is an essential component during CRISPR interference (Brouns et al. 2008). Cas3 is a large two-domain protein in type I-E systems comprising an N-terminal HD phosphohydrolase domain and a C-terminal Superfamily 2 helicase domain. Biochemical analyses of *Streptococcus thermophilus* Cas3 have shown that Cas3 displays Mg-dependent ATPase activity, which is coupled to unwinding of DNA/DNA and RNA/DNA duplexes in 3′ to 5′ direction. Cas3 also shows Mg-dependent nuclease activity by the HD domain with a preference for ssDNA substrates (Sinkunas et al. 2011). The Cas3 HD domain from *T. thermophilus* adopts a globular structure with a concave surface in which the active site resides comprising two metal-ion binding sites (Mulepati and Bailey 2011). Contrary to Cas3 from *S. thermophilus* and *M. jannaschii* (Beloglazova et al. 2011; Sinkunas et al. 2011), the endonuclease activity of the Cas3 HD domain from *T. thermophilus* was reported to be activated by transition metal ions, such as manganese and nickel. These activities are in line with a mechanism in which Cas3 inflicts permanent damage to target DNA once that target DNA has been identified by Cascade through R-loop formation.

This hypothesis has recently been confirmed for the *E. coli* type I-E system. Upon infection with a targeted phage, Cascade first localizes the target DNA independently from Cas3, and subsequently recruits Cas3 via the Cse1 subunit (Westra et al. 2012). Evidence supporting the Cas3–Cse1 interaction includes the presence of naturally occurring Cas3–Cse1 fusion proteins in several species with a type I-E system. After Cas3 is recruited to the Cascade-bound R-loop, Cas3 engages the target DNA by generating a nick; most likely in close proximity to the target sequence (Westra et al. 2012). Subsequently, based on in vitro analyses (Beloglazova et al. 2011; Sinkunas et al. 2011), Cas3 initiates progressive ATP-dependent unwinding of the target DNA at the nicked site using the helicase domain, while cleaving unwound single-stranded target DNA by the HD domain. This mode of single-stranded target DNA degradation from a DNA duplex may generate the strand-specific precursors used for new spacer acquisition events (Swarts et al. 2012) (Fig. 6.3). It is anticipated that Cascade dissociates from the target DNA during Cas3-mediated target DNA degradation, recycling the complex to find new target DNA molecules.

Most homologues of the Cas3 protein in the different type I systems share the same domain architecture: an N-terminal HD nuclease domain and a C-terminal DExH helicase domain. Next to the above-mentioned Cas3–Cse1 fusion proteins in some type I-E systems, there are several other noteworthy exceptions to this rule. In some type I-A systems, the two domains of Cas3 are encoded as separate proteins, while other cases show Cas3 proteins that have a switched domain organization: an N-terminal DExH helicase with an C-terminal HD nuclease domain. Lastly, the I-F system is known to express a Cas2–Cas3 fusion protein (Table 6.1).

6.3.3.4 Immune Escape

Using an M13 phage test system and a plasmid transformation assay, it was found that invaders escape type I-E CRISPR-Cas resistance by introducing point mutations in their protospacers at very specific positions in or directly adjacent to the protospacer (Semenova et al. 2011). The requirements for crRNA matching are strict only for a noncontiguous seven-nucleotide region of a protospacer (positions 1–5, 7, and 8 of the spacer) near the 5′ end of the crRNA. This critical seven-nucleotide region is called crRNA seed sequence, given its resemblance to the seed sequence in eukaryotic small RNAs in RNA interference. Mutations in the seed region abolish CRISPR-Cas-mediated immunity by reducing the binding affinity of the crRNA-guided Cascade complex for protospacer DNA. Invader DNA genomes may therefore remain undetected in a cell when mutations in the seed region of the protospacers have occurred, resulting in immune escape. Based on the higher binding affinity of ssDNA probes, a seed sequence of approximately eight nucleotides was also identified in for the type I-F Csy-complex from *P. aeruginosa* (Wiedenheft et al. 2011b), suggesting that seed sequences may be a common theme in type I systems. The crRNA seed sequence has been hypothesized to play a role in enhancing the efficiency of scanning the invader DNA for a protospacer match. In contrast to the strict base pairing requirements of the seed sequence, up to five mutations outside the seed region were tolerated without loss of resistance of the *E. coli* type I-E system. This property limits the possibilities for phages to escape and may allow a single crRNA to effectively target numerous related phages. The type I-F system from *P. aeruginosa* was recently also shown to be actively resisting virus infection when virus genome targeting spacers were present in the CRISPR arrays (Cady et al. 2012). Remarkably, up to four mismatches between the CRISPR spacer and virus protospacer still provided a level of resistance, but one additional mutation in the protospacer allowed the virus to escape immunity (Cady et al. 2012).

6.3.3.5 Protospacer Adjacent Motif

In addition to mutations in the seed region of the protospacer, invaders can also evade immunity by mutating a conserved trinucleotide motif just outside the protospacer. The importance of these short nucleotide sequences was originally shown in the type II system of *S. thermophilus* by sequencing phages that had overcome host immunity by mutating a single nucleotide of the motif (Deveau et al. 2008). The emerging picture of the PAM is one of ubiquitous occurrence, variation in sequence, and size (2–4 nt), as well as location on either side of the protospacer (Lillestøl et al. 2009; Mojica et al. 2009). For *E. coli,* it was shown that point mutations at each of the three positions in the PAM severely reduced the binding affinity of Cascade for these probes, again suggesting that these invading DNA molecules may remain undetected in host cells. Recent analyses showed that PAMs are checked by Cse1 only in the targeted strand of dsDNA molecules (Sashital et al. 2012; Westra et al. 2012). The functional importance of the PAM and seed sequence suggests that protospacer recognition initiates at the PAM from where the crRNA invades the dsDNA and attempts to base pair the seed sequence of the crRNA. This is then followed by progressive propagation of crRNA spacer–phage protospacer pairing in 3′ direction of the spacer, leading to local unwinding of the double-stranded protospacer and full R-loop formation (Semenova et al. 2011) (Figs. 6.1b and 6.3). The extension of pairing beyond the seed segment appears to be relatively insensitive to mismatches, as up to five mutations are allowed between the crRNA and the protospacer in this region, before immunity is lost. The stepwise recognition process of DNA targets may enable the Cascade complex to locate protospacers with greater efficiency.

6.3.3.6 Recognition of the Target DNA or CRISPR DNA

In addition to enhancing target localization, the PAM requirement has the additional benefit of preventing autoimmunity that results from targeting the genomic CRISPR array. It is interesting to note that the vast majority of protospacers with random sequence flanks are not targeted in PAM systems, (50/64, including the one flanked by the CRISPR repeat) and that only a few are targeted (4/64, those containing a PAM). In contrast, the nonPAM containing type III-A CRISPR-Cas system of *Staphylococcus epidermidis* targets almost all sequences containing a protospacer (63/64), while only one sequence (1/64, containing three nucleotides from the CRISPR repeat) is not targeted (Marraffini and Sontheimer 2010). Type III-B systems therefore appear to be pure self-DNA recognition systems, while the much more strict PAM-governed systems are best described as target recognition systems. These differences have implications for the ease with which invaders may escape the immunity in these two systems. While invaders can escape PAM-governed systems by point mutation of the PAM, the nonPAM system of type III-A would typically require multiple mutations to mimic self-DNA, reducing the chances of immune escape.

6.4 Outlook

Type I CRISPR-Cas systems are the largest and most diverse CRISPR-Cas type. Yet, only three of the six subtypes have been studied genetically and biochemically. It will be of interest to characterize all six types to understand and appreciate the mechanistic differences between the subtypes and to obtain insight into their evolution. Understanding the molecular mechanism of Cascade-like complexes is expected to come from atomic resolution structures of type I-A, I-C, I-E, or I-F systems. In addition, with experimental model systems for spacer acquisition now in place, the mechanism of spacer insertion and repeat duplication will also soon be deciphered.

References

Al-Attar S, Westra ER, van der Oost J, Brouns SJ (2011) Clustered regularly interspaced short palindromic repeats (CRISPRs): the hallmark of an ingenious antiviral defense mechanism in prokaryotes. Biol Chem 392:277–289

Babu M, Beloglazova N, Flick R, Graham C, Skarina T, Nocek B, Gagarinova A, Pogoutse O, Brown G, Binkowski A, Phanse S, Joachimiak A, Koonin EV, Savchenko A, Emili A, Greenblatt J, Edwards AM, Yakunin AF (2011) A dual function of the CRISPR-Cas system in bacterial antivirus immunity and DNA repair. Mol Microbiol 79:484–502

Barrangou R, Fremaux C, Deveau H, Richards M, Boyaval P, Moineau S, Romero DA, Horvath P (2007) CRISPR provides acquired resistance against viruses in prokaryotes. Science 315:1709–1712

Beloglazova N, Brown G, Zimmerman MD, Proudfoot M, Makarova KS, Kudritska M, Kochinyan S, Wang S, Chruszcz M, Minor W, Koonin EV, Edwards AM, Savchenko A, Yakunin AF (2008) A novel family of sequence-specific endo ribonucleases associated with the clustered regularly interspaced short palindromic repeats. J Biol Chem 283:20361–20371

Beloglazova N, Petit P, Flick R, Brown G, Savchenko A, Yakunin AF (2011) Structure and activity of the Cas3 HD nuclease MJ0384, an effector enzyme of the CRISPR interference. EMBO J 30:4616–4627

Bolotin A, Quinquis B, Sorokin A, Ehrlich SD (2005) Clustered regularly interspaced short palindrome repeats (CRISPRs) have spacers of extrachromosomal origin. Microbiology 151:2551–2561

Brouns SJ, Jore MM, Lundgren M, Westra ER, Slijkhuis RJ, Snijders AP, Dickman MJ, Makarova KS, Koonin EV, van der Oost J (2008) Small CRISPR RNAs guide antiviral defense in prokaryotes. Science 321:960–964

Cady KC, Bondy-Denomy J, Heussler GE, Davidson AR, O'Toole GA (2012) The CRISPR/Cas adaptive immune system of *Pseudomonas aeruginosa* mediates resistance to naturally occurring and engineered phages. J Bacteriol 194:5728–5738

Carte J, Wang R, Li H, Terns RM, Terns MP (2008) Cas6 is an endoribonuclease that generates guide RNAs for invader defense in prokaryotes. Genes Dev 22:3489–3496

Carte J, Pfister NT, Compton MM, Terns RM, Terns MP (2010) Binding and cleavage of CRISPR RNA by Cas6. RNA 16:2181–2188

Datsenko KA, Pougach K, Tikhonov A, Wanner BL, Severinov K, Semenova E (2012) Molecular memory of prior infections activates the CRISPR/Cas adaptive bacterial immunity system. Nat Commun 3:945

Deltcheva E, Chylinski K, Sharma CM, Gonzales K, Chao Y, Pirzada ZA, Eckert MR, Vogel J, Charpentier E (2011) CRISPR RNA maturation by trans-encoded small RNA and host factor RNase III. Nature 471:602–607

Deveau H, Barrangou R, Garneau JE, Labonte J, Fremaux C, Boyaval P, Romero DA, Horvath P, Moineau S (2008) Phage response to CRISPR-encoded resistance in *Streptococcus thermophilus*. J Bacteriol 190:1390–1400

Diez-Villasenor C, Almendros C, Garcia-Martinez J, Mojica FJ (2010) Diversity of CRISPR loci in *Escherichia coli*. Microbiology 156:1351–1361

Ebihara A, Yao M, Masui R, Tanaka I, Yokoyama S, Kuramitsu S (2006) Crystal structure of hypothetical protein TTHB192 from *Thermus thermophilus* HB8 reveals a new protein family with an RNA recognition motif-like domain. Protein Sci Publ Protein Soc 15:1494–1499

Erdmann S, Garrett RA (2012) Selective and hyperactive uptake of foreign DNA by adaptive immune systems of an archaeon via two distinct mechanisms. Mol Microbiol 85:1044–1056

Garneau JE, Dupuis ME, Villion M, Romero DA, Barrangou R, Boyaval P, Fremaux C, Horvath P, Magadan AH, Moineau S (2010) The CRISPR/Cas bacterial immune system cleaves bacteriophage and plasmid DNA. Nature 468:67–71

Gesner EM, Schellenberg MJ, Garside EL, George MM, Macmillan AM (2011) Recognition and maturation of effector RNAs in a CRISPR interference pathway. Nat Struct Mol Biol 18: 688–692

Goren MG, Yosef I, Auster O, Qimron U (2012) Experimental definition of a clustered regularly interspaced short palindromic duplicon in *Escherichia coli*. J Mol Biol 423:14–16

Gudbergsdottir S, Deng L, Chen Z, Jensen JV, Jensen LR, She Q, Garrett RA (2011) Dynamic properties of the Sulfolobus CRISPR/Cas and CRISPR/Cmr systems when challenged with vector-borne viral and plasmid genes and protospacers. Mol Microbiol 79:35–49

Hale CR, Zhao P, Olson S, Duff MO, Graveley BR, Wells L, Terns RM, Terns MP (2009) RNA-guided RNA cleavage by a CRISPR RNA-Cas protein complex. Cell 139:945–956

Hale CR, Majumdar S, Elmore J, Pfister N, Compton M, Olson S, Resch AM, Glover CV 3rd, Graveley BR, Terns RM, Terns MP (2012) Essential features and rational design of CRISPR RNAs that function with the Cas RAMP module complex to cleave RNAs. Mol Cell 45: 292–302

Haurwitz RE, Jinek M, Wiedenheft B, Zhou K, Doudna JA (2010) Sequence- and structure-specific RNA processing by a CRISPR endonuclease. Science 329:1355–1358

Hunter S, Jones P, Mitchell A, Apweiler R, Attwood TK, Bateman A, Bernard T, Binns D, Bork P, Burge S, de Castro E, Coggill P, Corbett M, Das U, Daugherty L, Duquenne L, Finn RD, Fraser M, Gough J, Haft D, Hulo N, Kahn D, Kelly E, Letunic I, Lonsdale D, Lopez R, Madera M, Maslen J, McAnulla C, McDowall J, McMenamin C, Mi H, Mutowo-Muellenet P, Mulder N, Natale D, Orengo C, Pesseat S, Punta M, Quinn AF, Rivoire C, Sangrador-Vegas A, Selengut JD, Sigrist CJ, Scheremetjew M, Tate J, Thimmajanarthanan M, Thomas PD, Wu CH, Yeats C, Yong SY (2012) InterPro in 2011: new developments in the family and domain prediction database. Nucleic Acids Res 40:D306–D312

Ishino Y, Shinagawa H, Makino K, Amemura M, Nakata A (1987) Nucleotide sequence of the iap gene, responsible for alkaline phosphatase isozyme conversion in *Escherichia coli*, and identification of the gene product. J Bacteriol 169:5429–5433

Ivančić-Baće I, Al Howard J, Bolt EL (2012) Tuning into interference: R-loops and Cascade complexes in CRISPR immunity. J Mol Biol 422:607–616

Jansen R, Embden JD, Gaastra W, Schouls LM (2002) Identification of genes that are associated with DNA repeats in prokaryotes. Mol Microbiol 43:1565–1575

Jinek M, Chylinski K, Fonfara I, Hauer M, Doudna JA, Charpentier E (2012) A programmable dual-RNA-guided DNA endonuclease in adaptive bacterial immunity. Science 337:816–821

Jore MM, Lundgren M, van Duijn E, Bultema JB, Westra ER, Waghmare SP, Wiedenheft B, Pul U, Wurm R, Wagner R, Beijer MR, Barendregt A, Zhou K, Snijders AP, Dickman MJ, Doudna JA, Boekema EJ, Heck AJ, van der Oost J, Brouns SJ (2011) Structural basis for CRISPR RNA-guided DNA recognition by Cascade. Nat Struct Mol Biol 18:529–536

Kunin V, Sorek R, Hugenholtz P (2007) Evolutionary conservation of sequence and secondary structures in CRISPR repeats. Genome Biol 8:R61

Lillestøl RK, Shah SA, Brugger K, Redder P, Phan H, Christiansen J, Garrett RA (2009) CRISPR families of the crenarchaeal genus *Sulfolobus*: bidirectional transcription and dynamic properties. Mol Microbiol 72:259–272

Lintner NG, Frankel KA, Tsutakawa SE, Alsbury DL, Copie V, Young MJ, Tainer JA, Lawrence CM (2011a) The structure of the CRISPR-associated protein Csa3 provides insight into the regulation of the CRISPR/Cas system. J Mol Biol 405:939–955

Lintner NG, Kerou M, Brumfield SK, Graham S, Liu H, Naismith JH, Sdano M, Peng N, She Q, Copie V, Young MJ, White MF, Lawrence CM (2011b) Structural and functional characterization of an archaeal clustered regularly interspaced short palindromic repeat (CRISPR)-associated complex for antiviral defense (CASCADE). J Biol Chem 286: 21643–21656

Lopez-Sanchez MJ, Sauvage E, Da Cunha V, Clermont D, Ratsima Hariniaina E, Gonzalez-Zorn B, Poyart C, Rosinski-Chupin I, Glaser P (2012) The highly dynamic CRISPR1 system of *Streptococcus agalactiae* controls the diversity of its mobilome. Mol Microbiol 85:1057–1071

Makarova KS, Haft DH, Barrangou R, Brouns SJ, Charpentier E, Horvath P, Moineau S, Mojica FJ, Wolf YI, Yakunin AF, van der Oost J, Koonin EV (2011) Evolution and classification of the CRISPR-Cas systems. Nat Rev Microbiol 9:467–477

Manica A, Zebec Z, Teichmann D, Schleper C (2011) In vivo activity of CRISPR-mediated virus defence in a hyperthermophilic archaeon. Mol Microbiol 80:481–491

Marraffini LA, Sontheimer EJ (2008) CRISPR interference limits horizontal gene transfer in staphylococci by targeting DNA. Science 322:1843–1845

Marraffini LA, Sontheimer EJ (2010) Self versus non-self discrimination during CRISPR RNA-directed immunity. Nature 463:568–571

Mojica FJ, Diez-Villasenor C, Garcia-Martinez J, Soria E (2005) Intervening sequences of regularly spaced prokaryotic repeats derive from foreign genetic elements. J Mol Evol 60:174–182

Mojica FJ, Diez-Villasenor C, Garcia-Martinez J, Almendros C (2009) Short motif sequences determine the targets of the prokaryotic CRISPR defence system. Microbiology 155:733–740

Mulepati S, Bailey S (2011) Structural and biochemical analysis of nuclease domain of clustered regularly interspaced short palindromic repeat (CRISPR)-associated protein 3 (Cas3). J Biol Chem 286:31896–31903

Mulepati S, Orr A, Bailey S (2012) Crystal structure of the largest subunit of a bacterial RNA-guided immune complex and its role in DNA target binding. J Biol Chem 287:22445–22449

Nam KH, Haitjema C, Liu X, Ding F, Wang H, Delisa MP, Ke A (2012) Cas5d protein processes pre-crRNA and assembles into a Cascade-like interference complex in subtype I-C/Dvulg CRISPR-cas system. Structure 20:1574–1584

Plagens A, Tjaden B, Hagemann A, Randau L, Hensel R (2012) Characterization of the CRISPR/Cas subtype I-A system of the hyperthermophilic crenarchaeon thermoproteus tenax. J Bacteriol 194:2491–2500

Pougach K, Semenova E, Bogdanova E, Datsenko KA, Djordjevic M, Wanner BL, Severinov K (2010) Transcription, processing and function of CRISPR cassettes in *Escherichia coli*. Mol Microbiol 77:1367–1379

Pourcel C, Salvignol G, Vergnaud G (2005) CRISPR elements in *Yersinia pestis* acquire new repeats by preferential uptake of bacteriophage DNA, and provide additional tools for evolutionary studies. Microbiology 151:653–663

Przybilski R, Richter C, Gristwood T, Clulow JS, Vercoe RB, Fineran PC (2011) Csy4 is responsible for CRISPR RNA processing in *Pectobacterium atrosepticum*. RNA Biol 8: 517–528

Pul U, Wurm R, Arslan Z, Geissen R, Hofmann N, Wagner R (2010) Identification and characterization of *E. coli* CRISPR-cas promoters and their silencing by H-NS. Mol Microbiol 75:1495–1512

Sapranauskas R, Gasiunas G, Fremaux C, Barrangou R, Horvath P, Siksnys V (2011) The *Streptococcus thermophilus* CRISPR/Cas system provides immunity in *Escherichia coli*. Nucleic Acids Res 39:9275–9282

Sashital DG, Jinek M, Doudna JA (2011) An RNA-induced conformational change required for CRISPR RNA cleavage by the endoribonuclease Cse3. Nat Struct Mol Biol 18:680–687

Sashital DG, Wiedenheft B, Doudna JA (2012) Mechanism of foreign DNA selection in a bacterial adaptive immune system. Mol Cell 46:606–615

Semenova E, Jore MM, Datsenko KA, Semenova A, Westra ER, Wanner B, van der Oost J, Brouns SJ, Severinov K (2011) Interference by clustered regularly interspaced short palindromic repeat (CRISPR) RNA is governed by a seed sequence. Proc Natl Acad Sci U S A 108:10098–10103

Sinkunas T, Gasiunas G, Fremaux C, Barrangou R, Horvath P, Siksnys V (2011) Cas3 is a single-stranded DNA nuclease and ATP-dependent helicase in the CRISPR/Cas immune system. EMBO J 30:1335–1342

Sternberg SH, Haurwitz RE, Doudna JA (2012) Mechanism of substrate selection by a highly specific CRISPR endoribonuclease. RNA 18:661–672

Swarts DC, Mosterd C, van Passel MW, Brouns SJ (2012) CRISPR interference directs strand specific spacer acquisition. PLoS ONE 7:e35888

Touchon M, Charpentier S, Clermont O, Rocha EP, Denamur E, Branger C (2011) CRISPR distribution within the *Escherichia coli* species is not suggestive of immunity-associated diversifying selection. J Bacteriol 193:2460–2467

Wang R, Preamplume G, Terns MP, Terns RM, Li H (2011) Interaction of the Cas6 riboendonuclease with CRISPR RNAs: recognition and cleavage. Structure 19:257–264

Westra ER, Brouns SJ (2012) The rise and fall of CRISPRs—dynamics of spacer acquisition and loss. Mol Microbiol 85:1021–1025

Westra ER, Pul U, Heidrich N, Jore MM, Lundgren M, Stratmann T, Wurm R, Raine A, Mescher M, Van Heereveld L, Mastop M, Wagner EG, Schnetz K, Van Der Oost J, Wagner R, Brouns SJ (2010) H-NS-mediated repression of CRISPR-based immunity in *Escherichia coli* K12 can be relieved by the transcription activator LeuO. Mol Microbiol 77:1380–1393

Westra ER, van Erp PB, Kunne T, Wong SP, Staals RH, Seegers CL, Bollen S, Jore MM, Semenova E, Severinov K, de Vos WM, Dame RT, de Vries R, Brouns SJ, van der Oost J (2012) CRISPR immunity relies on the consecutive binding and degradation of negatively supercoiled invader DNA by Cascade and Cas3. Mol Cell 46:595–605

Wiedenheft B, Zhou K, Jinek M, Coyle SM, Ma W, Doudna JA (2009) Structural basis for DNase activity of a conserved protein implicated in CRISPR-mediated genome defense. Structure 17:904–912

Wiedenheft B, Lander GC, Zhou K, Jore MM, Brouns SJ, van der Oost J, Doudna JA, Nogales E (2011a) Structures of the RNA-guided surveillance complex from a bacterial immune system. Nature 477:486–489

Wiedenheft B, van Duijn E, Bultema JB, Waghmare SP, Zhou K, Barendregt A, Westphal W, Heck AJ, Boekema EJ, Dickman MJ, Doudna JA (2011b) RNA-guided complex from a bacterial immune system enhances target recognition through seed sequence interactions. Proc Natl Acad Sci U S A 108:10092–10097

Yosef I, Goren MG, Kiro R, Edgar R, Qimron U (2011) High-temperature protein G is essential for activity of the *Escherichia coli* clustered regularly interspaced short palindromic repeats (CRISPR)/Cas system. Proc Natl Acad Sci U S A 108:20136–20141

Yosef I, Goren MG, Qimron U (2012) Proteins and DNA elements essential for the CRISPR adaptation process in *Escherichia coli*. Nucleic Acids Res 40:5569–5576

Zhang J, Rouillon C, Kerou M, Reeks J, Brugger K, Graham S, Reimann J, Cannone G, Liu H, Albers SV, Naismith JH, Spagnolo L, White MF (2012) Structure and mechanism of the CMR complex for CRISPR-mediated antiviral immunity. Mol Cell 45:303–313

Chapter 7
Type II: *Streptococcus thermophilus*

Marie-Ève Dupuis and Sylvain Moineau

Abstract *Streptococcus thermophilus* is an important industrial lactic acid bacterium that contains one of the best-studied models of the CRISPR-Cas system. Four CRISPR-Cas loci have been identified in this species, but only two have been demonstrated to be significantly active, i.e., new spacers can be naturally acquired following exposure to viruses and plasmids. CRISPR-Cas and CRISPR-Cas are classified as type II systems and are recognized by the presence of the signature gene *cas*9, in addition to the universal *cas*1 and *cas*2 genes. Two subgroups are currently distinguished due to the presence of another gene: subtype II-A (e.g., CRISPR-Cas and CRISPR-Cas of *S. thermophilus*) contains a *csn*2*-like* gene, whereas subtype II-B contains a *cas*4*-like* gene. For type II systems, crRNA biogenesis is directed by tracrRNA, implicating the Cas9 protein and the host non-Cas protein RNase III. Cas9 is also involved in the cleavage of plasmid and phage double-stranded DNA during the interference step.

Contents

M.-È. Dupuis · S. Moineau (✉)
Département de biochimie, de microbiologie et de bio-informatique,
Faculté des sciences et de génie, Université Laval, Québec G1V 0A6, Canada
e-mail: Sylvain.Moineau@bcm.ulaval.ca

M.-È. Dupuis
e-mail: marie-eve.dupuis.2@ulaval.ca

R. Barrangou and J. van der Oost (eds.), *CRISPR-Cas Systems*,
DOI: 10.1007/978-3-642-34657-6_7, © Springer-Verlag Berlin Heidelberg 2013

7.1 Historical Perspectives

CRISPR-Cas systems were first identified in the *S. thermophilus* species during genome sequence analysis of CNRZ1066 and LMG 18311 strains (Bolotin et al. 2004). The two bacterial genomes have in common more than 90 % of coding sequences, but one of the main differences occurred in the CRISPR loci. The first locus, called CRISPR1 (subtype II-A, but with a longer version of *csn*2 gene, see below), was present in both strains and an additional locus, called CRISPR2 (subtype III-A), was identified only in LMG 18311. A third *S. thermophilus* complete genome (LMD-9) was subsequently published (Makarova et al. 2006a), and an extra locus, CRISPR3 (subtype II-A, with a shorter *csn*2 gene), was revealed (Horvath et al. 2008). Overall, strain LMD-9 possesses at least 14 *cas* genes (Goh et al. 2011; Horvath et al. 2008). The fourth CRISPR locus (CRISPR4-Cas, subtype I-E) was recently identified in the industrial *S. thermophilus* strain DGCC7710 (Horvath and Barrangou 2010). Two other *S. thermophilus* genome sequences have since been made available: strain ND03 contains the first three CRISPR-Cas systems (Sun et al. 2011) while CRISPR data has not yet been published for strain JIM 8232 (Delorme et al. 2011). Comparative genomic analysis of JIM 8232 revealed nine hypervariable regions of strain-specific DNA compared to previously sequenced *S. thermophilus* strains, including CRISPR sequence regions. Figure 7.1 presents the currently known *S. thermophilus* CRISPR-Cas systems and the prevalence of *cas* genes among the sequenced genomes. Strain DGCC7710 is used as the model here because this strain has been heavily studied partly, because it is used in the dairy industry and partly because the main functional steps have been experimentally confirmed (see below).

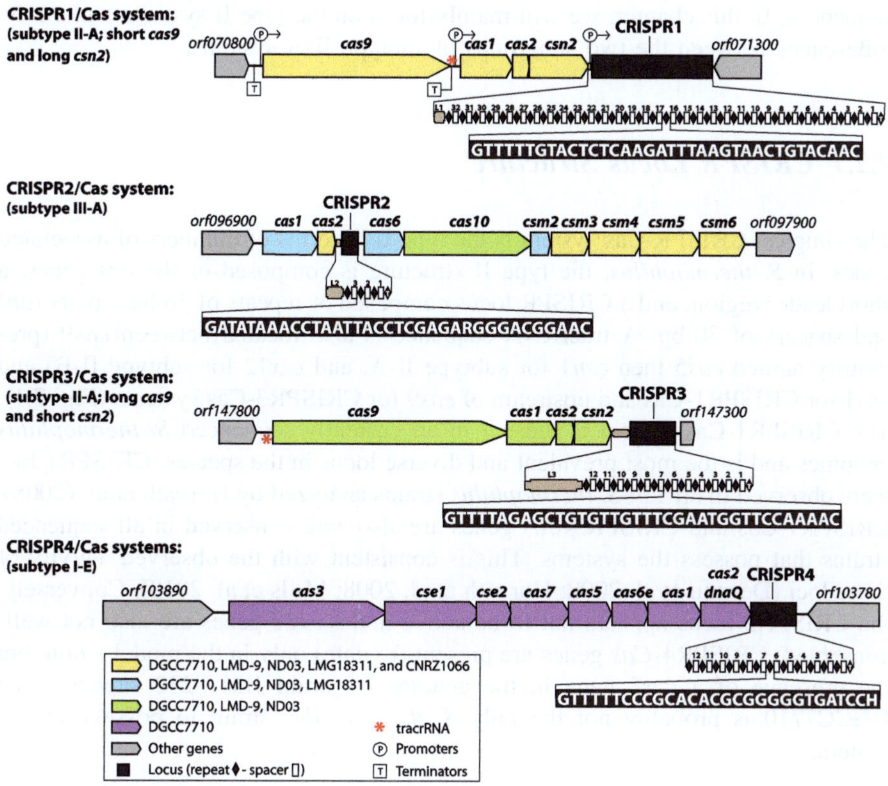

Fig. 7.1 CRISPR-Cas systems of *S. thermophilus* DGCC7710. Genetic organization is detailed for each locus of the model strain *S. thermophilus* DGCC7710. *Rectangles* and *arrows* represent repeat-spacer units and *cas* genes, respectively. The genes are colored according to their prevalence in *S. thermophilus* published genomes, as indicated in the box. Below the genomic organization, information on CRISPR-Cas elements is given: composition of repeat-spacer arrays (*black diamonds and white squares,* respectively), repeat sequences, and leader region (*L1, L2, L3, and L4 boxes*). TracrRNA sequences are located only for appropriate systems, more specifically for CRISPR1-Cas and CRISPR3-Cas systems, and are represented by a red asterisk. Available transcriptional information for CRISPR1-Cas systems is given, with "*P*" indicating the promoter and "*T*" indicating the terminator. Figure adapted from Horvath and Barrangou (2010)

7.2 Type II Systems of *S. thermophilus*

For *S. thermophilus* CRISPR1-Cas and CRISPR3-Cas (subtype II-A), four *cas* genes are associated with each of the locus [including the universal *cas* genes *cas*1 and *cas*2 (previously named *cas*6)], compared to nine and eight genes, respectively, for CRISPR2- and CRISPR4-Cas loci. The CRISPR1- and CRISPR3-Cas systems have historically been considered to be homologs, but differences do exist between these systems, notably in terms of repeat sequences and Cas protein sequences. In this chapter, we will mainly focus on the type II systems and on the differences between the two subgroups of subtype II-A systems.

7.2.1 CRISPR Locus Structure

The simplest CRISPR-Cas system is the type II in terms of numbers of associated genes. In *S. thermophilus,* the type II structure is composed of the *cas* genes, a short leader region, and a CRISPR locus composed of repeats of 36 base pairs (bp) and spacers of 30 bp. A tracrRNA sequence is also located between *cas*9 (previously named *cas*5 then *csn*1 for subtype II-A, and *csx*12 for subtype II-B) and *cas*1 for CRISPR1-Cas and upstream of *cas*9 for CRISPR3-Cas systems (Fig. 7.1). The CRISPR1-Cas system is present in all currently sequenced *S. thermophilus* genomes and is the most prevalent and diverse locus in the species. CRISPR1 loci were observed in all 124 *S. thermophilus* strains analyzed by Horvath et al. (2008). CRISPR1-*Cas* and CRISPR3-*Cas* genes are also well-conserved in all sequenced strains that possess the systems. This is consistent with the observed activity of those loci (Deveau et al. 2008; Horvath et al. 2008; Mills et al. 2010). Conversely, the CRISPR2 locus appears not to be active and its *cas* genes are also not well-conserved. CRISPR4-*Cas* genes are present (to date) only in the model strain, but the presence of a *cas*3 gene in the genome of strain JIM 8232 suggests that DGCC7710 is probably not the only *S. thermophilus* strain to possess such a system.

7.2.1.1 Repeats

For *S. thermophilus*, CRISPR1 locus is composed of identical repeats of a nearly perfect 36 bp palindrome interspaced by unique spacers, generally 30 bp in length. A similar observation can be made for CRISPR3 locus. In 2008, Horvath et al. (2008) analyzed the diversity and activity of CRISPR loci in 124 *S. thermophilus* strains and showed that CRISPR repeats and *cas* genes were locus specific and functionally coupled. Possible variants of repeat sequences and their frequency were noted. For the CRISPR1 locus, two variants and three terminal repeats (degenerated at the 3′ end) were identified. The typical repeat sequence (i.e., the most frequent sequence) was identified at a frequency of 99.7 % (for a total of 2,820 repeats). For the CRISPR3 locus, no degenerated terminal repeat was reported. To date, the maximum number of repeats in the same locus is 52 for the CRISPR1 locus (*S. thermophilus* JIM76) and 16 for the CRISPR3 locus (*S. thermophilus* SMQ-301). An alignment of DGCC7710 repeat sequences, with similar repeats from *S. thermophilus* and other genera, is presented in Fig. 7.2. Interestingly, the identical *S. thermophilus* CRISPR3-like repeats are found in *Streptococcus macacae*, *Streptococcus mitis*, *Streptococcus mutans*, and

A. Repeat of subtype II-A systems (with long *cas9* and short *csn2*)

Organism	Repeat	Mismatches
	`* **  *     *                    * *`	
Streptococcus thermophilus (CRISPR3)	GTTTTAGAGCTGTGTTGTTTCGAATGGTTCCAAAAC	
Streptococcus macacae, mitis, mutans, sanguinis	GTTTTAGAGCTGTGTTGTTTCGAATGGTTCCAAAAC	(0)
Streptococcus agalactiae, anginosus, bovis, equinus, gallolyticus	GTTTTAGAGCTGTGCTGTTTCGAATGGTTCCAAAAC	(1)
Streptococcus suis	GTTTTAGAGCTGTGCTGTTTCGAATGGTTTCAAAAC	(2)
Streptococcus dysgalactiae	GTTTTAGAGCTATGTTGTTTTGAATGGTCCCAAAAC	(3)
Streptococcus pyogenes	GTTTTAGAGCTATGCTGTTTTGAATGGTCCCAAAAC	(4)
Streptococcus pseudoporcinus	GTTTTAGAGCTATGTTATTTTGAATGATTCCAAAAC	(4)
Enterococcus italicus	GTTTTAGAGCTATGTTGAATCGAATGCTTCCAAAAC	(4)
Streptococcus equi	GTTTTAGAGTTATGCTGTATTGAATGGTTCCAAAAC	(5)
Enterococcus faecium	GTTTTAGAGCTATGCTGATTTGAATGCTTCCAAAAC	(5)
Enterococcus faecalis	GTTTTAGAGTCATGTTGTTTAGAATGGTACCAAAAC	(5)
Listeria monocytogenes	GTTTTAGAGCTATGTTATTTTGAATGCTACCAAAAC	(5)
Lactobacillus animalis	GTTTTAGAGCTATGTTGTTTTGTATGACTCCAAAAC	(5)
Treponema denticola	GTTTGAGAGTTGTGTAATTTAAGATGGATCTCAAAC	(10)
Olsenella uli	GTTTTGGGGCAGTGTCGTTTTGACTGGTAATCAAAC	(10)
Catenibacterium mitsuokai	GTTTTAGGGTTATGTTATTTTGAACTGAATTAAAAC	(11)
Coriobacterium glomerans	GTTTTGGAGCAGTGTCGTTCTGACTGGTAATCCAAC	(11)
Eggerthella sp.	GTTTTGGAGCAGTGGAATTCTGACTGGTAGTCAAAC	(12)
Acidaminococcus sp.	GTTTGAGAGATATGTAAATTCAAAGGATAATCAAAC	(13)
Finegoldia magna	GTTTGAGAATGATGTAATTTCATATAGGTATTAAAC	(14)
Pseudoramibacter alactolyticus	GTTTGAGAACCTTGTTATTTACGATACTTCTTTAGC	(14)
Anaerococcus tetradius	GTTTGAGAGTTATGTAATTTAAGTAGTAAGTAAAAC	(15)
Megasphaera sp.	GTTTGAGAGTAATGTAATTCTATAAATGTCTAAAAC	(15)
Lactobacillus iners	GTTTGAGAGTTATGTAATTTATGCACTAGGTAAAAC	(16)
Lactobacillus salivarius	GTTTCAGAAGTATGTTAAATCAATAAGGTTAAGACC	(16)
Ruminococcus lactaris	GTTTGAGTACCATACATTTTTACATCATTCTCAAAC	(16)
Dorea longicatena	GCTTGTGTACCATACATTTTTACATCATTCTCAAAC	(18)

B. Repeat of subtype II-A systems (with short *cas9* and long csn2)

Organism	Repeat	Mismatches
	` **** ***        * *           * ***`	
Streptococcus thermophilus (CRISPR1)	GTTTTTGTACTCTCAAGATTTAAGTAACTGTACAAC	
Streptococcus gordonii, mitis, oralis, vestibularis	GTTTTTGTACTCTCAAGATTTAAGTAACTGTACAAC	(0)
Streptococcus anginosus	GTTTTTGTACTCTCAAGATTTAAGTAACTGTAAAAC	(1)
Streptococcus macedonicus	GTTTTTGTACTCTCAAGATTTAAGTAACCGTACAAC	(1)
Streptococcus gallolyticus, infantarius, pasteurianus	GTTTTTGTACTCTCAAGATTTAAGTAACCGTAAAAC	(2)
Streptococcus suis	GTTTTTGTACTCTCAAGATTTAAGTAACAGTAAAAC	(2)
Enterococcus faecalis	GTTTTTGTACTCTCAATAATTTCTTATCAGTAAAAC	(8)
Lactobacillus farciminis	GTTTTTGTACCTTAAAGAATCTAGAAATAGTAAAAC	(10)
Butyrivibrio fibriosolvens	GTTTTAGTACCCGGGAAAATTAAGTGATTGGAAAAC	(11)
Enterococcus rectale	ATTTTAGTAACTGAATAATTTACGTGACTGTAAAAC	(12)
Eubacterium ventriosum	ATTTTAGTACCTGAAGAAATTAAGTTATCGTAAAAC	(13)

C. CRISPR1 and CRISPR3 repeats of *Streptococcus thermophilus*

Organism	Repeat	Mismatches
	`***** *  ** *   * **  * *   *  * ***`	
Streptococcus thermophilus (CRISPR3)	GTTTTAGAGCTGTGTTGTTTCGAATGGTTCCAAAAC	
Streptococcus thermophilus (CRISPR1)	GTTTTTGTACTCTCAAGATTTAAGTAACTGTACAAC	(17)

Fig. 7.2 Sequence alignments of *S. thermophilus* DGCC7710 repeats. Panel A. Comparison of the repeat sequence of CRISPR3 locus of *S. thermophilus* DGCC7710 with other repeats associated with subtype II-A systems, possessing a shorter version of the *csn*2 gene. Mismatches in repeat sequences of other organisms are underlined in gray and the number of different nucleotides is indicated in parenthesis. Panel B. Comparison of the repeat sequence of CRISPR1 locus of *S. thermophilus* DGCC7710 with other repeats associated with subtype II-A systems having a longer *csn*2 gene. Panel C. Comparison of the nucleotide sequences of CRISPR1 and CRISPR3 repeats

Streptococcus sanguinis, while only one mismatch is found in the repeats of *Streptococcus agalactiae*, *Streptococcus anginosus*, *Streptococcus bovis*, *Streptococcus equinus,* and *Streptococcus gallolyticus*. Similarly, highly identical repeats of *S. thermophilus* CRISPR1 are found in *Streptococcus gordonii*, *S. mitis*, *Streptococcus oralis*, and *Streptococcus vestibularis*. Clearly, the repeats of both subtype II-A groups are highly conserved in the CRISPR loci found in several other streptococci. The maximum number of observed mismatches in streptococcal sequences is five for subtype II-A St-CRISPR3 group and two for St-CRISPR1 group. Of note, it was recently found that the minimal number of repeats for spacer acquisition (for type I-E in *Escherichia coli*) is only one and that the added repeat was a duplication of that first repeat (Yosef et al. 2012).

7.2.1.2 Spacers

The field of *S. thermophilus* research has made significant strides in confirming the foreign origin of CRISPR spacers. Bolotin et al. (2005) analyzed CRISPR structures from *S. thermophilus* and *S. vestibularis* using bioinformatic tools, revealing homology between spacers and extrachromosomal elements. Over one-third (36 %) of the analyzed spacers had significant matches with available sequences; the best matches were with sequences from phage genomes (75 %), and streptococcal and lactococcal plasmids (20 %). These observations were confirmed experimentally by Barrangou et al. (2007). The latter study arguably launched the CRISPR-Cas field by demonstrating that *S. thermophilus* cells harboring a CRISPR-Cas system could acquire small pieces of DNA from the genomes of invading virulent phages and insert those as spacers at a CRISPR locus, providing immunity against subsequent infections with the same or related phage in a sequence-specific manner (Barrangou et al. 2007). Typically, spacer names are numerically associated with their "historical position" [i.e., spacer #1 is the oldest spacer and is situated upstream of the terminal repeat (Horvath and Barrangou 2010; Makarova et al. 2011)]. A comparative analysis demonstrated that newly acquired spacers perfectly matched the genomic sequences of phage proto-spacers (Deveau et al. 2008). In fact, this sequence identity is needed to confer the phage-resistant phenotype to these *S. thermophilus* strains termed BIMs, for "Bacteriophage-Insensitive Mutants".

Moreover, Horvath et al. (2008) analyzed more than 900 spacers and revealed homology to phage (77 %), plasmid (16 %), and *S. thermophilus* chromosomal sequences (7 %). Spacer polymorphisms among *S. thermophilus* strains suggested that the loci evolved via polarized spacer additions at the 5′-end of the locus, corresponding to the region near the leader region. The activity of a locus was defined as the capacity to integrate new spacers and this activity is correlated to the CRISPR diversity observed by in silico analysis (Horvath et al. 2008). The more active CRISPR-Cas locus is, the more spacers are present in a given strain. This in silico analyses calculated that almost 90 % of the spacers have a length of 30 bp, with others ranging from 29 to 31 bp, consistent with previous bioinformatic analysis (Bolotin et al. 2005) and in vivo experiments (Deveau et al. 2008; Garneau et al. 2010; Mills et al. 2010). The *S. thermophilus* CRISPR1-Cas system can also naturally acquire spacers from a rolling-circle replicating plasmids leading to plasmid loss (Garneau et al. 2010). Strains containing spacers derived from plasmids are named PIMs, for "Plasmid-Interfering Mutants".

Bolotin et al. (2005) have shown that one-third of their spacer sequences had no obvious extrachromosomal origin. In addition to phage and plasmid interference, it has also been proposed that CRISPR-Cas systems may act as a microbial regulatory system by controlling general mRNA transcript levels (Makarova et al. 2006b). For *S. thermophilus*, Horvath et al. (2008) have found four spacers (11 %), three CRISPR1 spacers and one CRISPR3 spacer, with 100 % identity to chromosomal sequences. In that case, the bacterial targets were the genes DtpT (a proton symporter), RexA (an ATP-dependent exonuclease), *Ster_0775* (a phage-associated DNA primase), and an intergenic region (between *Ster*_0810 and *Ster*_0811 genes). No information is available about a potential PAM near those putative bacterial proto-spacers or the potential interfering activity of those spacers against self-sequences.

7.2.2 Leader Regions

CRISPR-Cas systems contain a leader region upstream of its CRISPR locus. Leader regions are non-coding AT-rich sequences that act as a promoter for CRISPR arrays. It is generally less than 500 bp in length. Horvath et al. (2009) calculated that the average length of leader regions of lactic acid bacterium (LAB) genomes was (138 ± 94) bp. The promoter function has been demonstrated in vivo for some species, such as *E. coli* (Pul et al. 2010), *Pyrococcus furiosus* (Hale et al. 2008), *Streptococcus pyogenes* (Deltcheva et al. 2011), *S. thermophilus* (Garneau 2009), and *Sulfolobus solfataricus* (Lillestøl et al. 2006). The entire locus is transcribed for *S. solfataricus* P1, whereas the oldest spacers appear not transcribed for *S. thermophilus* DGCC7710. More specifically for *S. thermophilus* DGCC7710, the CRISPR1 and CRISPR3 leader regions are, respectively, 63 nucleotides (nt) and 321 nt in length. Interestingly, a stretch of 50 nt of the CRISPR1 leader shows about 60 % of identity with a part of the CRISPR3 leader (positions 204–250). The leader regions of lactic acid bacteria genomes have no particular sequence conservation, except for the CRISPR3 locus (St-CRISPR3 group) that possesses a 72 bp conserved section (Horvath et al. 2008). That region is particularly well-conserved in the *Streptococcus* genus (as the strains included in Table 7.1 and others) at the extremity adjacent to the first repeat. In *E. coli*, it was shown that only 60 bp of the leader sequence of subtype I-E systems is needed to provide acquisition, specially the 20 bp near the first repeat (Yosef et al. 2012).

7.2.3 cas *Genes and Related Proteins*

Type II systems are genetically organized as an operon with the typical gene order "*cas*9-*cas*1-*cas*2-*csn*2/*cas*4", and their gene orientation is consistent with the direction of adjacent repeats. The fourth *cas* gene of type II systems is now named *csn*2 or *cas*4, depending on the subtype to which it belongs, although some type II systems lack the fourth genes (see Sect. 7.3.1). The name of *cas* genes has been a matter of debate in recent years and some confusion may have resulted. For example, CRISPR-Cas of *S. thermophilus* DGCC7710, contains a *csn*2 gene, but previous publications have used the name *cas*7 for this gene.

Table 7.1 Subtype II A systems (with short *csn2*) sharing amino acid homology with CRISPR3-Cas of *S thermophilus* DGCC7710

Bacterial species and strain name		*cas* genes positions[a]		Protein length (aa)[b]				Number of spacers
		Start	End	Cas9	Cas1	Cas2	Csn2 (short)	
Streptococcus thermophilus								
DGCC7710	[CRISPR3]	–	–	1,388	289	114	219	12
LMD-9	[CRISPR3]	1,379,975	1,378,775	1,388	289	114	219	8
ND03	[CRISPR3]	1,366,596	1,365,396	1,388	289	57	219	20
Streptococcus agalactiae								
2,603 V/R		902,090	908,063	1,370	289	108	221	25
A909		980,303	986,276	1,370	289	113	221	15
NEM316		160,813	166,807	1,377	289	113	221	14
Streptococcus dysgalactiae								
ATCC12394		1,239,336	1,238,139	1,371	289	113	242	8
GGS_124		1,176,755	1,175,624	1,371	289	113	242	18
Streptococcus equi								
MGCS10565		1,369,339	1,368,136	1,348	289	107	224	17
Streptococcus gallolyticus								
ATCC BAA-2069		1,520,905	1,519,708	1,370	288	114	221	26
UNC34		1,518,984	1,517,790	1,371	288	114	221	12
Streptococcus mutans								
NN2025		737,258	743,154	1,345	288	107	220	70
UA159		1,330,942	1,329,655	1,345	288	109	190	6
Streptococcus pyogenes								
M1 GAS (SF370)		854,757	860,722	1,368	289	113	220	6
MGAS315		743,040	749,005	1,368	289	108	220	0
MGAS2096		813,084	819,049	1,368	289	113	242	2
MGAS5005		773,340	779,305	1,368	289	113	220	3
MGAS6180		771.231	777.196	1,368	289	113	242	4

(continued)

Table 7.1 (continued)

Bacterial species and strain name	*cas* genes positions[a]		Protein length (aa)[b]				Number of spacers
	Start	End	Cas9	Cas1	Cas2	Csn2 (short)	
MGAS9429	852,508	858,473	1,368	289	113	242	2
MGAS10270	844,446	850,255	1,368	289	113	190	2
NZ131	821,210	827,175	1,368	289	113	220	4
SSI-1	1,149,610	1,148,413	1,368	289	113	220	0
Streptococcus suis							
ST1	1,293,105	1,292,242	1,381	288	110	221	3
Coriobacterium glomerans							
PW2	2,036,091	2,042,122	1,384	292	102	226	9
Eggerthella sp.							
YY7918	2,684,425	2,683,228	1,380	292	111	227	43
Finegoldia magna							
ATCC 29328	73,918	79,790	1,348	290	101	215	14
Lactobacillus salivarius							
UCC118	114,740	120,808	1,149	301	101	223	27
Listeria monocytogenes							
10403S	2,641,981	2,640,786	1,334	288	113	220	30
J0161	2,735,374	2,734,179	1,334	288	113	220	19
Olsenella uli							
DSM 7084	1,407,777	1,406,580	1,399	292	111	226	30

[a] Positions are given only when the complete genome sequence is available.

7.2.3.1 csn2

In *S. thermophilus*, CRISPR1-Cas and CRISPR3-Cas systems show some differences for the *csn*2 gene, mainly its length, 350 and 219 nt, respectively. The two systems are classified within the subtype II-A, but they may still be distinguished according to which version of the *csn*2 gene they carry. When this gene is inactivated, spacer acquisition is no longer possible (Deveau et al. 2008). No details are known about the exact function of this protein during the spacer acquisition stage but some protein structures are available.

Csn2 of CRISPR subtype II-A [previously named CASS4 (Makarova et al. 2006b)], contains no known conserved domain. This protein is a member of the protein family called the "SPy1049 family" because it is composed of Csn-like protein similar to the one from *S. pyogenes* M1 GAS. The first structure of a small Csn2 protein came from *Enterococcus faecalis* ATCC 4200 (PDB: 3S5U; Ef.Csn2) (Nam et al. 2011). Two other structures of the small version of that protein were recently made available: from *S. agalactiae* ATCC 13813 (PDB: 3QHQ; Sag.Csn2) (Ellinger et al. 2012) and from *S. pyogenes* SF370 (PDB: 3TOC; Spy.Csn2) (Koo et al. 2012). Overall, all three structures share a diamond-shaped ring structure from homotetramerization. All monomer contains two main domains: the head, comprising N-terminal and C-terminal parts, and the tail, forming by the middle part. Two flexible regions linked the two domains. Two protomers are linked together by the α/β-domain of their head, and two dimers interact through the α-helical domain to form the final and stable ring. Some differences between the two protomers of the dimer were observed for Sag.Csn2 and Ef.Csn2. Nam et al. (2011) determined that the inner face of the ring is likely important for function because of one non-specific dsDNA and four Ca^{2+} binding sites. Calcium ions are important for the stability of the ring for all studied small Cns2 proteins, for their oligomerization, and for DNA cleavage. These binding sites were confirmed by others (Ellinger et al. 2012). But it was also shown that only linear dsDNA may enter into the positively charged channel and that the protein is stable at pH 7.0–9.0. For Sag.Csn2, the overall dimension is 70 Å in width, 55 Å in length, and with an inner channel of 30 Å.

Regarding the larger version of Csn2, there is currently only structure available and from the CRISPR1 locus of *S. thermophilus* LMG 18311 (PDB: 3ZTH) (Lee et al. 2012). It shares only 11 % of identity and 19 % of similarity with the sequence of the Ef.Csn2 protein, but the diamond-shaped ring structure is conserved and is likely universal for subtype II-A systems. However, no Ca^{2+}ion is needed for tetramerization but the C-terminal region is needed.

7.2.3.2 *cas*4

The *csn*2 gene was first predicted to be a functional analog of *cas*4 genes (COG1468, TIGR00372) (Jansen et al. 2002). These genes are found among the subtype II-B members [previously named CASS4a (Makarova et al. 2006b)]. No domain with a known function was found within subtype II-B *cas*4 genes. Interestingly, *cas*4 genes are also present in type I systems: I-A (previously included in the Apern and CASS5 groups), I-B (Treap-Hmari and CASS7), I-C (Dvulg and CASS1), or I-D (newly identified subtype) (Makarova et al. 2011). These Cas4 proteins contain a RecB exonuclease motif. It was suggested that three cysteine residues play a role in a DNA binding process, and a tyrosine residue permits the covalent bond between the Cas4 protein and the cleaved DNA (Jansen et al. 2002). It is believed that *cas*4 also plays a role in the spacer acquisition stage.

7.2.3.3 *cas*1 and *cas*2

Little is known about the universal *cas*1 and *cas*2 genes except their ubiquitous property, but they were suspected for a long time to act during the adaptation step (Beloglazova et al. 2008; Ebihara et al. 2006; Wiedenheft et al. 2009). Only recently, Yosef et al. (2012) demonstrated that Cas1 and Cas2 are essential in spacer acquisition of *E. coli* subtype I-E CRISPR-Cas system. Takeuchi et al. (2012) suggested that Cas1 and Cas2 protein sequences evolve more slowly than other Cas proteins as they possess strong evolutionary conservation. Currently, six and five protein structures are available for Cas1 and Cas2, respectively. None of those structures include streptococcal proteins. Cas1 protein structures include published structures from *E. coli* (PDB: 3NKD and 3NKE; Babu et al. 2011) and *Pseudomonas aeruginosa* [PDB: 3GOD (Wiedenheft et al. 2009)], as well as unpublished structures from *Pyrococcus horikoshii* [PDB: 3PV9 (Petit et al. 2010)], *Thermotoga maritima* [PDB: 3LFX (Beloglazova et al. 2010)], and *Aquifex aeolicus* [PDB: 2YZS (Ebihara et al. 2007)]. Some non-structural information is also available for Cas1 (SSO1450) from a *S. solfataricus* strain (Han et al. 2009). These data show that Cas1-like proteins are high-affinity nucleic acid binding proteins, metal-dependant, often without sequence specificity, and they contain four strictly conserved residues (Makarova et al. 2002). Cas1 was suspected to have DNA binding properties (Han et al. 2009; Babu et al. 2011) and it was also demonstrated that it produced dsDNA fragments of 80 bp in length in *P. aeruginosa* strain 14 (subtype I-F system, formerly pest and CASS3) (Wiedenheft et al. 2009). Its potential roles are then related to processing foreign nucleic acids, such as recognition, cleavage, and/or integration of sequences.

Cas2 (previously named *cas*6 in *S. thermophilus*) structures include the published Cas2 from *Desulfovibrio vulgaris* (PDB: 3OQ2; Samai et al. 2010), *S. solfataricus* [PDB: 2IVY (Oke et al. 2010), and 2I8E (Beloglazova et al. 2008)] and the unpublished Cas2 from *S. solfataricus* [PDB: 3EXC (Proudfoot et al. 2008)] and *Thermus thermophilus* [PDB: 1ZPW (Ihsanawati et al. 2005)]. Cas2 proteins exhibit good structural conservation (91–100 %) and even if their role remains unclear, they are known to be small proteins (80–120 amino acid [aa]), metal-dependent, and putative endoribonucleases. Specificity for uracil-rich regions was seen for *Sulfolobus* Cas2 (Beloglazova et al. 2008) and recently, a relationship with VapD (Virulent Associated Protein) was described, with a shared ferredoxin-like fold (Kwon et al. 2012). Cas2 may facilitate spacer selection and/or integration of new spacers (Makarova et al. 2011). It could also degrade phage transcripts or inhibit global transcription through RNA cleavage (Beloglazova et al. 2008).

7.2.3.4 *cas*9

The signature *cas*9 gene of *S. thermophilus* (previously named *cas*5 and *csn*1 for subtype II-A) codes for a large protein possessing two distinct domains: McrA/HNH and RuvC/RNaseH. The HNH motif is characteristic of many nucleases that act on dsDNA, whereas RuvC/RNaseH includes proteins that show a wide spectrum of nucleolytic functions, acting on RNA and DNA molecules. Those domains are situated in the middle and at the N-terminus of the Cas9 protein, respectively. Unfortunately, no structures are available yet for any of the Cas9 protein.

Studies with *S. thermophilus* DGCC7710 have shown that inactivating the *cas*9 gene of the CRISPR1-Cas system results in loss of the phage resistance phenotype (Deveau et al. 2008; Garneau et al. 2010). It has also been shown that the *cas*9 gene of the *S. thermophilus* DGCC7710 CRISPR3 locus, when expressed in *E. coli*, is the only gene necessary for the interference phenotype (Sapranauskas et al. 2011). Mutation analysis using alanine replacement outlined the importance of the two domains in the interference function. It was suggested that the HNH-domain of Cas9 proteins can be involved in dsDNA degradation or that the RuvC/RNaseH domain may be responsible for dsDNA breaks. It was also suggested that the RuvC/RNaseH domains may also act during crRNA maturation, which was supported by the discovery that Cas9 helped to generate mature crRNA (Deltcheva et al. 2011). Another function suggested that Cas9 concerns a role as a "molecular anchor" to permit the pairing of tracrRNA with pre-crRNA. Finally, Cas9 may be involved in the second cleavage during crRNA processing and/or protect tracrRNA and pre-crRNA against other host RNases. Recently, Jinek et al. (2012) have confirmed for the first time, in *S. pyogenes* SF370, that each domain is responsible for the cleavage of the dsDNA during the interference step. Indeed, Spy.Cas9 was shown to cleave linear and supercoiled plasmids and they confirmed that the HNH domain cleaves the complementary strand while the RuvC part cleaves the noncomplementary one.

For subtype II-B systems, *cas*9 was previously named *csx*12, as those found in *Wolinella succinogenes* (e.g., strain DSM 1740), *Francisella tularensis* (e.g., strain SCHU S4), *Legionella pneumophila* (e.g., strain Paris), and *Burkholderiales bacterium* (e.g., strain 1_1_47). HNH domains are also present in these proteins but share no similarity with *Streptococcus* CRISPR-Cas systems. Since NCBI suggested that those systems are likely degenerated CRISPR-Cas systems, they will not be discussed further.

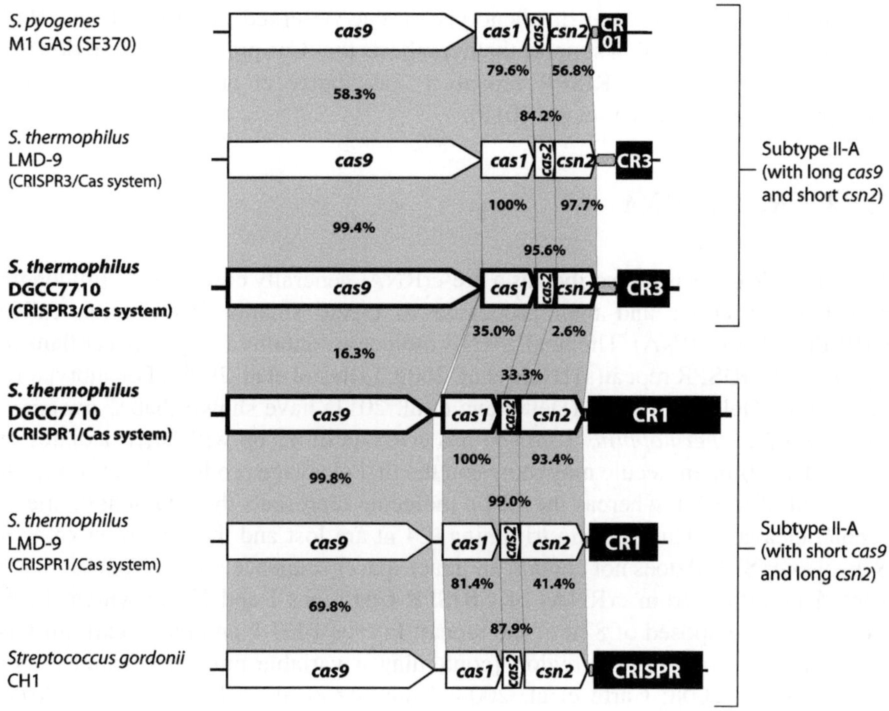

Fig. 7.3 Comparative analyses of Cas proteins of *S. thermophilus* CRISPR-Cas systems of subtype II-A and related streptococci. The *cas* genes are shown as *white arrows* and CRISPR loci arrays are shown as *black rectangles*. Loci of the model strain *S. thermophilus* DGCC7710 are presented in the *middle* of the figure, and compared with the Cas proteins of both type II subtypes. Multi-alignment comparisons were calculated with the national center for biotechnology information (NCBI) "Protein Blast" tool. The numbers in gray shading indicate percent identity between homologous Cas protein sequences. Percentages were calculated by dividing the number of identical amino acids by highest number of amino acids of the two proteins compared

7.2.3.5 Comparison of Cas Proteins of Different Groups of Subtype II-A Systems

The paralogs proteins of CRISPR1-Cas and CRISPR3-Cas of *S. thermophilus* DGCC7710 have limited identity between each other (respectively 16, 35, 33, and 3 % for Cas9, Cas1, Cas2, and Csn2 proteins). There are also significant differences in protein size, mainly for genes *cas*9 and *csn*2. In fact, *S. thermophilus* CRISPR1-Cas proteins are closer to other CRISPR1-Cas proteins from other streptococci than to *S. thermophilus* CRISPR3-Cas proteins (Fig. 7.3). Overall, Cas proteins are very well-conserved among *S. thermophilus* strains. There are more than 93 % identity for all Cas proteins from strain DGCC7710 compared to LMD-9. This is in line with an ancient gene duplication, and suggests that each CRISPR system has its own dedicated set of co-evolved Cas proteins. It is consistent with the absence of phenotype rescue observed in KO mutants (Barrangou et al. 2007), which support the hypothesis that Cas proteins are functionally paired with particular CRISPR repeats (Chakraborty et al. 2010; Godde and Bickerton 2006; Takeuchi et al. 2012).

7.2.4 CRISPR-RNA

The CRISPR locus is transcribed as a pre-crRNA, generally covering the full length repeat-spacer array, and then processed to create smaller RNA units, called CRISPR-RNA (crRNA). The final crRNA molecule contains a single spacer flanked by a partial CRISPR repeat(s) (Hale et al. 2009; Lillestøl et al. 2009). For subtype II-A systems, Deltcheva et al. (Deltcheva et al. 2011) have shown that *S. pyogenes* SF370 (and *S. thermophilus* LMD-9) has crRNAs of 42 bp with intermediates of 66 bp. The 66 bp molecule may represent the first cleavage product (30 nt of spacer and 36 nt of repeat), whereas the 42 bp molecule represents the final length after a second cleavage. Thus, during trimming, 24 nt are lost and the mature crRNA of *S. pyogenes* SF370 does not contain an intact spacer sequence: only 20 out of 30 are kept. This differs from crRNAs of CRISPR-Cas types I and III, in which the 5′ extremity is composed of 8 nt of the repeat; In type I-E/I-F and the 3′ extremity is also composed of a repeat region [containing a variable number of nucleotides (Brouns et al. 2008; Carte et al. 2008; Haurwitz et al. 2010; Jore et al. 2011; Wiedenheft et al. 2011)]. In type III, trimming occurs at the 3′ end of the crRNA, resulting in two major species that generally differ 6 nt (for details, see Chap. 5).

7.2.5 tracrRNA Sequences

Trans-encoded small CRISPR-RNA (called tracrRNA) is transcribed from a genomic region, near the type II CRISPR-Cas systems, which possesses a stretch of partial complementarity to the repeat (and to their associated portions on the crRNA precursor transcripts). Recent differential RNA sequencing of *S. pyogenes* SF370 revealed that tracrRNAs are implicated in crRNA maturation, along with endogenous RNase III, and Cas9 protein activities (Deltcheva et al. 2011). TracrRNA was found in many other bacterial strains, including *S. thermophilus* LMD-9. More precisely, according to the putative promoter and terminator, the tracrRNA of the latter strain should be processed to a final length of 70 nt for both the CRISPR1-Cas and CRISPR3-Cas systems. For the other studied strains (containing a subtype II-A system from St-CRISPR3 group), the entire tracrRNA putative length was evaluated to be 100 or 110 nt, and the processed form to be 90, 80, 70, or 65 nt. In each case, the final form involves a loss of nucleotides in the stretch region corresponding to the repeat sequence.

Our sequence alignment analysis also revealed some differences for tracrRNA depending on the St-CRISPR group (Fig. 7.4). First, the two tracrRNA of

CRISPR1 and CRISPR3 loci are homologs with their associated repeat sequence (at 75 and at 83 %, respectively). Second, the tracrRNA sequence is situated after the CRISPR1 *cas*9 gene, whereas the CRISPR3 tracrRNA is located between the *cas*9 and the *cas*1 genes. Finally, tracrRNA orientation varied by depending on the subtype: it is in the inversed orientation with respect to the *cas* genes and the locus for CRISPR3 systems as opposed to the conventional orientation of CRISPR1 systems. Interestingly, even if repeat sequences are different in other bacterial species of the same subtype II-A (as compared to the model sequence of DGCC7710), the differential nucleotide(s) of the repeat are matching their associated tracr-stretch sequence. This comparison highlights a phenomenon of co-evolution that signifies that some nucleotide positions seem to be more critical than others.

7.2.6 Proto-Spacer-Adjacent Motifs

In addition to the proto-spacer identity necessary for CRISPR interference, a proto-spacer-adjacent motif (termed PAM) is flanking the proto-spacer region (Makarova et al. 2011), downstream in the case of type II systems in *S. thermophilus* (Deveau et al. 2008; Horvath et al. 2008). The PAM is implicated in the *S. thermophilus* CRISPR1-Cas resistance phenotype as a perfect match to the sequence 5′-NNAGAAW-3′ is needed for interference. Phages can escape CRISPR-Cas interference with a single mutation in the PAM region. A different PAM (5′-NGGNG-3′) was discovered for CRISPR3-Cas systems of *S. thermophilus* (Horvath et al. 2008). In the orthologs system of *S. pyogenes*, the motif is smaller and included only the 5′-NGG-3′ sequence (Jinek et al. 2012). Recently, PIM experiments revealed that PAM identity may be less conserved than previously believed (Garneau et al. 2010). Many *S. thermophilus* PIM strains prevented plasmid transformation even if the PAM sequence flanking the proto-spacer was not a perfect match (single or double nucleotide mismatches) in the incoming plasmid. This lower stringency for PAM conservation could be due to the lower selective pressure for plasmid as compared to phage infection. The importance of PAM to distinguish self and non-self sequences during interference (i.e., spacers and proto-spacers) and for spacer acquisition was recently confirmed for another subtype II-A CRISPR-Cas systems (*S. agalactiae*) (Lopez-Sanchez et al. 2012).

7.3 Prevalence of Subtype II-A Systems

7.3.1 Complete Subtype II-A Systems

A recent comparative analysis revealed that at least 103 genomes contain the type II signature *cas*9 gene (Makarova et al. 2011). However, of these, only a few strains contain a complete operon including all four type II genes. Based on our

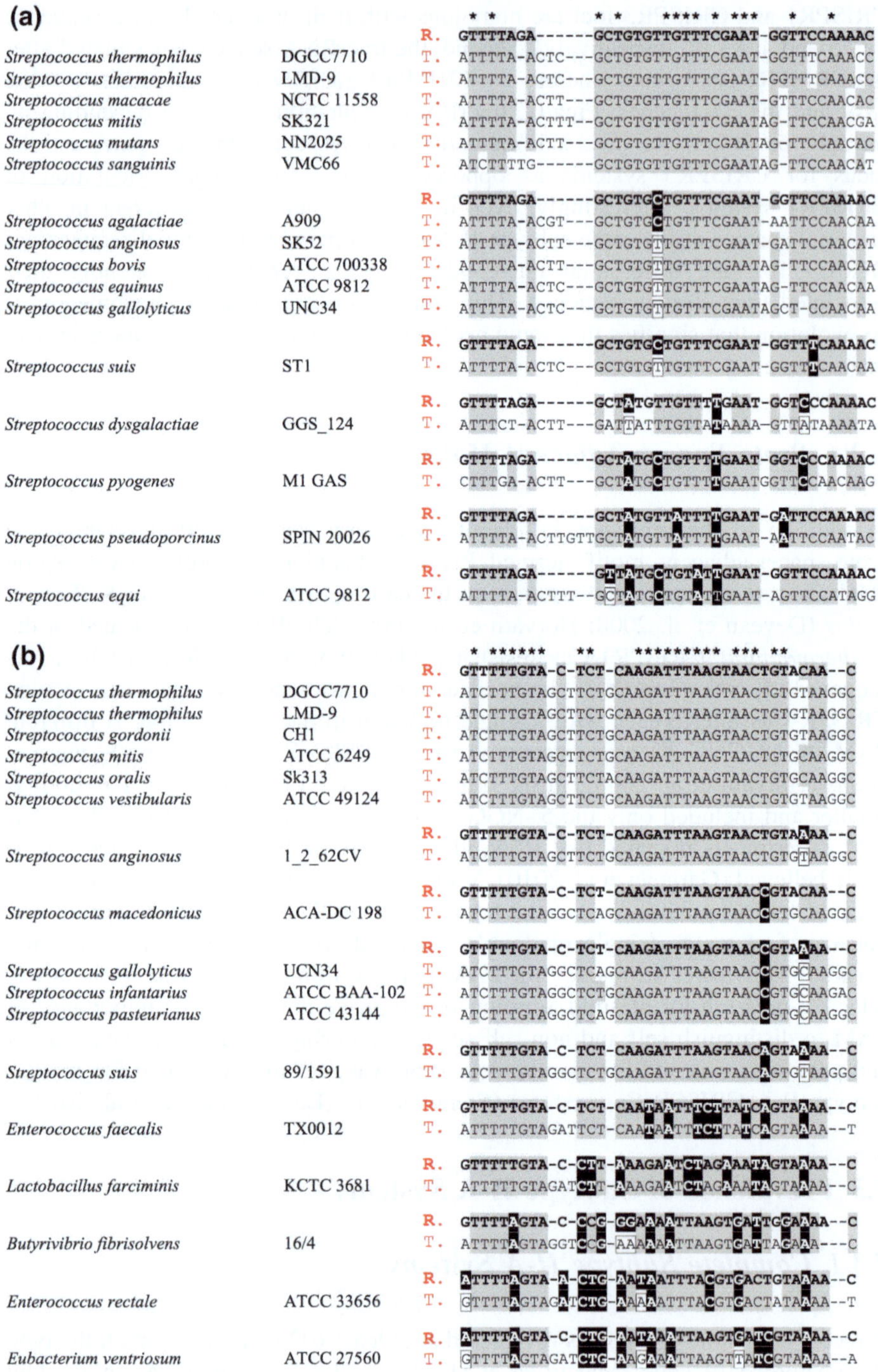
(a)
* * * * **** * *** * *
R. GTTTTAGA------GCTGTGTTGTTTCGAAT-GGTTCCAAAAC
Streptococcus thermophilus DGCC7710 T. ATTTTA-ACTC---GCTGTGTTGTTTCGAAT-GGTTTCAAACC
Streptococcus thermophilus LMD-9 T. ATTTTA-ACTC---GCTGTGTTGTTTCGAAT-GGTTTCAAACC
Streptococcus macacae NCTC 11558 T. ATTTTA-ACTT---GCTGTGTTGTTTCGAAT-GTTTCCAACAC
Streptococcus mitis SK321 T. ATTTTA-ACTTT--GCTGTGTTGTTTCGAATAG-TTCCAACGA
Streptococcus mutans NN2025 T. ATTTTA-ACTT---GCTGTGTTGTTTCGAATAG-TTCCAACAC
Streptococcus sanguinis VMC66 T. ATCTTTTG------GCTGTGTTGTTTCGAATAG-TTCCAACAT
R. GTTTTAGA------GCTGTGCTGTTTCGAAT-GGTTCCAAAAC
Streptococcus agalactiae A909 T. ATTTTA-ACGT---GCTGTGCTGTTTCGAAT-AATTCCAACAA
Streptococcus anginosus SK52 T. ATTTTA-ACTT---GCTGTGTTGTTTCGAAT-GATTCCAACAT
Streptococcus bovis ATCC 700338 T. ATTTTA-ACTT---GCTGTGTTGTTTCGAATAG-TTCCAACAA
Streptococcus equinus ATCC 9812 T. ATTTTA-ACTC---GCTGTGTTGTTTCGAATAG-TTCCAACAA
Streptococcus gallolyticus UNC34 T. ATTTTA-ACTC---GCTGTGTTGTTTCGAATAGCT-CCAACAA
R. GTTTTAGA------GCTGTGCTGTTTCGAAT-GGTTTCAAAAC
Streptococcus suis ST1 T. ATTTTA-ACTC---GCTGTGTTGTTTCGAAT-GGTTTCAACAA
R. GTTTTAGA------GCTATGTTGTTTTGAAT-GGTCCCAAAAC
Streptococcus dysgalactiae GGS_124 T. ATTTCT-ACTT---GATTATTTGTTATAAAA–GTTATAAAATA
R. GTTTTAGA------GCTATGCTGTTTTGAAT-GGTCCCAAAAC
Streptococcus pyogenes M1 GAS T. CTTTGA-ACTT---GCTATGCTGTTTTGAATGGTTCCAACAAG
R. GTTTTAGA------GCTATGTTATTTTGAAT-GATTCCAAAAC
Streptococcus pseudoporcinus SPIN 20026 T. ATTTTA-ACTTGTTGCTATGTTATTTTGAAT-AATTCCAATAC
R. GTTTTAGA------GTTATGCTGTATTGAAT-GGTTCCAAAAC
Streptococcus equi ATCC 9812 T. ATTTTA-ACTTT--GCTATGCTGTATTGAAT-AGTTCCATAGG
(b)
* ****** ** ********** *** **
R. GTTTTTGTA-C-TCT-CAAGATTTAAGTAACTGTACAA--C
Streptococcus thermophilus DGCC7710 T. ATCTTTGTAGCTTCTGCAAGATTTAAGTAACTGTGTAAGGC
Streptococcus thermophilus LMD-9 T. ATCTTTGTAGCTTCTGCAAGATTTAAGTAACTGTGTAAGGC
Streptococcus gordonii CH1 T. ATCTTTGTAGCTTCTGCAAGATTTAAGTAACTGTGTAAGGC
Streptococcus mitis ATCC 6249 T. ATCTTTGTAGCTTCTGCAAGATTTAAGTAACTGTGCAAGGC
Streptococcus oralis Sk313 T. ATCTTTGTAGCTTCTACAAGATTTAAGTAACTGTGCAAGGC
Streptococcus vestibularis ATCC 49124 T. ATCTTTGTAGCTTCTGCAAGATTTAAGTAACTGTGTAAGGC
R. GTTTTTGTA-C-TCT-CAAGATTTAAGTAACTGTAAAA--C
Streptococcus anginosus 1_2_62CV T. ATCTTTGTAGCTTCTGCAAGATTTAAGTAACTGTGTAAGGC
R. GTTTTTGTA-C-TCT-CAAGATTTAAGTAACCGTACAA--C
Streptococcus macedonicus ACA-DC 198 T. ATCTTTGTAGGCTCAGCAAGATTTAAGTAACCGTGCAAGGC
R. GTTTTTGTA-C-TCT-CAAGATTTAAGTAACCGTAAAA--C
Streptococcus gallolyticus UCN34 T. ATCTTTGTAGGCTCAGCAAGATTTAAGTAACCGTGCAAGGC
Streptococcus infantarius ATCC BAA-102 T. ATCTTTGTAGGCTCTGCAAGATTTAAGTAACCGTGCAAGAC
Streptococcus pasteurianus ATCC 43144 T. ATCTTTGTAGGCTCAGCAAGATTTAAGTAACCGTGCAAGGC
R. GTTTTTGTA-C-TCT-CAAGATTTAAGTAACAGTAAAA--C
Streptococcus suis 89/1591 T. ATCTTTGTAGGCTCTGCAAGATTTAAGTAACAGTGTAAGGC
R. GTTTTTGTA-C-TCT-CAATAATTTCTTATCAGTAAAA--C
Enterococcus faecalis TX0012 T. ATTTTTGTAGATTCT-CAATAATTTCTTATCAGTAAAA--T
R. GTTTTTGTA-C-CTT-AAAGAATCTAGAAATAGTAAAA--C
Lactobacillus farciminis KCTC 3681 T. ATTTTTGTAGATCTT-AAAGAATCTAGAAATAGTAAAA--C
R. GTTTTAGTA-C-CCG-GGAAAATTAAGTGATTGCAAAA--C
Butyrivibrio fibrisolvens 16/4 T. ATTTTAGTAGGTCCG-AAAAAATTAAGTGATTAGAAA---C
R. ATTTTAGTA-A-CTG-AATAATTTACGTGACTGTAAAA--C
Enterococcus rectale ATCC 33656 T. GTTTTAGTAGATCTG-AAAAATTTACGTGACTATAAAA--T
R. ATTTTAGTA-C-CTG-AATAAATTAAGTGATCGTAAAA--C
Eubacterium ventriosum ATCC 27560 T. GTTTTAGTAGATCTG-AAGAAATTAAGTTATCGTAAAA--A

◄ **Fig. 7.4** Comparative nucleotide analyses between the repeat and the tracrRNA region on the chromosomal DNA. The parts of tracrRNA that share identity with the repeat sequence are shown for *S. thermophilus* and other subtype II-A CRISPR-Cas systems. *Panel A* presentation of St-CRISPR3 group of subtype II-A systems with short *csn*2 genes. *Panel B* presentation of St-CRISPR1 group of subtype II-A systems with longer *csn*2 genes. For both panels, the sequence of the repeat (*R*) is presented above the corresponding stretch of the tracrRNA sequence (*T*). Alignments were done to fit with the position of the *S. thermophilus* repeat, and perfect matches are underlined in *gray*. Nucleotides in *black squares* represent mismatches with the repeat sequence of the model strain *S. thermophilus* DGCC7710. The corresponding nucleotide on the tracrRNA sequence is also included in a *black square* if the mismatch is "conserved" on the tracrRNA sequence (*as shown by the covariance phenomenon*). Contrarily, the nucleotide is included in an *empty square* if there is a difference between compared sequences for the same system. *Asterisks* denote the nucleotide positions that are always identical between the repeat and the tracrRNA sequences of the same system

bioinformatic analysis with the NCBI "Protein Blast" tool and the GenBank database, 26 and 83 strains have systems homologs to the *S. thermophilus* DGCC7710 CRISPR1-Cas and CRISPR3-Cas systems, respectively. However, no spacer acquisition or interference activity has yet been demonstrated for mainly all these systems. Some of the above strains/species with CRISPR-Cas type II elements lack a fourth gene, such as for *Neisseria meningitidis, Campylobacter jejuni*, and *Treponema denticola*. However, these systems have yet to be studied in detail and to our knowledge, no evidence of functional adaptation and/or interference activity is available (Tasaki et al. 2012; Deshpande et al. 2011; Merchant-Patel et al. 2010; Price et al. 2007; Schouls et al. 2003). It is tempting to speculate that they may lack the spacer acquisition function due to the lack of the fourth gene.

Thirty subtype II-A systems similar to CRISPR3-Cas of DGCC7710 are described in Table 7.1. It seems that protein sizes are generally well-conserved in this subtype and that it may be possible to have up to 70 spacers in the same II-A locus (e.g., *S. mutans* NN2025). The majority of these systems belong to the *Streptococcus* genus (divided into eight different species) and even some non-related species groups within that subtype (*Coriobacterium glomerans*, *Eggerthella* sp., *Finegoldia magna*, *Listeria monocytogenes*, and *Olsenella uli*).

Many others subtype II-A systems, associated with CRISPR1-Cas systems of *S. thermophilus*, are found in other strains/species (Table 7.2). The majority of these systems are found in eleven streptococci species. Other organisms include *Butyrivibrio fibrisolvens*, *E. faecalis*, *Eubacterium rectale*, *Eubacterium ventriosum*, and *Lactobacillus farciminis*. The protein sizes of these systems are also similar. The Csn2 protein is larger than Cas1 in these systems and, contrary to its St-CRISPR3 homolog with short *csn*2 genes, their Cas9 protein is nearly 200 AA smaller. The larger CRISPR array of St-CRISPR1 group, presented in Table 7.2, contains 44 spacers. Globally, we observe that subtype II-A systems with shorter *csn*2 (St-CRISPR3 group) are more prevalent and more conserved than subtype II-A with long *csn*2 (St-CRISPR1 group). On the other hand, it is interesting to note that the St-CRISPR1 (with long *csn*2 and short *cas*9 genes) is ubiquitous in the *S. thermophilus* species.

Table 7.2 Subtype II B systems (with long *csn*2) sharing amino acid homology with CRISPR1-Cas of *S thermophilus* DGCC7710

Bacterial species and strain name		*cas* genes positions[a]		Protein length (aa)[b]				Number of spacers
		Start	End	Cas9	Cas1	Cas2	Csn2 (long)	
Streptococcus thermophilus								
CNCM I-1630	–	–	–	595/302*	167/113*	107	350	12
CNRZ1066	[CRISPR1]	619,189	625,037	1,128	303	107	350	41
DGCC7710	[CRISPR1]	–	–	1,121	303	107	350	32
JIM 8232		706,443	712,270	1,121	303	107	350	42
JIM8777		708,034	713,879	1,127	303	107	350	26
LMD-9	[CRISPR1]	643,235	649,062	1,121	303	107	350	16
LMG18311	[CRISPR1]	624,007	629,838	1,122	303	107	350	33
ND03	[CRISPR1]	633,621	639,449	1,121	303	44	350	36
Streptococcus anginosus								
1_2_62CV		–	–	1,125	303	102	347	4
Streptococcus gallolyticus								
UCN34	1,511,433	1,510,027	1,130	303	92	349	14	
Streptococcus gordonii								
CH1		1,426,750	1,425,339	1,136	307	107	345	26
Streptococcus infantarius								
ATCC BAA-102		–	–	1,129	304	107	349	30
Streptococcus macedonicus								
ACA-DC 198		1,025,472	1,029,177	1,044	303	107	349	17
Streptococcus mitis								
ATCC 6249		–	–	1,134	355	107	347	47
Streptococcus oralis								
Sk313		–	–	1,134	302	98	196	44

(continued)

Table 7.2 (continued)

Bacterial species and strain name	*cas* genes positions[a]		Protein length (aa)[b]				Number of spacers
	Start	End	Cas9	Cas1	Cas2	Csn2 (long)	
Streptococcus pasteurianus							
ATCC 43144	1,400,035	1,398,629	1,130	303	107	349	37
Streptococcus sp.							
C150	–	–	1,139	303	107	350	12
Streptococcus suis							
89/1591	–	–	1,122	304	107	348	6
Streptococcus vestibularis							
ATCC 49124	–	–	1,128	303	107	350	9
F0396	–	–	1,038	303	44	350	5
Butyrivibrio fibrisolvens							
16/4	309,663	308,957	765 / 177*	302	107	241	22
Enterococcus faecalis							
Fly1	–	–	1,150	304	107	69 / 263*	8
T11	–	–	1,150	304	107	352	21
TX0012	–	–	1,150	306	99	352	2
Eubacterium rectale							
ATCC 33656	1,591,112	1,596,816	1,114	302	108	327	44
Eubacterium ventriosum							
ATCC 27560	–	–	1,107	305	108	331	14
Lactobacillus farciminis							
KCTC 3681	–	–	1,126	302	97	251#	5

[a] Positions are given only when the complete genome sequence is available .For draft genomes, the CRISPR-Cas locus was found on one or many contigs
[b] *Two genes (ORF) are associated with that putative protein, # positioned at one extremity of a contig. In both cases, no precise protein lengths can be given

7.3.2 *Unique Properties of Each Group of Subtype II-A Systems*

Overall, despite some similarities, CRISPR3-Cas and CRISPR3-Cas systems are clearly distinct. First, sequence similarities are low even at the protein level. Second, the repeat sequences as well as the tracrRNA and the crRNA are also significantly distinct. Third, the operon architecture, the position of the tracrRNA regions and the number transcriptional promoters also differ. The CRISPR1 leader region is not conserved whereas the CRISPR3 leader is. It is very interesting, from a bacterial point of view, to carry two distinct, independent, and active CRISPR-Cas systems as it is the case for *S. thermophilus* DGCC7710 (Magadán et al. 2012).

7.4 The Three Stages of Type II CRISPR-Cas Systems

Three main stages have been identified in the CRISPR-Cas system: (1) acquisition of new spacers (also called the adaptation or the immunization step), (2) biogenesis of RNAs (the expression step), and (3) targeting and cleavage of invading foreign proto-spacers (the interference or the immunity step). Type II systems have distinctive features, which will be detailed in the next sections.

7.4.1 *CRISPR-Adaptation*

The immunization step of a cell involves the insertion of one or several new spacers in combination with a new repeat, generally at the leader end. It has been shown that this insertion is generally polarized to the 5′-end (Deveau et al. 2008; Horvath et al. 2008), near the leader region, and that it is occurring in less than 1 % of the bacterial cells of a population (Yosef et al. 2012), but the exact mechanism remains unclear. The polarization allows to identify a timeline of interactions based on the spacer order (Barrangou et al. 2007). The Cas1 and Cas2 proteins, in addition to Csn2 for type II systems, have been linked to that stage (Barrangou et al. 2007; Beloglazova et al. 2008; Ebihara et al. 2006; Garneau et al. 2010; Wiedenheft et al. 2009).

In the first comparative studies, it was assumed that the selection of new spacers was random because no particular regions of the foreign genetic element (i.e., phage genome in numerous studies) were overrepresented. Indeed, no noticeable preference was observed for a genome strand, a transcriptional module, a coding or non-coding regions, or a phage groups (Deveau et al. 2008). Later, it was suggested that the PAMs located near the proto-spacer are implicated in spacer selection (Deveau et al. 2008; Horvath et al. 2008). Of note, the presence of a PAM provides a way to distinguish between the spacer (no PAM) in the CRISPR locus and the proto-spacer flanking sequence (with a PAM) in the foreign DNA. In *E. coli* subtype I-E system, it was shown that the spacer acquisition is strand-specific if a previous spacer allows recognition but not interference (Swarts et al. 2012; Datsenko et al. 2012). No biological evidence allows transferring that priming phenomenon principle to type II system mechanisms.

The most likely scenario for adaptation is that the foreign dsDNA (plasmid or phage) is recognized through a PAM, then a CRISPR protein complex cleaves the DNA to obtain a pre-spacer. It is not known that if DNA is already 30 bp in length, but we assume that it is not since spacer orientation is conserved when it is inserted in a locus. The PAM is most likely still involved at this point and must be present in the pre-spacer fragment. As dsDNA binding properties have been reported for Cas1 and Csn2 (subtype II-A) proteins (Nam et al. 2011; Wiedenheft et al. 2009; Yosef et al. 2012), it is probable that Cas1, Csn2, and Cas4 (subtype II-B analog of Csn2) act on DNA at the moment of the pre-spacer selection and/or the spacer insertion. The fragment is then incorporated into the CRISPR locus. The spacer length will be between 28 and 31 nt, with a majority of 30 (Bolotin et al. 2005; Deveau et al. 2008; Garneau et al. 2010; Mills et al. 2010). Since insertion is polarized, the leader region may also be implicated in the insertion event in addition to the PAM sequence. Of note, among the *S. thermophilus* phage genomic sequences available, there are more than 200 CRISPR1 and 400 CRISPR3 PAMs per phage genome (averaging 274 ± 30 and 464 ± 25, respectively). Thus in *S. thermophilus*, there are well over 600 potential spacers that can be acquired by subtype II-A CRISPR-Cas systems from a single-phage genome.

During the spacer acquisition stage, a new repeat is also added by an unknown mechanism. Deletions of one or many (up to 17) old spacers have been (rarely) detected during the spacer insertion by *S. thermophilus* type II systems (Deveau et al. 2008; Mills et al. 2010). Interestingly, a study on population dynamics predicted that spacer deletion can occur when the CRISPR array is longer than 30 repeat-spacer units, and the deleted spacer will be randomly selected with probability proportional to its distance to the leader sequence (He and Deem 2010). Others have also suggested that it may be a way to control the length of the CRISPR locus (Al-Attar et al. 2011; Mills et al. 2010; Sorek et al. 2008). Perhaps carrying several spacers have a fitness cost for the strain or it is less stable. Internal spacer duplications have also been observed in both groups of subtype II-A systems (Bolotin et al. 2005). That kind of duplication was observed in both CRISPR1 and CRISPR3 loci of *S. thermophilus* ND03, and it is also found in the CRISPR3 of DGCC7710. Recently, analyses of CRISPR arrays in *S. agalactiae* stains have shown that spacer deletions appear to occur to keep mobile genetic elements into the bacterial population and to create heterogeneity of the mobilome (Lopez-Sanchez et al. 2012). Interestingly, some strains show differences among older spacers, which is different with leader-polarized insertions in *S. thermophilus* loci.

There is no data to suggest at which moment of the phage infection or the plasmid replication the pre-spacer is generated. It is possible that only defective molecules, non-infectious phage, or non-replicative plasmid, may be used as templates (Abedon 2011).

7.4.1.1 BIMs and PIMs Production

Although the precise acquisition mechanism is not yet known, it is now technically relatively easy in *S. thermophilus* to get spontaneous and naturally occurring BIMs, by challenging a strain with virulent phage lysate under various conditions. BIMs were reported for many strains, such as DGCC7710 [infected by phages 2972 or 858 (Barrangou et al. 2007; Deveau et al. 2008; Garneau et al. 2010; Horvath et al. 2008)], SMQ-301 [infected by DT1 (Deveau et al. 2008)], CSK938 [infected by 5000, 5002, 5102, and 5077 (Mills et al. 2010)], CSK939 [by 5027 and 5093 (Mills et al. 2010)], and CSK944 [infected by 5196 (Mills et al. 2010)]. An efficient way to obtain *S. thermophilus* BIMs is to mix phages and bacteria on solid media at a relatively low multiplicity of infection (ratio of 0.01–1 phage/cell) and incubate until single colonies appear. PCR reactions can be used then to rapidly confirm possible spacer insertions in these colonies. If a new repeat-spacer unit has been added at the 5′ (leader) end of a given CRISPR locus, one should observe a shift in the PCR fragment size visualized by electrophoresis through an agarose gel. DNA sequencing can provide additional detail about the spacer sequence and the characteristics of the associated proto-spacer. Since it is possible to have different CRISPR BIMs within a single colony, these colonies need to be purified through streaking. Thus, it is possible to create, with successive phage challenges, a multi-resistant BIM (Deveau et al. 2008), which will have acquired several new and different interfering spacers targeting different phages that have been encountered. Phage cross-resistance has also been reported (i.e., resistance to different phages) with acquisition of a unique new spacer (Deveau et al. 2008; Mills et al. 2010). Furthermore, numerous distinct phage resistant derivatives can be generated from the same wild-type strain (Deveau et al. 2008; Mills et al. 2010). Similarly, PIMs can be obtained as described recently (Garneau et al. 2010). First, a plasmid containing a selection marker (for example antibiotic resistance) can be introduced into a CRISPR+ plasmid-free strain. A representative transformant is then grown in liquid medium for several generations in the absence of selection pressure and aliquots are screened for antibiotic-sensitive colonies and spacer acquisition. Recently, it was suggested that CRISPR-escape mutant (CEM) phages may drive the acquisition stage with a primed adaptation in type I-E (Swarts et al. 2012; Datsenko et al. 2012).

7.4.2 Expression of CRISPR-Cas System

Three main genetic elements are transcribed constitutively and simultaneously. First, *cas* genes are transcribed from one (St-CRISPR3 group) or two (St-CRISPR1 group) promoters situated upstream of the *cas*9 genes (for both groups) and between the *cas*9 and *cas*1 genes (for St-CRISPR1 group only). For strain *S. thermophilus* LMD-9, it was shown that the expression of some *cas* genes increased significantly at the beginning of infection by phage DT1 and a peak was observed after 5 min (Goh et al. 2011). Gene expression was most differently regulated at the *cas*1 and *cas*2 location of the CRISPR1-Cas locus, while protein overexpression of CRISPR1.Cas9 and CRISPR3.Cas9 (5–7-fold) was observed in strain DGCC7710 (Young et al. 2012). Second, pre-crRNA transcripts are obtained from the repeat-spacer units, mainly from a third promoter located in the CRISPR leader. It was shown that, due to read-through, the transcript from the *cas*1-*cas*2-*csn*2 promoter of CRISPR1 of *S. thermophilus* DGC7710 comprised also part of the CRISPR array but not all native 32 spacers (Garneau 2009). This CRISPR expression from two promoters would lead to overrepresentation of crRNAs coming from the newly acquired spacer. The processed spacer-containing transcript will provide the mature crRNAs (39–42 bp for subtype II-A system) necessary for the interference stage (Deltcheva et al. 2011) after intermediate production of a 66 nt form, which represents the first cleavage in each repeat of the transcript. Finally, the tracrRNA region is also transcribed as a long transcript (100–200 nt). Deltcheva et al. (2011) have shown that co-processing occurs for the two RNA molecules, involving the activity of the endogenous RNase III and the Cas9 protein. The tracrRNA will be processed to a smaller molecule ($\approx$89 nt), then to a final form of $\approx$75 nt. The smaller molecule contains only part of the repeat-like stretch to bind later with the crRNA. More precisely, the sizes for tracrRNA were found to be 100-80-70 nt for the CRISPR1-Cas system and 110-80-70 nt for the CRISPR3-Cas system of *S. thermophilus* LMD-9. Any of Cas1, Cas2, or Csn2 proteins are implicated in that process as the respective deletion mutants revealed. As mentioned before, it was also proposed that Cas9 acts as an anchor for the base-pairing of tracrRNA and pre-crRNA.

7.4.3 CRISPR Interference

Because of the expression described in the previous section, the CRISPR-Cas system containing cells are always ready to efficiently combat a foreign element that contains a matching proto-spacer and PAM. The cleavage of the invading foreign dsDNA has been demonstrated in vivo on phage genome and plasmid for the CRISPR1-Cas system in *S. thermophilus* DGCC7710 (Garneau et al. 2010). Cleavage occurs within the proto-spacer, specifically 3 nt upstream of the PAM sequence in all cases tested, that was confirmed later for the CRISPR3 system of the same strain (Magadán et al. 2012) and for *S. pyogenes* CRISPR-Cas system (Jinek et al. 2012). In some cases with CRISPR1 of *S. thermophilus*, a second cleavage site was observed in the proto-spacer 19 or 20 nt upstream of the PAM sequence (Garneau et al. 2010). Analysis of different BIMs and PIMs revealed that the CRISPR1-Cas system probably acts in a 3′-end, ruler-anchored manner. A *S. thermophilus* mutated strain that could not produce a Cas9 protein, did not show a CRISPR endonuclease activity. So, the dsDNA cleavage activity of the CRISPR-Cas system may explain the natural scarcity of plasmids in *S. thermophilus*.

The Cas9 protein of the CRISPR3-Cas system in *S. thermophilus* DGCC7710 has been expressed in *E. coli* (Sapranauskas et al. 2011). The authors demonstrated interference with plasmid transformation and phage infection but DNA cleavage was not studied. However, this later study showed that the CRISPR3-Cas system of *S. thermophilus* may be exported successfully to another unrelated bacterial species, in this case *E. coli*. It is unknown whether the transfer would be as successful with the CRISPR1-Cas system because it seems to be more *Streptococcus*-related, but it should be. In the interference stage, it is assumed that the crRNA and its associated Cas proteins are guided to the matching incoming dsDNA, thereby producing a R-loop between the DNA and the crRNA (Westra and Brouns 2012; Howard et al. 2011; Jore et al. 2011). Consequently, some specific domains of Cas9 recognize this hybrid structure and cleave the target dsDNA (Jinek et al. 2012). Finally, it appears that the cleaved DNA is not totally degraded (Garneau et al. 2010; Jinek et al. 2012), but the cleavage is still sufficient to block phage DNA replication, transcription, and/or packaging, as well as plasmid replication.

7.4.3.1 Interference-Escaping Phage Mutants

The interference phenotype of type II CRISPR-Cas systems is visible as phage resistance and reduced transformation frequency for plasmids. It is known, however, that some mutant phages may still cause infection and some plasmid molecules may still be transformed (Deveau et al. 2008; Garneau et al. 2010; Mills et al. 2010; Sapranauskas et al. 2011). These CRISPR-resistant invaders generally have a mutation in the proto-spacers or in the PAM. The most frequent mutation is a single nucleotide polymorphism (SNP). Thus, even a single nucleotide change is sufficient to circumvent the interference phenotype conferred by a specific spacer. Other mutations observed included additional nucleotide mismatches or deletions. On the other hand, if a BIM contains several interfering spacers, it significantly reduces the possibility of mutation for the invader at each proto-spacer or PAM regions. It has already been proven in vivo that the phage efficiency of plaquing significantly decreases with the increasing number of new spacers (Deveau et al. 2008). It was also hypothesized that further the mutations are separated from the PAM, the more they are tolerated by the interference machinery (Sapranauskas et al. 2011). This phenomenon is referred as the seed sequence, which is a minimal sequence that is complementary between the crRNA and the target, and needed to provide interference of a specific target (Semenova et al. 2011; Wiedenheft et al. 2011). More precisely, mutations at positions 25 and 28 of the proto-spacers (i.e., positions −3 and −6 from the PAM) affect plasmid transformation, but mutations at positions 2, 11, 18, and 23 do not. The phage escape-mutants described by Deveau et al. (2008) had nucleotide mutations at the proto-spacer positions 23, 25, 26, 27, and 28. The importance of these positions is in agreement with dsDNA cleavage of CRISPR1-Cas occurring after the 27th nucleotide within the proto-spacers (Garneau et al. 2010; Magadán et al. 2012).

7.5 Applications of the Type II CRISPR-Cas Systems

Besides the obvious scientific interest in understanding this fascinating biological system, CRISPR-Cas systems have several potential applications. The two most currently used applications will be discussed below.

7.5.1 Typing Application for S. thermophilus

The polymorphism of the spacer content has rapidly found an application as a genotyping technique, called spoligotyping. The polarized spacer insertion can be used to differentiate between bacterial strains by using spacer sequences. PCR reactions can amplify the variable spacer content and sequence analyses will reveal relatedness between strains. Previously, this technique was used to characterize bacterial agents from an epidemiological perspective, such as *L. pneumophila* (Ginevra et al. 2012), *Mycobacterium tuberculosis* (Comas et al. 2009; Zhang et al. 2010), *Corynebacterium diptheriae* (Mokrousov et al. 2009), *C. jejuni* (Price et al. 2007), and *Yersinia pestis* (Cui et al. 2008). This interesting value-added strain-typing tool can also be used to compare industrial strains for patent issues. Spoligotyping has been used to differentiate food industrial bacteria, such as *Lactobacillus acidophilus* (Russell et al. 2006), and *S. thermophilus* strains (Horvath et al. 2008).

7.5.2 Industrial Application of S. thermophilus BIMs

The possibility of obtaining a natural BIM with a high number of new interfering spacers is even more exciting from an industrial point of view. *S. thermophilus* is a LAB that is generally recognized as safe and it is used for the manufacture of several fermented dairy products such as yogurt and specialized cheeses. The manufacture of these products requires the inoculation of 10^7 carefully selected bacterial cells (known as starter culture) per ml of pasteurized milk to control the fermentation and to obtain high-quality products. The starter culture is a combination of LAB among which *S. thermophilus* is one of the most important species. Considering that 10^{11} bacterial cells are needed to produce 1 kg of cheese, it is clear that LAB are of considerable interest to the cheese industry (Quiberoni et al. 2010). It is widely acknowledged that an increased productivity within existing cheese manufacturing facilities will lead to milk fermentation failures due to virulent phages. The added LAB cells will always come into contact with phages present in the non-sterile but pasteurized milk. Although phage levels are usually low, a specific phage population can increase very rapidly if phage-sensitive cells are present in the starter culture. The ensuing lysis of a large number of sensitive bacterial cells will delay or even halt the fermentation process, leading to low-quality products or, in worse cases, discarded unfermented milk. For decades, the dairy industry has relied on an array of strategies to control this natural phenomenon. But in spite of these extensive efforts, "phage attacks" remain today the most common cause of slow or incomplete milk fermentation. The availability of BIMs obtained through the exploitation of the CRISPR-Cas system will add to our arsenal to control phages.

7.6 Conclusion

For *S. thermophilus*, information on active subtype II-A CRISPR-Cas systems was obtained mainly from microbiological experiments (such as phage infections) and includes only general functional descriptions. The precise roles of each Cas protein are not yet fully defined, but other protein characterizations and future mechanistic studies will help to define the molecular details of this fascinating bacterial immune system.

Acknowledgments We thank B.D. Conway for editorial assistance, M. Villion for commenting on the chapter, and scientists at Danisco/DuPont for discussions and support over the years. S.M. acknowledges funding from the Natural Sciences and Engineering Research Council (NSERC) of Canada (Discovery program). S.M. holds a Tier 1 Canada Research Chair on Bacteriophages.

References

Abedon ST (2011) Facilation of CRISPR adapation. Bacteriophages 1(3):179–181

Al-Attar S, Westra ER, van der Oost J, Brouns SJ (2011) Clustered regularly interspaced short palindromic repeats (CRISPRs): the hallmark of an ingenious antiviral defense mechanism in prokaryotes. Biol Chem 392(4):277–289

Babu M, Beloglazova N, Flick R, Graham C, Skarina T, Nocek B, Gagarinova A, Pogoutse O, Brown G, Binkowski A, Phanse S, Joachimiak A, Koonin EV, Savchenko A, Emili A, Greenblatt J, Edwards AM, Yakunin AF (2011) A dual function of the CRISPR-Cas system in bacterial antivirus immunity and DNA repair. Mol Microbiol 79(2):484–502

Barrangou R, Fremaux C, Deveau H, Richards M, Boyaval P, Moineau S, Romero DA, Horvath P (2007) CRISPR provides acquired resistance against viruses in prokaryotes. Science 315(5819):1709–1712

Beloglazova N, Brown G, Zimmerman MD, Proudfoot M, Makarova KS, Kudritska M, Kochinyan S, Wang S, Chruszcz M, Minor W, Koonin EV, Edwards AM, Savchenko A, Yakunin AF (2008) A novel family of sequence-specific endoribonucleases associated with the clustered regularly interspaced short palindromic repeats. J Biol Chem 283(29):20361–20371

Beloglazova N, Skarina T, Petit P, Flick R, Brown G, Savchenko A, Yakunin AF (2010) Crystal structure and nuclease activity of tm1797, a Cas1 protein from *Thermotoga maritima*. Protein database number: 3LFX

Bolotin A, Quinquis B, Renault P, Sorokin A, Ehrlich SD, Kulakauskas S, Lapidus A, Goltsman E, Mazur M, Pusch GD, Fonstein M, Overbeek R, Kyprides N, Purnelle B, Prozzi D, Ngui K, Masuy D, Hancy F, Burteau S, Boutry M, Delcour J, Goffeau A, Hols P (2004) Complete sequence and comparative genome analysis of the dairy bacterium *Streptococcus thermophilus*. Nat Biotechnol 22(12):1554–1558

Bolotin A, Quinquis B, Sorokin A, Ehrlich SD (2005) Clustered regularly interspaced short palindrome repeats (CRISPRs) have spacers of extrachromosomal origin. Microbiology 151(8):2551–2561

Brouns SJ, Jore MM, Lundgren M, Westra ER, Slijkhuis RJ, Snijders AP, Dickman MJ, Makarova KS, Koonin EV, van der Oost J (2008) Small CRISPR RNAs guide antiviral defense in prokaryotes. Science 321(5891):960–964

Carte J, Wang R, Li H, Terns RM, Terns MP (2008) Cas6 is an endoribonuclease that generates guide RNAs for invader defense in prokaryotes. Genes Dev 22(24):3489–3496

Chakraborty S, Snijders AP, Chakravorty R, Ahmed M, Tarek AM, Hossain MA (2010) Comparative network clustering of direct repeats (DRs) and *cas* genes confirms the possibility of the horizontal transfer of CRISPR locus among bacteria. Mol Phylogenet Evol 56(3):878–887

Comas I, Homolka S, Niemann S, Gagneux S (2009) Genotyping of genetically monomorphic bacteria: DNA sequencing in *Mycobacterium tuberculosis* highlights the limitations of current methodologies. PLoS ONE 4(11):e7815

Cui Y, Li Y, Gorgé O, Platonov ME, Yan Y, Guo Z, Pourcel C, Dentovskaya SV, Balakhonov SV, Wang X, Song Y, Anisimov AP, Vergnaud G, Yang R (2008) Insight into microevolution of *Yersinia pestis* by clustered regularly interspaced short palindromic repeats. PLoS ONE 3(7):e2652

Datsenko KA, Pougach K, Tikhonov A, Wanner BL, Severinov K, Semenova E (2012) Molecular memory of prior infections activates the CRISPR-Cas adaptive bacterial immunity system. Nat Commun 10(3):945

Delorme C, Bartholini C, Luraschi M, Pons N, Loux V, Almeida M, Guédon E, Gibrat JF, Renault P (2011) Complete genome sequence of the pigmented *Streptococcus thermophilus* strain JIM8232. J Bacteriol 193(19):5581–5582

Deltcheva E, Chylinski K, Sharma CM, Gonzales K, Chao Y, Pirzada ZA, Eckert MR, Vogel J, Charpentier E (2011) CRISPR RNA maturation by trans-encoded small RNA and host factor RNase III. Nature 471(7340):602–607

Deshpande NP, Kaakoush NO, Mitchell H, Janitz K, Raftery MJ, Li SS, Wilkins MR (2011) Sequencing and validation of the genome of a *Campylobacter concisus* reveals intra-species diversity. PLoS ONE 6(7)

Deveau H, Barrangou R, Garneau JE, Labonté J, Fremaux C, Boyaval P, Romero DA, Horvath P, Moineau S (2008) Phage response to CRISPR-encoded resistance in *Streptococcus thermophilus*. J Bacteriol 190(4):1390–1400

Ebihara A, Yao M, Masui R, Tanaka I, Yokoyama S, Kuramitsu S (2006) Crystal structure of hypothetical protein TTHB192 from *Thermus thermophilus* HB8 reveals a new protein family with an RNA recognition motif-like domain. Protein Sci 15(6):1494–1499

Ebihara A, Yokoyama S, Kuramitsu S (2007) Crystal structure of uncharacterized conserved protein from *Aquifex aeolicus*. Protein database number: 2YZS

Ellinger P, Arslan Z, Wurm R, Tschapek B, MacKenzie C, Pfeffer K, Panjikar S, Wagner R, Schmitt L, Gohlke H, Pul Ü, Smits SH (2012) The crystal structure of the CRISPR-associated protein Csn2 from *Streptococcus agalactiae*. J Struct Biol 178(3):350–362

Garneau JE (2009) Caractérisation du système CRISPR-cas chez *Streptococcus thermophilus*. Université Laval master thesis, Département de biochimie, de microbiologie et de bio-informatique. Québec, p 109

Garneau JE, Dupuis M-È, Villion M, Romero DA, Barrangou R, Boyaval P, Fremaux C, Horvath P, Magadán AH, Moineau S (2010) The CRISPR/Cas bacterial immune system cleaves bacteriophage and plasmid DNA. Nature 468(7320):67–71

Ginevra C, Jacotin N, Diancourt L, Guigon G, Arquilliere R, Meugnier H, Descours G, Vandenesch F, Etienne J, Lina G, Caro V, Jarraud S (2012) *Legionella pneumophila* ST1/Paris-Pulsotype subtyping by spoligotyping. J Clin Microbiol (Epub ahead of print)

Godde JS, Bickerton A (2006) The repetitive DNA elements called CRISPRs and their associated genes: evidence of horizontal transfer among prokaryotes. J Mol Evol 62:718–729

Goh YJ, Goin C, O'Flaherty S, Altermann E, Hutkins R (2011) Specialized adaptation of a lactic acid bacterium to the milk environment: the comparative genomics of *Streptococcus thermophilus* LMD-9. Microb Cell Fact 10(Suppl 1):S22

Hale C, Kleppe K, Terns RM, Terns MP (2008) Prokaryotic silencing (psi)RNAs in *Pyrococcus furiosus*. RNA 14(12):2572–2579

Hale CR, Zhao P, Olson S, Duff MO, Graveley BR, Wells L, Terns RM, Terns MP (2009) RNA-guided RNA cleavage by a CRISPR RNA-Cas protein complex. Cell 139(5):945–956

Han D, Lehmann K, Krauss G (2009) SSO1450–a CAS1 protein from *Sulfolobus solfataricus* P2 with high affinity for RNA and DNA. FEBS Lett 583(12):1928–1932

Haurwitz RE, Jinek M, Wiedenheft B, Zhou K, Doudna JA (2010) Sequence- and structure-specific RNA processing by a CRISPR endonuclease. Science 329(5997):1355–1358

He J, Deem MW (2010) Heterogeneous diversity of spacers within CRISPR (clustered regularly interspaced short palindromic repeats). Phys Rev Lett 105(12):128102

Horvath P, Barrangou R (2010) CRISPR/Cas, the immune system of bacteria and archaea. Science 327(5962):167–170

Horvath P, Coûté-Monvoisin AC, Romero DA, Boyaval P, Fremaux C, Barrangou R (2009) Comparative analysis of CRISPR loci in lactic acid bacteria genomes. Int J Food Microbiol 131(1):62–70

Horvath P, Romero DA, Coûté-Monvoisin AC, Richards M, Deveau H, Moineau S, Boyaval P, Fremaux C, Barrangou R (2008) Diversity, activity, and evolution of CRISPR loci in *Streptococcus thermophilus*. J Bacteriol 190(4):1401–1412

Howard JA, Delmas S, Ivančić-Baće I, Bolt EL (2011) Helicase dissociation and annealing of RNA-DNA hybrids by *Escherichia coli* Cas3 protein. Biochem J 439(1):85–95

Ihsanawati S, Murayama K, Shirouzu M, Yokoyama S (2005) Crystal structure of a hypothetical protein TT1823 from *Thermus thermophilus*. Protein database number: 1ZPW

Jansen R, van Embden JD, Gaastra W, Schouls LM (2002) Identification of genes that are associated with DNA repeats in prokaryotes. Mol Microbiol 43(6):1565–1575

Jinek M, Chylinski K, Fonfara I, Hauer M, Doudna JA, Charpentier E (2012) A programmable dual-RNA-Guided DNA endonuclease in adaptive bacterial immunity. Science 337:816–821

Jore MM, Brouns SJ, van der Oost J (2011) RNA in defense: CRISPRs protect prokaryotes against mobile genetic elements. Cold Spring Harb Perspect Biol. doi:10.1101/cshperspect.a003657

Koo Y, Jung DK, Bae E (2012) Crystal structure of *Streptococcus pyogenes* csn2 reveals calcium-dependent conformational changes in its tertiary and quaternary structure. PLoS ONE 7(3):e33401

Kwon AR, Kim JH, Park SJ, Lee KY, Min YH, Im H, Lee I, Lee KY, Lee BJ (2012) Structural and biochemical characterization of HP0315 from *Helicobacter pylori* as a VapD protein with an endoribonuclease activity. Nucleic Acids Res (Epub ahead of print)

Lee KH, Lee SG, Eun Lee K, Jeon H, Robinson H, Oh BH (2012) Identification, structural, and biochemical characterization of a group of large Csn2 proteins involved in CRISPR-mediated bacterial immunity. Proteins. doi:10.1002/prot.24138 (Epub ahead of print)

Lillestøl RK, Redder P, Garrett RA, Brügger K (2006) A putative viral defence mechanism in archaeal cells. Archaea 2(1):59–72

Lillestøl RK, Shah SA, Brügger K, Redder P, Phan H, Christiansen J, Garrett RA (2009) CRISPR families of the crenarchaeal genus *Sulfolobus*: bidirectional transcription and dynamic properties. Mol Microbiol 72(1):259–272

Lopez-Sanchez MJ, Sauvage E, Da Cunha V, Clermont D, Ratsima Hariniaina E, Gonzalez-Zorn B, Poyart C, Rosinski-Chupin I, Glaser P (2012) The highly dynamic CRISPR1 system of *Streptococcus agalactiae* controls the diversity of its mobilome. Mol Microbiol. doi:10.1111/j.1365-2958.2012.08172.x

Magadán AH, Dupuis M-È, Villion M, Moineau S (2012) Cleavage of phage DNA by the *Streptococcus thermophilus* CRISPR3-Cas system. PLoS ONE 7(7):e40913

Makarova KS, Aravind L, Grishin NV, Rogozin IB, Koonin EV (2002) A DNA repair system specific for thermophilic Archaea and bacteria predicted by genomic context analysis. Nucleic Acids Res 30(2):482–496

Makarova KS, Grishin NV, Shabalina SA, Wolf YI, Koonin EV (2006a) A putative RNA-interference-based immune system in prokaryotes: computational analysis of the predicted enzymatic machinery, functional analogies with eukaryotic RNAi, and hypothetical mechanisms of action. Biol Direct 1:7

Makarova KS, Haft DH, Barrangou R, Brouns SJ, Charpentier E, Horvath P, Moineau S, Mojica FJ, Wolf YI, Yakunin AF, van der Oost J, Koonin EV (2011) Evolution and classification of the CRISPR-Cas systems. Nat Rev Microbiol 9(6):467–477

Makarova KS, Slesarev A, Wolf Y, Sorokin A, Mirkin B, Koonin EV, Pavlov A, Pavlova N, Karamychev V, Polouchine N, Shakhova V, Grigoriev I, Lou Y, Rohksar D, Lucas S, Huang K, Goodstein DM, Hawkins T, Plengvidhya V, Welker D, Hughes J, Goh Y, Benson A, Baldwin K, Lee JH, Díaz-Muñiz I, Dosti B, Smeianov V, Wechter W, Barabote R, Lorca G, Altermann E, Barrangou R, Ganesan B, Xie Y, Rawsthorne H, Tamir D, Parker C, Breidt F, Broadbent J, Hutkins R, O'Sullivan D, Steele J, Unlu G, Saier M, Klaenhammer T, Richardson P, Kozyavkin S, Weimer B, Mills D (2006b) Comparative genomics of the lactic acid bacteria. Proc Natl Acad Sci USA 103(42):15611–15616

Merchant-Patel S, Blackall PJ, Templeton J, Price EP, Tong SY, Huygens F, Giffard PM (2010) *Campylobacter jejuni* and *Campylobacter coli* genotyping by high-resolution melting analysis of a flaA fragment. Appl Environ Microbiol 76(2):493–499

Mills S, Griffin C, Coffey A, Meijer WC, Hafkamp B, Ross RP (2010) CRISPR analysis of bacteriophage-insensitive mutants (BIMs) of industrial *Streptococcus thermophilus* -implications for starter design. J Appl Microbiol 108(3):945–955

Mokrousov I, Vyazovaya A, Kolodkina V, Limeschenko E, Titov L, Narvskaya O (2009) Novel macroarray-based method of *Corynebacterium diphtheriae* genotyping: evaluation in a field study in Belarus. Eur J Clin Microbiol Infect Dis 28(6):701–703

Nam KH, Kurinov I, Ke A (2011) Crystal structure of clustered regularly interspaced short palindromic repeats (CRISPR)-associated Csn2 protein revealed Ca^{2+}-dependent double-stranded DNA binding activity. J Biol Chem 286(35):30759–30768

Oke M, Carter LG, Johnson KA, Liu H, McMahon SA, Yan X, Kerou M, Weikart ND, Kadi N, Sheikh MA, Schmelz S, Dorward M, Zawadzki M, Cozens C, Falconer H, Powers H, Overton IM, Van Niekerk CAJ, Peng X, Patel P, Garrett RA, Prangishvili D, Botting CH, Coote PJ, Dryden DTF, Barton GJ, Schwarz-Linek U, Challis GL, Taylor GL, White MF, Naismith JH (2010) The scottish structural proteomics facility: targets, methods and outputs. J Struct Funct Genomics 11:167–180

Petit P, Brown G, Savchenko A, Yakunin 2010 Structure determination and overall comparison of three Cas1 proteins. Protein database number: 3PV9

Price EP, Smith H, Huygens F, Giffard PM (2007) High-resolution DNA melt curve analysis of the clustered, regularly interspaced short-palindromic-repeat locus of *Campylobacter jejuni*. Appl Environ Microbiol 71(10):3431–3436

Proudfoot M, Brown M, Singer AU, Skarina T, Tan K, Kagan O, Edwards AM, Joachimiak A, Savchenko A, Yakunin AF 2008 Structure of the RNase SSO8090 from *Sulfolobus solfataricus*. Protein database number: 3EXC

Pul U, Wurm R, Arslan Z, Geissen R, Hofmann N, Wagner R (2010) Identification and characterization of *E. coli* CRISPR-cas promoters and their silencing by H-NS. Mol Microbiol 75(6):1495–1512

Quiberoni A, Moineau S, Rousseau GM, Reinheimer J, Ackermann H-W (2010) *Streptococcus thermophilus* bacteriophages. Int Dairy J 20(10):1–8

Russell WM, Barrangou R, Horvath P (2006) Detection and typing of bacterial strains. US Patent Application 20060199190

Samai P, Smith P, Shuman S (2010) Structure of a CRISPR-associated protein Cas2 from *Desulfovibrio vulgaris*. Acta Crystallogr Sect F Struct Biol Cryst Commun 66:1552–1556

Sapranauskas R, Gasiunas G, Fremaux C, Barrangou R, Horvath P, Siksnys V (2011) The *Streptococcus thermophilus* CRISPR/Cas system provides immunity in *Escherichia coli.* Nucleic Acids Res 39(21):9275–9282

Schouls LM, Reulen S, Duim B, Wagenaar JA, Willems RJ, Dingle KE, Colles FM, Van Embden JD (2003) Comparative genotyping of *Campylobacter jejuni* by amplified fragment length polymorphism, multilocus sequence typing, and short repeat sequencing: strain diversity, host range, and recombination. J Clin Microbiol 41(1):15–26

Semenova E, Jore MM, Datsenko KA, Semenova A, Westra ER, Wanner B, van der Oost J, Brouns SJ, Severinov K (2011) Interference by clustered regularly interspaced short palindromic repeat (CRISPR) RNA is governed by a seed sequence. Proc Natl Acad Sci USA 108(25):10098–10103

Sorek R, Kunin V, Hugenholtz P (2008) CRISPR–a widespread system that provides acquired resistance against phages in bacteria and archaea. Nat Rev Microbiol 6(3):181–186

Sun Z, Chen X, Wang J, Zhao W, Shao Y, Wu L, Zhou Z, Sun T, Wang L, Meng H, Zhang H, Chen W (2011) Complete genome sequence of *Streptococcus thermophilus* strain ND03. J Bacteriol 193(3):793–794

Swarts DC, Mosterd C, van Passel MW, Brouns SJ (2012) CRISPR interference directs strand specific spacer acquisition. PLoS ONE 7(4):e35888

Takeuchi N, Wolf YI, Makarova KS, Koonin EV (2012) Nature and intensity of selection pressure on CRISPR-associated genes. J Bact (Epub ahead of print)

Tasaki E, Hirayama J, Tazumi A, Hayashi K, Hara Y, Ueno H, Moore JE, Millar BC, Matsuda M (2012) Molecular identification and characterization of clustered regularly interspaced short palindromic repeats (CRISPRs) in a urease-positive thermophilic *Campylobacter* sp. (UPTC). World J Microbiol Biotechnol 28(2):713–20

Westra ER, Brouns SJ (2012) The rise and fall of CRISPRs—dynamics of spacer acquisition and loss. Mol Microbiol. doi:10.1111/j.1365-2958.2012.08170.x

Wiedenheft B, van Duijn E, Bultema JB, Waghmare SP, Zhou K, Barendregt A, Westphal W, Heck AJ, Boekema EJ, Dickman MJ, Doudna JA (2011) RNA-guided complex from a bacterial immune system enhances target recognition through seed sequence interactions. Proc Natl Acad Sci USA 108(25):10092–10097

Wiedenheft B, Zhou K, Jinek M, Coyle SM, Ma W, Doudna JA (2009) Structural basis for DNase activity of a conserved protein implicated in CRISPR-mediated genome defense. Structure 17(6):904–912

Yosef I, Goren MG, Qimron U (2012) Proteins and DNA elements essential for the CRISPR adaptation process in *Escherichia coli.* Nucleic Acids Res 40(12):5569–76

Young JC, Dill BD, Pan C, Hettich RL, Banfield JF, Shah M, Fremaux C, Horvath P, Barrangou R, Verberkmoes NC (2012) Phage-induced expression of CRISPR-associated proteins is revealed by shotgun proteomics in *Streptococcus thermophilus.* PLoS ONE 7(5):e38077

Zhang J, Abadia E, Refregier G, Tafaj S, Boschiroli ML, Guillard B, Andremont A, Ruimy R, Sola C (2010) *Mycobacterium tuberculosis* complex CRISPR genotyping: improving efficiency, throughput and discriminative power of 'spoligotyping' with new spacers and a microbead-based hybridization assay. J Med Microbiol 59:285–294

Chapter 8
Type III CRISPR-Cas Systems and the Roles of CRISPR-Cas in Bacterial Virulence

Asma Hatoum-Aslan, Kelli L. Palmer, Michael S. Gilmore and Luciano A. Marraffini

Abstract Type III CRISPR-Cas systems constitute nearly a quarter of all known CRISPR systems and have been found to reside in both archaea and bacteria, including important bacterial pathogens such as staphylococci and mycobacteria. By blocking the horizontal transfer of bacteriophage and conjugative plasmids, CRISPR-Cas systems not only protect against foreign invaders, but also prevent the acquisition of virulence factors and antibiotic resistance cassettes that are encoded in these mobile genetic elements. For this reason, these systems can have a broad impact on the evolution of bacterial pathogens. This chapter explores in-depth our current understanding of the molecular mechanisms of CRISPR interference in the type III systems in particular, and more generally, the existing data supporting the role of CRISPR-Cas systems in the emergence of bacterial pathogens.

A. Hatoum-Aslan · L. A. Marraffini (✉)
Laboratory of Bacteriology, The Rockefeller University, 1230 York Avenue,
New York, NY 10065, USA
e-mail: marraffini@rockefeller.edu

A. Hatoum-Aslan
e-mail: ahatoum@rockefeller.edu

K. L. Palmer · M. S. Gilmore
Department of Ophthalmology, Harvard Medical School, 243 Charles Street,
Boston, MA 02114, USA
e-mail: kelli.palmer@utdallas.edu

M. S. Gilmore
e-mail: michael_gilmore@meei.harvard.edu

K. L. Palmer · M. S. Gilmore
Department of Microbiology and Molecular Genetics, Harvard Medical School,
200 Longwood Avenue, Boston, MA 02115, USA

K. L. Palmer
Department of Molecular and Cell Biology, University of Texas at Dallas,
800 W Campbell Road, Richardson, TX 75080, USA

R. Barrangou and J. van der Oost (eds.), *CRISPR-Cas Systems*,
DOI: 10.1007/978-3-642-34657-6_8, © Springer-Verlag Berlin Heidelberg 2013

Contents

8.1 Introduction to Type III Systems

CRISPR loci have been found in 80 % archaeal genomes and about 40 % of bacterial genomes sequenced to date (Haft et al. 2005; Makarova et al. 2006). Consistent with this trend, a survey of the taxonomic distribution of all CRISPR types (I–III) has revealed that type III systems are more commonly found in archaea, occurring in ~75 % of the organisms containing CRISPR-Cas loci (Makarova et al. 2011b). On the other hand, about 40 % of the bacterial genomes harboring CRISPR contain type III CRISPR systems. The biological significance of this overrepresentation of CRISPR in archaea is yet to be determined.

Type III systems comprise nearly a quarter of all CRISPR-Cas systems identified in archaea and bacteria (Makarova et al. 2011b), and have been found to reside in important bacterial pathogens such as staphylococci and mycobacteria (Makarova et al. 2011b; Marraffini and Sontheimer 2008). Furthermore, a type III system found in a clinical *Staphylococcus epidermidis* isolate can prevent the uptake of antibiotic resistance genes (Marraffini and Sontheimer 2008), thus providing a novel pathway that can be engineered to limit the spread of antibiotic resistance in clinical settings. This chapter will be dedicated to type III CRISPR systems and what is known of the mechanisms by which they attack target nucleic acids. We will also review the role of CRISPR interference in the emergence and evolution of bacterial pathogens.

What distinguishes type III systems from the others? While nearly all CRISPR types possess the universal *cas*1 and *cas*2 genes thought to be involved in the adaptation phase of CRISPR interference (Deveau et al. 2010; Horvath and Barrangou 2010; Marraffini and Sontheimer 2010a), the hallmark of type III systems is the presence of the signature gene *cas*10, which encodes a large protein homologous to palm-domain polymerases (Makarova et al. 2011b) (see Chap. 3). Type III systems also harbor multiple genes that encode repeat associated mysterious proteins (RAMPs), a superfamily of Cas proteins that possess one or more RNA recognition motifs and a characteristic glycine-rich loop (Makarova et al. 2011a). Emerging evidence (described in more detail later) has shown that both

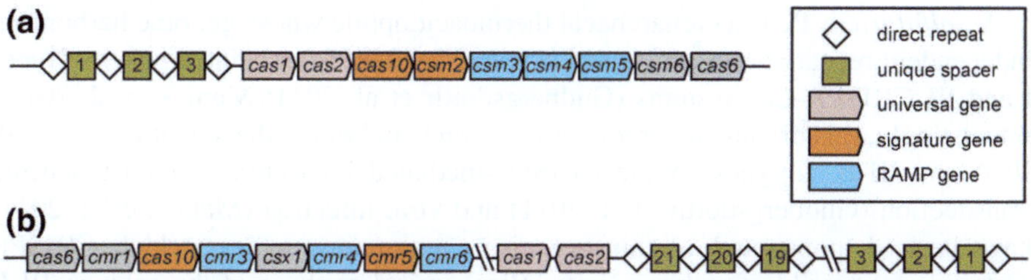

Fig. 8.1 Two model type III CRISPR systems. **a** The *S. epidermidis* type III-A system has two arrays, with three and two unique spacers. Nine *cas* genes are found in-between these arrays. **b** *P. furiosus* has seven repeat-spacer arrays and 29 potential *cas* genes. One genomic locus is known that harbors *cas* genes from the type III-B system and the adjacent repeat-spacer array. Unique spacers (*green boxes*), direct repeats (*white diamonds*), universal genes (*brown arrows*), signature genes (*orange arrows*), and RAMP genes (*blue arrows*) are indicated for each locus. Genes labeled with *gray arrows* do not fall into any of the above categories

Cas10 and RAMPs play roles in the defense phase of CRISPR interference, particularly during crRNA biogenesis (Carte et al. 2008; Hatoum-Aslan et al. 2011) and targeting (Hale et al. 2009, 2012; Makarova et al. 2011b).

Type III systems can be further classified into subtypes III-A and III-B based upon the phylogeny of their *cas*1 gene, as well as the complement of RAMPs they possess (Makarova et al. 2011b) (see Chap. 3) (Fig. 8.1). In addition, while III-A systems have the signature *csm*2 gene, III-B systems harbor the signature *cmr*5 gene. The best characterized type III CRISPR systems reside in the bacterium *S. epidermidis* (III-A), and the archaea *Pyrococcus furiosus* (III-B) and *Sulfolobus solfataricus* (III-A and III-B). Data generated from these systems have helped lay the foundation for what is known today about CRISPR interference.

S. epidermidis RP62A is a clinical isolate that harbors a single III-A CRISPR locus with nine *cas* genes flanked by two repeat-spacer arrays (Fig. 8.1). Early experiments have established the in vivo functionality of this system (Marraffini and Sontheimer 2008). The first spacer (*spc*1) matches a region of the *nickase* gene found on all staphylococcal conjugative plasmids, and can prevent the conjugative transfer of a plasmid harboring antibiotic resistance genes. The second spacer (*spc*2) matches a structural gene of the *S. epidermidis* bacteriophage CNPH82 (Daniel et al. 2007) and can prevent infection (our unpublished results).

P. furiosus is a hyperthermophilic euryarchaeon whose genome harbors seven independent repeat-spacer arrays and 29 potential CRISPR-associated genes that belong to types I and III CRISPR-Cas systems (Hale et al. 2008, 2009). Included among these genes are *cas*10 and *cmr*5, the hallmarks of III-B CRISPR-Cas systems (Makarova et al. 2011b) (Fig. 8.1). Evidence suggesting the in vivo functionality of this system has recently been reported (Hale et al. 2012), and in vitro studies of purified Cas and Cmr proteins in complex with crRNAs have led to a detailed understanding of the mechanisms underlying crRNA biogenesis and targeting in type III systems.

S. solfataricus P2 is a crenarchaeal thermoacidophile whose genome harbors six independent repeat-spacer arrays and at least 21 CRISPR-associated genes of types I and III CRISPR-Cas systems (Gudbergsdottir et al. 2011; Manica et al. 2011). Interestingly, this organism harbors *cas*10, *csm*2, and *csm*5, the signatures for both III-A and III-B subtypes. While CRISPR-mediated interference against plasmid transduction (Gudbergsdottir et al. 2011) and viral infection (Manica et al. 2011) have been demonstrated in this organism, it remains unclear which CRISPR system is responsible for this activity. Biochemical analysis of the subtype III-B machinery of *S. solfataricus* P2 revealed an alternative targeting mechanism (Zhang et al. 2012).

Studies of these model systems have uncovered many similarities as well as some striking differences in the ways they generate crRNAs and target their respective nucleic acids. The remainder of this section will focus on CRISPR RNA biogenesis and targeting in these type III systems.

8.2 CRISPR RNA Biogenesis

CrRNAs have been detected by northern blot analysis under optimal growth conditions in both bacteria and archaea that harbor type III systems (Deng et al. 2011; Gudbergsdottir et al. 2011; Hale et al. 2008; Marraffini and Sontheimer 2010b; Tang et al. 2005) (see Chap. 4). This observation suggests constitutive expression of crRNAs. However, a DNA binding protein in *Sulfolobales* called Cbp1 has recently been reported to bind CRISPR repeats and enhance transcription of the repeat-spacer array in a yet unknown mechanism (Deng et al. 2011; Peng et al. 2003). It is unclear whether a similar form of transcription regulation exists in bacteria (see Chap. 4).

In type III CRISPR-Cas systems, crRNA biogenesis begins with the transcription of the repeat-spacer array into a long precursor (pre-crRNA). This precursor is subsequently cleaved to liberate mature crRNAs in two distinct stages: primary processing and maturation (Fig. 8.2a). During primary processing, the pre-crRNA is cut within each direct repeat to liberate intermediate crRNAs that consist of individual spacers flanked on both ends by partial repeats (Hale et al. 2008; Hatoum-Aslan et al. 2011). The process of maturation involves additional degradation of the 3′ end of the intermediate, eliminating repeat sequences at this end and reducing the size of the intermediate crRNA (see Chap. 5). In both systems, this results in the generation of two mature crRNA species that are composed of a full or partial spacer sequence flanked by eight nucleotides of repeat sequence on its 5′ end. This "5′-tag" is the only repeat sequence that remains associated with the mature crRNA. Mature crRNAs are easily detected by northern analysis and are the most abundant small RNA species, indicating that maturation is a highly robust and efficient process.

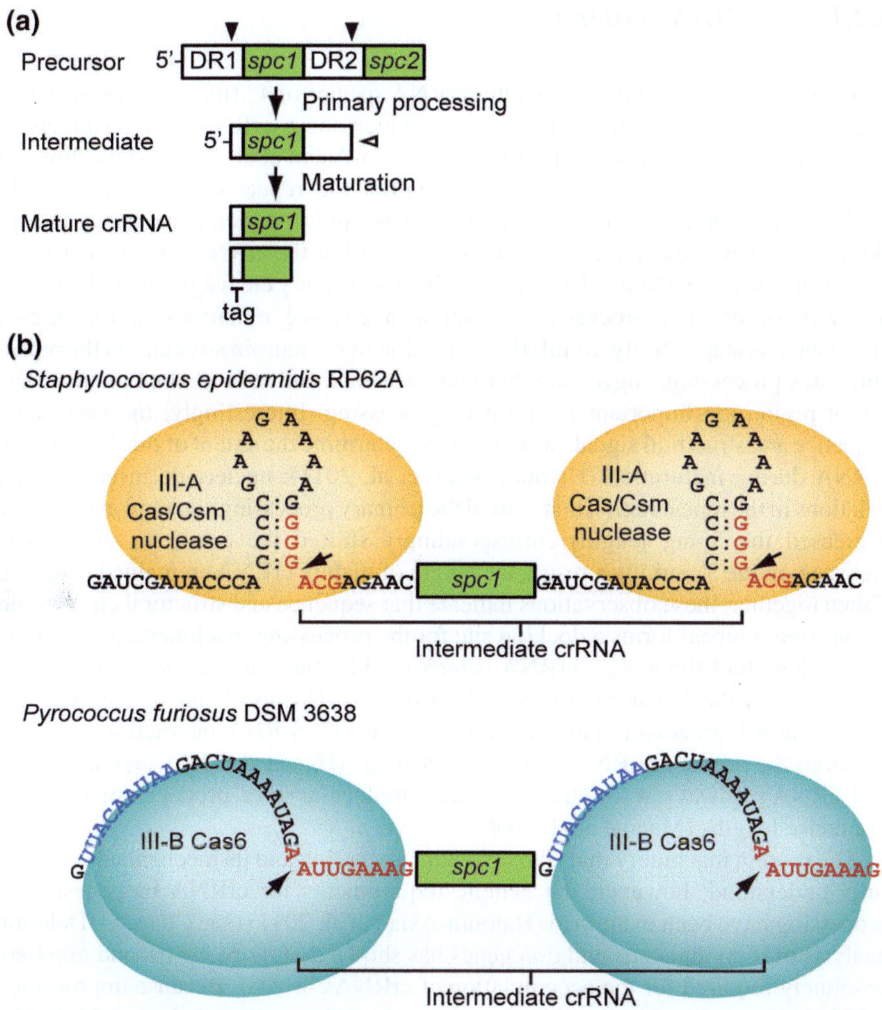

Fig. 8.2 CRISPR RNA biogenesis in type III systems. **a** Mature crRNAs are generated from a precursor repeat-spacer array in two distinct stages: primary processing and maturation. During primary processing, the precursor is cleaved within consecutive direct repeats (*solid arrowheads*) to generate crRNA intermediates. During maturation, these intermediates undergo further nucleolytic cleavage on the 3′ end (*open arrowhead*) to yield mature crRNAs. Mature crRNAs consist of a whole or partial unique spacer flanked by eight nucleotides of repeat sequence on the 5′ end, known as the 5′ "tag" or "handle". **b** Primary processing by the *S. epidermidis* type III-A system requires a consensus sequence surrounding the cleavage site (the residues required for efficient cleavage are shown in *red*, the *arrow* marks the precise cleavage site) as well as a hairpin structure within the repeat. Precursor processing requires Cas6, Csm4, and/or Csm1 (*yellow oval*) in vivo. Csm2, Csm3 and Csm5 are required for maturation of the intermediate crRNA and may also belong to a Cas/Csm nuclease complex. Primary processing by the *P. furiosus* Type III-B system is performed by the Cas6 endoribonuclease (*blue oval*), which may act as a dimer. Upstream repeat sequences (*blue*) provide a binding site for Cas6, whereas the cleavage site (*red*) is in the downstream region of the repeat

8.2.1 Type III-A Systems

In *S. epidermidis,* maturation generates crRNA species of 43 (major species) and 37 nucleotides in length. The mechanism by which these crRNAs are so precisely measured has been recently determined (Hatoum-Aslan et al. 2011; Marraffini and Sontheimer 2010b). Primary processing within the repeat occurs exactly eight nucleotides upstream of the spacer at the base of a predicted hairpin structure (Fig. 8.2b). This cleavage event defines the 5′ end of the crRNA. A six nucleotide consensus sequence (GGGACG) surrounding the primary cleavage site is absolutely required for primary processing, as single nucleotide mutations in this region eliminate this step entirely. Similarly, disruption of the hairpin structure in the repeat abrogates processing, suggesting that both sequence and structural elements in the repeat region are important for primary processing. Interestingly, the very same sequence and structural signals were found to determine the extent of the 3′ end of the crRNA during maturation (Hatoum-Aslan et al. 2011): nucleotide insertions and deletions in the repeat region that shifted the primary processing site (and extended or decreased the 5′-tag length) correspondingly shifted the extent of nucleolytic cleavage on the 3′ end. As a result, the length of mature crRNAs remained constant. Taken together, these observations indicate that sequence and structural cues within the upstream repeat forms a docking site for the processing machinery, and that the precise length of the mature crRNA is measured by the Cas processing machinery rather than by the distance between crRNA repeats. This mechanism is reminiscent of the miRNA processing pathway in which a DGCR8-RNA interaction precisely positions the primary miRNA for Drosha cleavage (Han et al. 2006), and of miRNA and siRNA pathways in which Dicer, itself a molecular ruler, produces small RNAs of precise lengths (Macrae et al. 2006).

The protein machinery that is involved in processing and its mechanism is not yet fully understood; however, the genetic requirements for crRNA biogenesis and processing have been examined (Hatoum-Aslan et al. 2011) (see Chap. 5). Deletion analysis of individual *cas* and *csm* genes has shown that *cas*6, *cas*10, and *csm*4 are absolutely required for the accumulation of crRNAs in vivo. Because unprocessed crRNA precursors are susceptible to nuclease degradation (Pougach et al. 2010), the absence of crRNAs in these deletion mutants likely reflects a lack of primary processing. Cas6 and Csm4 belong to the RAMP superfamily of proteins that harbor an RNA-recognition motif and catalytic histidine, which suggest the potential for ribonuclease activity (Makarova et al. 2011b). Because Cas6 homologs in other CRISPR systems have been shown to carry out primary processing (Carte et al. 2008; Haurwitz et al. 2010), Cas6 is the most likely candidate to perform this function in *S. epidermidis*. While the precise roles of Cas10 and Csm4 during crRNA accumulation are yet to be determined, possible functions may include the activation of Cas6 or protection of the crRNAs from cellular nucleases.

Genetic analysis implicated *csm*2, *csm*3, and *csm*5 in crRNA maturation (Hatoum-Aslan et al. 2011). Remarkably, mutants lacking these genes retained the ability to catalyze the primary cleavage event, but exhibited a complete deficiency in maturation and remained in the intermediate stage. This demonstrates that while primary cleavage and maturation are anchored to the same primary processing site on upstream repeat regions, they are carried out by different protein machinery and catalytic mechanisms. Supporting their direct role in maturation, Csm3 and Csm5 are also RAMP proteins with a predicted catalytic histidine capable of nucleolytic activity (Makarova et al. 2011a). Alternatively, Csm2, Csm3, and/or Csm5 could recruit cellular ribonucleases or trigger a rearrangement of a processing complex that presents the 3′ end of the intermediate crRNA to cellular nucleases.

8.2.2 Type III-B Systems

The *P. furiosus* DSM 3638 genome contains two islands of *cas* genes, some of which are specific to type III-B, and seven independent repeat-spacer arrays (Hale et al. 2008) (Fig. 8.1). Sequencing of small RNAs cloned from this organism detected crRNAs from all seven loci, thus demonstrating their constitutive expression. Interestingly, promoter-proximal spacers were more highly represented in sequencing reads, suggesting a mechanism of polar expression. Northern analysis of crRNAs from two of these loci revealed mature species of 45 and 39 nucleotides in length, both of which are derived from a 69 nucleotide intermediate. While the latter is reminiscent of the crRNA species generated by the type III-A model system, there are notable differences in the processing pathways that lead to these similarly sized mature crRNAs.

A screen of recombinant Cas proteins identified Cas6 as the endoribonuclease responsible for primary processing in this type III-B system (Carte et al. 2008). In vitro data generated using purified components revealed that this cleavage occurs eight nucleotides upstream of the spacer, giving rise to an eight-nucleotide tag. The majority (~70 %) of crRNAs produced from all seven CRISPR loci in vivo have an identical eight-nucleotide tag (Hale et al. 2009, 2012), suggesting it plays an important role in CRISPR interference. Because crRNA repeats in this system are largely unstructured (Wang et al. 2011), cues that direct primary processing are limited to sequence elements within repeats. Indeed, a series of structural and functional studies in vitro identified and characterized two distinct sites on the crRNA repeat that are important for efficient and accurate primary processing (Carte et al. 2008, 2010; Wang et al. 2011) (Fig. 8.2b). The first site, located within the 5′ half of the repeat (specifically, nucleotides 2–10), was shown to form a binding region for Cas6; single nucleotide substitutions in this region eliminate Cas6 binding and cleavage. The second site, located in the 3′ half of the repeat (specifically, nucleotides 22–30), presumably binds and positions the cleavage site within the active site of the enzyme. Mutants in the latter retain the ability to bind Cas6 but exhibit diminished/absent cleavage, or a shift in the cleavage site. A linker region between these two sites of at least eight nucleotides is required to bridge the Cas6 binding and cleavage sites (Wang et al. 2011).

The co-crystal structure of Cas6 in complex with repeat RNA revealed a putative dimerization interface (Wang et al. 2011), indicating it may act as a dimer in solution. Mutational analysis revealed a trio of conserved amino acids (Tyr31, His46, and Lys52) that are important for cleavage activity and likely define an active site (Carte et al. 2010). Additionally, a glycine-rich loop adjacent to the putative active site forms a basic patch on the enzyme surface that enhances crRNA binding (Wang et al. 2011). These findings predict a mechanism of primary cleavage in III-B systems in which the Cas6 binding site tethers the 5′ end of the crRNA, resulting in a wrapping of the repeat around the enzyme and precise positioning of the crRNA cleavage site within the active site of the enzyme (Fig. 8.2b). This cleavage is independent of divalent metal ions and results in the generation of 5′-hydroxyl and 2′,3′ cyclic phosphate ends (Carte et al. 2008). The active site architecture and reaction characteristics of Cas6 predict a mechanism of cleavage similar to that catalyzed by the archaeal tRNA splicing endonuclease (Calvin and Li 2008; Carte et al. 2010; Xue et al. 2006). Interestingly, native *P. furiosus* Cas6 isolated from cell extracts was found stably associated with intermediate crRNAs, suggesting it may influence downstream events in the CRISPR pathway (Carte et al. 2010).

The genes involved in crRNA maturation in this system are yet to be established, yet it is tempting to speculate about their identity based upon similarities in both type III systems. A number of RAMPs have been grouped into three broad categories according to their sequence similarities and structural features (Makarova et al. 2011a). In this classification scheme, the Type III-A RAMPs Csm3, Csm4, and Csm5 are orthologs of the Type III-B RAMPs Cmr6, Cmr3, and Cmr4, respectively. We speculate that the demonstrated roles of Csm3 and Csm5 in Type III-A crRNA maturation (Hatoum-Aslan et al. 2011) are most likely played by Cmr6 and Cmr4 during Type III-B crRNA maturation. Interestingly, deep sequencing revealed a clear distinction between crRNAs found in whole cell extracts versus those bound to Cmr1-Cas10/Cmr3-6 complexes (Hale et al. 2012). CrRNAs isolated from the complex had more precisely defined mature lengths of 45 and 39 nt when compared to the less-defined crRNA lengths found in whole cell extracts. This observed bias of mature crRNAs inhabiting the complex supports the notion that the complex itself might be involved in actively trimming the crRNA.

The *S. solfataricus* genome harbors both types III-A and III-B CRISPR systems, as well as six repeat-spacer arrays (Gudbergsdottir et al. 2011; Manica et al. 2011). Deep sequencing analysis of crRNAs associated with the type III-B protein complex (Cmr1, Cas10, Cmr3-7) revealed that all six CRISPR loci are expressed to variable degrees (Zhang et al. 2012). Unlike the sequencing results in *P. furiosus*, there was no enrichment of promoter-proximal spacers observed in *S. solfataricus*. Northern analysis revealed the mean size of crRNAs associated with the complex centered around 45 nt. The genes involved in primary processing and maturation are yet to be determined; however, it is speculated that *cas6*, of which there are four copies, is most likely responsible for primary processing in this organism.

8.3 Targeting

Mature crRNAs act as guides for a targeting complex (Gesner et al. 2011; Hale et al. 2009; Jore et al. 2011; Sashital et al. 2011; Semenova et al. 2011; Wiedenheft et al. 2011) that patrols the cellular milieu for invasive genetic elements and cleaves their genome once a matching sequence (protospacer) is detected (Garneau et al. 2010). The targeting mechanism in type III systems is not well understood; however, the available information continues to underscore the differences CRISPR-Cas systems can exhibit, even when classified in the same subtype.

8.3.1 Type III-A Systems

The type III-A system in *S. epidermidis* prevents plasmid conjugation and bacteriophage infection by *spc*1- and *spc*2-mediated targeting (Marraffini and Sontheimer 2008) (our unpublished results, Fig. 8.3). The proto-spacer DNA, rather than its corresponding mRNA, is targeted in this system. The latter was demonstrated by disrupting the proto-spacer with a self-splicing intron, and showing that this modification renders the plasmid resistant to *spc*1-mediated interference. How do targeting complexes avoid cleaving the spacer DNA within the chromosomal CRISPR

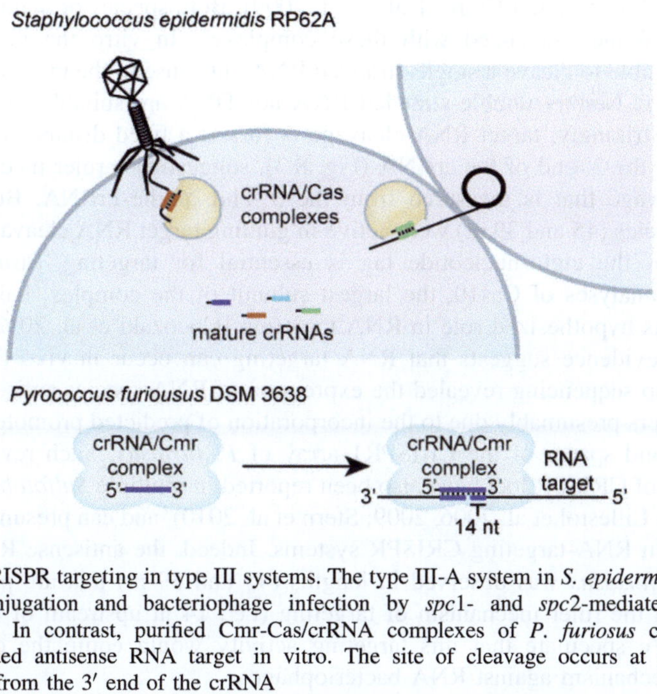

Fig. 8.3 CRISPR targeting in type III systems. The type III-A system in *S. epidermidis* prevents plasmid conjugation and bacteriophage infection by *spc*1- and *spc*2-mediated targeting, respectively. In contrast, purified Cmr-Cas/crRNA complexes of *P. furiosus* can cleave a single-stranded antisense RNA target in vitro. The site of cleavage occurs at precisely 14 nucleotides from the 3′ end of the crRNA

locus? In the type III-A system of *S. epidermidis*, prevention of autoimmunity was narrowed down to the complementarity between the 5′-tag on the crRNA and the 3′ region adjacent to the protospacer DNA: perfect complementarity between these two regions in the CRISPR array prevents targeting, while mismatches result in the targeting of the foreign nucleic acid (Marraffini and Sontheimer 2010b).

In contrast to type III systems, the target of type I and type II systems contain a proto-spacer adjacent motif (PAM) that is required for targeting (Mojica et al. Microbiology, 2009; Sapranauskas et al. Nucleic Acids Research, 2011; Westra et al. Mol Cell, 2012) and also provides the basis for self versus. non-self-discrimination. In these systems the PAM, and a minimum of seven crRNA-complementing nucleotides in the proto-spacer (the so-called 'seed sequence') allow for CRISPR interference to take place (Semenova et al. PNAS, 2011; Wiedenheft et al. PNAS, 2011). While the existence of a PAM in type III-A systems is yet to be determined (the low number of known targets prevents a significant alignement of flanking sequences), the 5′-end sequence of the crRNAs of *S. epidermidis* is essential for targeting and most likely constitutes a seed sequence (Inbal Maniv and Luciano Marraffini, unpublished results).

8.3.2 Type III-B Systems

Mass spectrometry analysis of purified *P. furiosus* targeting complexes revealed the presence of the signature proteins Cas10 and Cmr5, as well as the RAMPs Cmr1, Cmr3, Cmr4, and Cmr6 (Hale et al. 2009). Both species of mature crRNAs were also found associated with these complexes. In vitro the purified Cmr complex is able to cleave a single-stranded RNA antisense to the mature crRNA in the complex. Neither double-stranded RNA nor DNA are suitable substrates for targeting. Strikingly, target RNA cleavage occurs at a fixed distance (14 nucleotides) from the 3′ end of the crRNA (Fig. 8.3), suggesting a ruler mechanism for target cleavage that is measured from the 3′ end of the crRNA. Both mature crRNA species (45 and 39 nt) were active in guiding target RNA cleavage, and the presence of the eight-nucleotide tag is essential for targeting. Structural and functional analyses of Cas10, the largest subunit of the complex, have recently ruled out its hypothesized role in RNA targeting (Cocozaki et al. 2012).

Recent evidence suggests that RNA targeting can occur in vivo (Hale et al. 2012). Deep sequencing revealed the expression of RNA species antisense to one of the spacers presumably due to the incorporation of predicted promoter elements in the second spacer of the CRISPR1 array of *P. furiosus*. Such reverse-strand transcripts of CRISPR loci have also been reported in multiple *Sulfolobales* (Deng et al. 2011; Lillestøl et al. 2006, 2009; Stern et al. 2010), and can presumably serve as targets in RNA-targeting CRISPR systems. Indeed, the antisense RNA generated in *P. furiosus* was observed in lengths (45 and 39 nt) that are predicted to result from the ruler mechanism of targeting (i.e., 14 nt upstream of the crRNA 3′ end). We speculate that this targeting activity would equip the cell with a defense mechanism against RNA bacteriophages.

Because *S. solfataricus* harbors both III-A and III-B CRISPR systems, data generated by in vivo studies cannot distinguish the effects of one from the other. However, a recent study has been able to characterize the III-B targeting mechanism using in vitro analysis (Zhang et al. 2012). In this study, a recombinant complex composed of Cmr1, Cas10, Cmr3-7, and crRNAs was reconstituted with an antisense target, and cleavage of the target RNA was observed. Interestingly, cleavage occurred in a number of locations, regardless of the crRNA 3′-end length, thus ruling out a mechanism of targeting that measures cleavage from the 3′ end of the crRNA. Instead, the latter observation suggests a sequence-dependent model of targeting. Indeed, mutational analysis of the target and the crRNA revealed that cleavage occurs between "UA" sites. Surprisingly, cleavage of the crRNA was also detected in the same locations, but in a less efficient manner, indicating that a single crRNA is capable of multiple-turnover catalysis. While RNA cleavage requires an intact tag sequence and a region of non-complementarity between the 5′-tag and the 3′ end of the target, RNA cleavage did not require strict complementarity between the crRNA and target. This tolerance for spacer–proto-spacer mismatches during targeting was also observed in *S. solfataricus* in vivo (Gudbergsdottir et al. 2011; Manica et al. 2011).

CRISPR targeting against invading plasmid and bacteriophage DNA has been demonstrated in numerous systems in vivo. As investigations continue to illuminate the mechanistic details underlying CRISPR interference in distinct systems, generalizations about different interference pathways become harder to make. A number of questions remain unanswered for type III CRISPR systems, particularly about the adaptation phase in which new spacers are acquired in response to plasmid and phage invasion. It is speculated that adaptation in all CRISPR types occurs by a similar mechanism since the genes most likely carrying out this process (*cas*1 and *cas*2) are universally present in all types, and are the most highly conserved (Makarova et al. 2011b; Takeuchi et al. 2011). However, the majority of *cas* genes, particularly the RAMPs, exhibit high sequence variability and likely high functional diversity (Takeuchi et al. 2011). In addition, little is known about in vivo targeting in type III systems. Is the DNA target cleaved by the type III-A machinery? Can RNA targeting by type III-B systems defend the cell from RNA bacteriophages or regulate gene expression? Further exploration into the model type III systems is expected to answer these outstanding questions as well as to reveal mechanistic details about CRISPR interference.

8.4 Role of CRISPR-Cas Systems in the Evolution of Bacterial Pathogens

Lateral or horizontal gene transfer, the exchange of genetic material between organisms, is the major source of genetic variability for bacterial evolution (Nakamura et al. 2004). HGT occurs by uptake of environmental DNA (transformation) or by the incorporation of heterologous DNA carried on mobile genetic elements such as plasmids (conjugation) and lysogenic phages (transduction) (Thomas and Nielsen 2005). There is experimental proof that CRISPR-Cas systems limit phage infection (Barrangou et al. 2007), plasmid conjugation (Marraffini and Sontheimer 2008), and natural transformation (Bikard et al. 2012). Therefore, CRISPR-Cas systems interfere with all routes of HGT and must play an important role in bacterial evolution.

HGT mechanisms and their natural barriers have been given increasing attention in the biomedical sciences. Pathogens have not only become more virulent but have also acquired resistance to virtually all known antibiotics. A critical health care issue is the rise of hospital- and community-associated meticillin (formerly methicillin)-resistant *Staphylococcus aureus* (MRSA) and vancomycin-resistant *S. aureus* (VRSA) (Furuya and Lowy 2006). The genesis of MRSA and VRSA strains is directly linked to the horizontal transfer of antibiotic resistance genes by plasmid conjugation. Sequencing of the first VRSA isolate (Weigel et al. 2003) revealed the presence of a multidrug resistance plasmid that was also present in an *Enterococcus faecalis* strain co-isolated from the same patient, suggesting that resistance was transferred between these species by conjugation. Likewise, sequencing of the highly virulent MRSA strain USA300 indicated that HGT has allowed the acquisition of elements that encode resistance and virulence determinants that enhance fitness and pathogenicity (Diep et al. 2006). *S. aureus* and *S. epidermidis* strains are the most common causes of nosocomial infections (Lim and Webb 2005; Lowy 1998; von Eiff et al. 2002), and mobile genetic elements can spread from one species to the other (Diep et al. 2006). The type III CRISPR-Cas system of *S. epidermidis* has been found to limit conjugation of pG0400 plasmid from *S. aureus* in the laboratory (Marraffini and Sontheimer 2008) and possibly constitute a natural barrier to the spread of antimicrobial resistance in natural environments. Interestingly, only one of the 29 *S. aureus* genomes completed so far harbor CRISPR loci (Grissa et al. 2007), an observation that suggests the absence of CRISPR may be a determining element in the rapid acquisition of virulence factors and antibiotic resistance by the most aggressive strains of this pathogen (Diep et al. 2006).

Consistent with a role for CRISPR-Cas systems in the control of antibiotic resistance gene dissemination (see also Chaps. 7 and 11), genomic analyses of *E. faecalis* and *E. faecium* have linked type II CRISPR-Cas absence with the evolution of multidrug resistance in high-risk lineages of those species (Palmer and Gilmore 2010). Unlike type III systems, type II systems are characterized by the type-specific gene *cas*9, and crRNA processing mediated by Cas9, RNase III, and the *trans*-encoded small RNA tracrRNA (Deltcheva et al. 2011; Makarova et al. 2011b), among other characteristics, discussed further in Chap. 8 of this book. *E. faecalis* and *E. faecium* are gastrointestinal tract bacteria in a wide range of hosts, but have also emerged as leading opportunistic pathogens. Certain lineages of these species are associated with horizontally acquired multidrug resistance and hospital-acquired infections, and are considered to be high-risk (Leavis et al. 2007; Willems et al. 2005). Analysis of antibiotic resistance, CRISPR, and *cas* gene content among 48 *E. faecalis* strains isolated from the early twentieth to the early twenty-first centuries revealed a statistically significant inverse relationship between *E. faecalis* CRISPR-Cas loci and antibiotic resistance acquired by HGT, with strains lacking CRISPR-Cas possessing more acquired antibiotic resistance traits (Palmer and Gilmore 2010). CRISPR-Cas loci were uniformly absent from each *E. faecalis* high-risk lineage examined, and were additionally absent from high-risk *E. faecium* strains with acquired resistance to vancomycin (Palmer and Gilmore 2010). In support of a role for CRISPR-Cas in regulating dissemination of antibiotic resistance genes in enterococci, CRISPR spacer sequences in *E. faecalis* have identity to sequences from antibiotic resistance plasmids (Palmer and Gilmore 2010). The variable distribution of CRISPR-Cas among *E. faecalis* isolates was independently confirmed by a second group which additionally noted a significant relationship between the presence of *cas* genes and the absence of certain plasmid-encoded virulence genes (Lindenstrauss et al. 2011). Collectively these results indicate that CRISPR-Cas loci are barriers to the uptake of antibiotic resistance genes in enterococci. It appears that enterococcal strains lacking CRISPR-Cas and concomitantly possessing multiple acquired antibiotic resistances have been selected for antibiotic therapy, contributing to the dominance of CRISPR-Cas-deficient phylogenetic lineages in the hospital environment. These results also suggest that CRISPR-Cas-deficient strains could be more efficient donors of antibiotic resistance genes to MRSA, as these strains possess more mobile content than strains with CRISPR-Cas (Palmer and Gilmore 2010) and may be more likely to encounter MRSA in clinical environments.

Experimental proof that acquisition of antibiotic resistance and virulence genes leads to the inactivation of CRISPR-Cas systems targeting these genes was recently obtained in the human pathogen *Streptococcus pneumoniae*. Natural transformation is fundamental for pneumococcal pathogenesis, as it provides the genetic diversity required for the rapid adaptation of pneumococci to clinical interventions such as the introduction of antibiotic therapy and anti-capsule vaccines (Croucher et al. 2011). A CRISPR-Cas system was engineered into *S. pneumoniae* (none of the sequenced strains harbor CRISPR loci) to target antibiotic resistance genes or capsule genes (Bikard et al. 2012). In vitro, CRISPR prevented the transformation and thus the acquisition of antibiotic resistance genes. However, CRISPR "escapers" were also obtained and their analysis revealed the presence of spacer deletions or the complete loss of the CRISPR locus. Moreover, in vivo, during pneumococcal infection of mice, CRISPR prevented the transfer of capsule genes, and therefore the transformation of non-encapsulated, avirulent pneumococci into encapsulated, virulent bacteria. In these experiments CRISPR "escapers" were also detected in mice that succumbed to pneumococcal infection in spite of the presence of CRISPR interference. Genetic analysis of these escapers showed mutations in the *cas* genes that inactivated CRISPR interference. Therefore these results support the notion that CRISPR loci can be detrimental to the evolvability of bacterial pathogens and, as explained above, could explain the absence of these barriers to horizontal transfer in pathogenic staphylococci, enterococci, and pneumococci.

Upon infection of the bacterial host, phages can undergo either lytic or lysogenic replication cycles. The lytic cycle results in the death of the host and release of the phage progeny into the environment. In the lysogenic cycle, a temperate phage integrates its genome into the bacterial chromosome, becoming an inheritable prophage. It has long been known that prophage-encoded genes play an important role in the virulence of pathogenic strains (Brussow et al. 2004). For example, many bacterial toxins reside in prophages found in the genomes of *Corynebacterium diphteriae*, *Clostridium botulinum*, *Vibrio cholerae*, *Escherichia coli*, *Streptococcus pyogenes,* and *S. aureus*. The contribution of prophages to the virulence of *Streptococcus pyogenes* (group A *Streptococcus* or GAS) is very well studied (Banks et al. 2002). It is hypothesized that the increase in the frequency and severity of infection, as well as the complex array of GAS clinical presentations, is the consequence of the acquisition and transfer of phage-encoded virulence factors such as streptococcal pyrogenic exotoxins (Spe) and *S. pyogenes* DNases (Spd). Of the 13 sequenced strains, 8 contain CRISPR systems and contain none or few prophages (Nozawa et al. 2011). On the other hand, strains lacking CRISPR are polylysogens. Moreover, many CRISPR spacers match sequences of prophages integrated into other strains. That is, there is a mutually exclusive relationship between CRISPR spacers and prophages, suggesting that CRISPR immunity can prevent not only phage lysis, but also lysogenesis. Such a mutually exclusive relationship has also been reported for *Sulfolobus islandicus* spacers that match prophages and plasmids (Held and Whitaker 2009). Therefore, CRISPR immunity against lysogenic bacteriophages interferes with the spread of virulence factors among pathogens.

Many of the virulence plasmids required to establish a successful infection by a number of bacterial pathogens are believed to have diverged from conjugative plasmids (Hu et al. 2009). Also, pathogenicity islands are flanked by transposable elements and therefore can transfer between different species by "hitch-hiking" on conjugative plasmids and temperate phages (Hacker and Kaper 2000). Analysis of the genomes of all sequenced isolates of *Salmonella enterica* revealed the presence of CRISPR loci in most of them (Fricke et al. 2011). While most spacers match prophage sequences, one spacer sequence was found that targets the *spv*-type virulence plasmids of this species. The spacer is found in a strain that lacks the plasmid, Saintpaul SARA23, but not in the *spv*-type plasmid-containing strain Typhimurium. Since phylogenetic analysis indicates that both strains share a common ancestor, it is possible that the acquisition of CRISPR interference against the *spv* virulence plasmid prevented the Saintpaul SARA23 ancestor from adapting to the same pathogenic lifestyle as the ancestor of Typhimurium and this could have led to the emergence of a new phylogenetic sublineage within the *S. enterica* species. Therefore, while direct experimental proof may be difficult to obtain, the prevention of conjugation and phage infection by CRISPR suggest an important role for these loci in the modulation and emergence of bacterial pathogens.

Finally, there is intriguing evidence that a CRISPR transcript regulates virulence in the intracellular pathogen *Listeria monocytogenes* via interaction with chromosomally encoded mRNAs. A small RNA possessing CRISPR-like properties (*rliB*) was detected in a study that identified several previously unknown, non-coding RNAs in *L. monocytogenes* EGD-e (Mandin et al. 2007). *rliB* possesses 5 tandem repeats of 29 nucleotides interspersed by sequences of 35–36 nucleotides, and is transcribed from a housekeeping (RpoD) promoter (Mandin et al. 2007). *rliB* was computationally predicted to interact with the *feoA* (ferrous iron acquisition) transcript, and *rliB* overexpression resulted in an increase in *feoA* transcript levels (as well as levels of the co-transcribed *feoB*) (Mandin et al. 2007), suggesting that *rliB* stabilized the *feoAB* transcript. Deletion of *rliB* resulted in more robust colonization of the mouse liver by *L. monocytogenes* EGD-e (Toledo-Arana et al. 2009). Taken together, these results suggest that *rliB* modulates *L. monocytogenes* virulence by interaction with *feoAB* transcripts; however, a direct link between these two observations has not been reported. Notably, *cas* genes were not detected near *rliB* in *L. monocytogenes* EGD-e (Mandin et al. 2007; Mraheil et al. 2011), although CRISPR-Cas loci are present in other *Listeria* sp. strains (Deltcheva et al. 2011; den Bakker et al. 2010). Further investigation is required to determine whether *rliB* is in fact associated with *cas* genes in *L. monocytogenes* EGD-e, and if not, whether *rliB* may have originated from an ancestral CRISPR-Cas locus and was co-opted for a secondary cellular function.

References

Banks DJ, Beres SB, Musser JM (2002) The fundamental contribution of phages to GAS evolution, genome diversification and strain emergence. Trends Microbiol 10(11):515–521

Barrangou R, Fremaux C, Deveau H, Richards M, Boyaval P, Moineau S, Romero DA, Horvath P (2007) CRISPR provides acquired resistance against viruses in prokaryotes. Science 315(5819):1709–1712

Bikard D, Hatoum-Aslan A, Mucida D, Marraffini LA (2012) CRISPR interference can prevent natural transformation and virulence acquisition during in vivo bacterial infection. Cell Host Microbe 12(2):177–186. doi:10.1016/j.chom.2012.06.003

Brussow H, Canchaya C, Hardt WD (2004) Phages and the evolution of bacterial pathogens: from genomic rearrangements to lysogenic conversion. Microbiol Mol Biol Rev 68(3):560–602

Calvin K, Li H (2008) RNA-splicing endonuclease structure and function. Cell Mol Life Sci 65(7–8):1176–1185. doi:10.1007/s00018-008-7393-y

Carte J, Pfister NT, Compton MM, Terns RM, Terns MP (2010) Binding and cleavage of CRISPR RNA by Cas6. RNA 16(11):2181–2188. doi:10.1261/rna.2230110, rna.2230110 [pii]

Carte J, Wang R, Li H, Terns RM, Terns MP (2008) Cas6 is an endoribonuclease that generates guide RNAs for invader defense in prokaryotes. Genes Dev 22(24):3489–3496

Cocozaki AI, Ramia NF, Shao Y, Hale CR, Terns RM, Terns MP, Li H (2012) Structure of the Cmr2 subunit of the CRISPR-Cas RNA silencing complex. Structure 20(3):545–553. doi:10.1016/j.str.2012.01.018

Croucher NJ, Harris SR, Fraser C, Quail MA, Burton J, van der Linden M, McGee L, von Gottberg A, Song JH, Ko KS, Pichon B, Baker S, Parry CM, Lambertsen LM, Shahinas D, Pillai DR, Mitchell TJ, Dougan G, Tomasz A, Klugman KP, Parkhill J, Hanage WP, Bentley SD (2011) Rapid pneumococcal evolution in response to clinical interventions. Science 331(6016):430–434

Daniel A, Bonnen PE, Fischetti VA (2007) First complete genome sequence of two *Staphylococcus epidermidis* bacteriophages. J Bacteriol 189(5):2086–2100

Deltcheva E, Chylinski K, Sharma CM, Gonzales K, Chao Y, Pirzada ZA, Eckert MR, Vogel J, Charpentier E (2011) CRISPR RNA maturation by trans-encoded small RNA and host factor RNase III. Nature 471(7340):602–607

den Bakker HC, Cummings CA, Ferreira V, Vatta P, Orsi RH, Degoricija L, Barker M, Petrauskene O, Furtado MR, Wiedmann M (2010) Comparative genomics of the bacterial genus *Listeria*: genome evolution is characterized by limited gene acquisition and limited gene loss. BMC Genomics 11:688. doi:10.1186/1471-2164-11-688

Deng L, Kenchappa CS, Peng X, She Q, Garrett RA (2011) Modulation of CRISPR locus transcription by the repeat-binding protein Cbp1 in Sulfolobus. Nucleic Acids Res. doi:10.1093/nar/gkr1111, gkr1111 [pii]

Deveau H, Garneau JE, Moineau S (2010) CRISPR/Cas system and its role in phage-bacteria interactions. Annu Rev Microbiol 64:475–493

Diep BA, Gill SR, Chang RF, Phan TH, Chen JH, Davidson MG, Lin F, Lin J, Carleton HA, Mongodin EF, Sensabaugh GF, Perdreau-Remington F (2006) Complete genome sequence of USA300, an epidemic clone of community-acquired meticillin-resistant *Staphylococcus aureus*. Lancet 367(9512):731–739

Fricke WF, Mammel MK, McDermott PF, Tartera C, White DG, Leclerc JE, Ravel J, Cebula TA (2011) Comparative genomics of 28 *Salmonella enterica* isolates: evidence for CRISPR-mediated adaptive sublineage evolution. J Bacteriol 193(14):3556–3568. doi:10.1128/JB.00297-11

Furuya EY, Lowy FD (2006) Antimicrobial-resistant bacteria in the community setting. Nat Rev Microbiol 4(1):36–45

Garneau JE, Dupuis ME, Villion M, Romero DA, Barrangou R, Boyaval P, Fremaux C, Horvath P, Magadan AH, Moineau S (2010) The CRISPR/Cas bacterial immune system cleaves bacteriophage and plasmid DNA. Nature 468(7320):67–71

Gesner EM, Schellenberg MJ, Garside EL, George MM, Macmillan AM (2011) Recognition and maturation of effector RNAs in a CRISPR interference pathway. Nat Struct Mol Biol 18(6):688–692

Grissa I, Vergnaud G, Pourcel C (2007) The CRISPRdb database and tools to display CRISPRs and to generate dictionaries of spacers and repeats. BMC Bioinformatics 8:172

Gudbergsdottir S, Deng L, Chen Z, Jensen JV, Jensen LR, She Q, Garrett RA (2011) Dynamic properties of the *Sulfolobus* CRISPR/Cas and CRISPR/Cmr systems when challenged with vector-borne viral and plasmid genes and protospacers. Mol Microbiol 79(1):35–49. doi:10.1111/j.1365-2958.2010.07452.x

Hacker J, Kaper JB (2000) Pathogenicity islands and the evolution of microbes. Annu Rev Microbiol 54:641–679

Haft DH, Selengut J, Mongodin EF, Nelson KE (2005) A guild of 45 CRISPR-associated (Cas) protein families and multiple CRISPR/Cas subtypes exist in prokaryotic genomes. PLoS Comput Biol 1(6):e60

Hale C, Kleppe K, Terns RM, Terns MP (2008) Prokaryotic silencing (psi)RNAs in *Pyrococcus furiosus*. RNA 14(12):2572–2579

Hale CR, Zhao P, Olson S, Duff MO, Graveley BR, Wells L, Terns RM, Terns MP (2009) RNA-guided RNA cleavage by a CRISPR RNA-Cas protein complex. Cell 139(5):945–956

Hale CR, Majumdar S, Elmore J, Pfister N, Compton M, Olson S, Resch AM, Glover CV, 3rd, Graveley BR, Terns RM, Terns MP (2012) Essential Features and Rational Design of CRISPR RNAs that Function with the Cas RAMP Module Complex to Cleave RNAs. Mol Cell. doi:10.1016/j.molcel.2011.10.023, S1097-2765(11)00955-5 [pii]

Han J, Lee Y, Yeom KH, Nam JW, Heo I, Rhee JK, Sohn SY, Cho Y, Zhang BT, Kim VN (2006) Molecular basis for the recognition of primary microRNAs by the Drosha-DGCR8 complex. Cell 125(5):887–901. doi:10.1016/j.cell.2006.03.043

Hatoum-Aslan A, Maniv I, Marraffini LA (2011) Mature clustered, regularly interspaced, short palindromic repeats RNA (crRNA) length is measured by a ruler mechanism anchored at the precursor processing site. Proc Natl Acad Sci USA 108(52):21218–21222. doi:10.1073/pnas.1112832108

Haurwitz RE, Jinek M, Wiedenheft B, Zhou K, Doudna JA (2010) Sequence- and structure-specific RNA processing by a CRISPR endonuclease. Science 329(5997):1355–1358

Held NL, Whitaker RJ (2009) Viral biogeography revealed by signatures in *Sulfolobus islandicus* genomes. Environ Microbiol 11(2):457–466

Horvath P, Barrangou R (2010) CRISPR/Cas, the immune system of bacteria and archaea. Science 327(5962):167–170

Hu X, Van der Auwera G, Timmery S, Zhu L, Mahillon J (2009) Distribution, diversity, and potential mobility of extrachromosomal elements related to the *Bacillus anthracis* pXO1 and pXO2 virulence plasmids. Appl Environ Microbiol 75(10):3016–3028

Jore MM, Lundgren M, van Duijn E, Bultema JB, Westra ER, Waghmare SP, Wiedenheft B, Pul U, Wurm R, Wagner R, Beijer MR, Barendregt A, Zhou K, Snijders AP, Dickman MJ, Doudna JA, Boekema EJ, Heck AJ, van der Oost J, Brouns SJ (2011) Structural basis for CRISPR RNA-guided DNA recognition by Cascade. Nat Struct Mol Biol 18(5):529–536

Leavis HL, Willems RJ, van Wamel WJ, Schuren FH, Caspers MP, Bonten MJ (2007) Insertion sequence-driven diversification creates a globally dispersed emerging multiresistant subspecies of *E. faecium*. PLoS Pathog 3(1):e7. doi:10.1371/journal.ppat.0030007

Lillestøl RK, Redder P, Garrett RA, Brugger K (2006) A putative viral defence mechanism in archaeal cells. Archaea 2(1):59–72

Lillestøl RK, Shah SA, Brugger K, Redder P, Phan H, Christiansen J, Garrett RA (2009) CRISPR families of the crenarchaeal genus *Sulfolobus*: bidirectional transcription and dynamic properties. Mol Microbiol 72(1):259–272

Lim SM, Webb SA (2005) Nosocomial bacterial infections in Intensive Care Units. I: organisms and mechanisms of antibiotic resistance. Anaesthesia 60(9):887–902

Lindenstrauss AG, Pavlovic M, Bringmann A, Behr J, Ehrmann MA, Vogel RF (2011) Comparison of genotypic and phenotypic cluster analyses of virulence determinants and possible role of CRISPR elements towards their incidence in *Enterococcus faecalis* and *Enterococcus faecium*. Syst Appl Microbiol 34(8):553–560. doi:10.1016/j.syapm.2011.05.002

Lowy FD (1998) *Staphylococcus aureus* infections. N Engl J Med 339(8):520–532

Macrae IJ, Zhou K, Li F, Repic A, Brooks AN, Cande WZ, Adams PD, Doudna JA (2006) Structural basis for double-stranded RNA processing by Dicer. Science 311(5758):195–198. doi:10.1126/science.1121638, 311/5758/195 [pii]

Makarova KS, Grishin NV, Shabalina SA, Wolf YI, Koonin EV (2006) A putative RNA-interference-based immune system in prokaryotes: computational analysis of the predicted enzymatic machinery, functional analogies with eukaryotic RNAi, and hypothetical mechanisms of action. Biol Direct 1:7

Makarova KS, Aravind L, Wolf YI, Koonin EV (2011a) Unification of Cas protein families and a simple scenario for the origin and evolution of CRISPR-Cas systems. Biol Direct 6:38. doi:10.1186/1745-6150-6-38, 1745-6150-6-38 [pii]

Makarova KS, Haft DH, Barrangou R, Brouns SJ, Charpentier E, Horvath P, Moineau S, Mojica FJ, Wolf YI, Yakunin AF, van der Oost J, Koonin EV (2011b) Evolution and classification of the CRISPR-Cas systems. Nat Rev Microbiol 9(6):467–477

Mandin P, Repoila F, Vergassola M, Geissmann T, Cossart P (2007) Identification of new noncoding RNAs in *Listeria monocytogenes* and prediction of mRNA targets. Nucleic Acids Res 35(3):962–974. doi:10.1093/nar/gkl1096

Manica A, Zebec Z, Teichmann D, Schleper C (2011) In vivo activity of CRISPR-mediated virus defence in a hyperthermophilic archaeon. Mol Microbiol 80(2):481–491

Marraffini LA, Sontheimer EJ (2008) CRISPR interference limits horizontal gene transfer in staphylococci by targeting DNA. Science 322(5909):1843–1845

Marraffini LA, Sontheimer EJ (2010a) CRISPR interference: RNA-directed adaptive immunity in bacteria and archaea. Nat Rev Genet 11(3):181–190

Marraffini LA, Sontheimer EJ (2010b) Self versus non-self discrimination during CRISPR RNA-directed immunity. Nature 463(7280):568–571

Mojica FJ, Diez-Villasenor C, Garcia-Martinez J, Almendros C (2009) Short motif sequences determine the targets of the prokaryotic CRISPR defence system. Microbiology 155 (Pt 3): 733–740

Mraheil MA, Billion A, Mohamed W, Mukherjee K, Kuenne C, Pischimarov J, Krawitz C, Retey J, Hartsch T, Chakraborty T, Hain T (2011) The intracellular sRNA transcriptome of *Listeria monocytogenes* during growth in macrophages. Nucleic Acids Res 39(10): 4235–4248. doi:10.1093/nar/gkr033

Nakamura Y, Itoh T, Matsuda H, Gojobori T (2004) Biased biological functions of horizontally transferred genes in prokaryotic genomes. Nat Genet 36(7):760–766

Nozawa T, Furukawa N, Aikawa C, Watanabe T, Haobam B, Kurokawa K, Maruyama F, Nakagawa I (2011) CRISPR inhibition of prophage acquisition in *Streptococcus pyogenes*. PLoS One 6(5):e19543

Palmer KL, Gilmore MS (2010) Multidrug-resistant enterococci lack CRISPR-cas. MBio 1(4)

Peng X, Brugger K, Shen B, Chen L, She Q, Garrett RA (2003) Genus-specific protein binding to the large clusters of DNA repeats (short regularly spaced repeats) present in *Sulfolobus genomes*. J Bacteriol 185(8):2410–2417

Pougach K, Semenova E, Bogdanova E, Datsenko KA, Djordjevic M, Wanner BL, Severinov K (2010) Transcription, processing and function of CRISPR cassettes in *Escherichia coli*. Mol Microbiol 77(6):1367–1379. doi:10.1111/j.1365-2958.2010.07265.x MMI7265 [pii]

Sapranauskas R, Gasiunas G, Fremaux C, Barrangou R, Horvath P, Siksnys V (2011) The Streptococcus thermophilus CRISPR/Cas system provides immunity in Escherichia coli. Nucleic Acids Res. doi:gkr606[pii]10.1093/nar/gkr606

Sashital DG, Jinek M, Doudna JA (2011) An RNA-induced conformational change required for CRISPR RNA cleavage by the endoribonuclease Cse3. Nat Struct Mol Biol 18(6):680–687

Semenova E, Jore MM, Datsenko KA, Semenova A, Westra ER, Wanner B, van der Oost J, Brouns SJ, Severinov K (2011) Interference by clustered regularly interspaced short palindromic repeat (CRISPR) RNA is governed by a seed sequence. Proc Natl Acad Sci USA 108(25):10098–100103

Stern A, Keren L, Wurtzel O, Amitai G, Sorek R (2010) Self-targeting by CRISPR: gene regulation or autoimmunity? Trends Genet 26(8):335–340

Takeuchi N, Wolf YI, Makarova KS, Koonin EV (2011) Nature and Intensity of Selection Pressure on CRISPR-Associated Genes. J Bacteriol. doi:10.1128/JB.06521-11, JB.06521-11 [pii]

Tang TH, Polacek N, Zywicki M, Huber H, Brugger K, Garrett R, Bachellerie JP, Huttenhofer A (2005) Identification of novel non-coding RNAs as potential antisense regulators in the archaeon *Sulfolobus solfataricus*. Mol Microbiol 55(2):469–481

Thomas CM, Nielsen KM (2005) Mechanisms of, and barriers to, horizontal gene transfer between bacteria. Nat Rev Microbiol 3(9):711–721

Toledo-Arana A, Dussurget O, Nikitas G, Sesto N, Guet-Revillet H, Balestrino D, Loh E, Gripenland J, Tiensuu T, Vaitkevicius K, Barthelemy M, Vergassola M, Nahori MA, Soubigou G, Regnault B, Coppee JY, Lecuit M, Johansson J, Cossart P (2009) The *Listeria* transcriptional landscape from saprophytism to virulence. Nature 459(7249):950–956. doi:10.1038/nature08080

von Eiff C, Peters G, Heilmann C (2002) Pathogenesis of infections due to coagulase-negative staphylococci. Lancet Infect Dis 2(11):677–685

Wang R, Preamplume G, Terns MP, Terns RM, Li H (2011) Interaction of the Cas6 riboendonuclease with CRISPR RNAs: recognition and cleavage. Structure 19(2):257–264

Weigel LM, Clewell DB, Gill SR, Clark NC, McDougal LK, Flannagan SE, Kolonay JF, Shetty J, Killgore GE, Tenover FC (2003) Genetic analysis of a high-level vancomycin-resistant isolate of *Staphylococcus aureus*. Science 302(5650):1569–1571

Westra ER, van Erp PB, Kunne T, Wong SP, Staals RH, Seegers CL, Bollen S, Jore MM, Semenova E, Severinov K, de Vos WM, Dame RT, de Vries R, Brouns SJ, van der Oost J (2012) CRISPR immunity relies on the consecutive binding and degradation of negatively supercoiled invader DNA by Cascade and Cas3. Mol Cell 46(5):595–605. doi:10.1016/j.molcel.2012.03.018

Wiedenheft B, Lander GC, Zhou K, Jore MM, Brouns SJ, van der Oost J, Doudna JA, Nogales E (2011) Structures of the RNA-guided surveillance complex from a bacterial immune system. Nature 477(7365):486–489. doi:10.1038/nature10402, nature10402 [pii]

Willems RJ, Top J, van Santen M, Robinson DA, Coque TM, Baquero F, Grundmann H, Bonten MJ (2005) Global spread of vancomycin-resistant *Enterococcus faecium* from distinct nosocomial genetic complex. Emerg Infect Dis 11(6):821–828

Xue S, Calvin K, Li H (2006) RNA recognition and cleavage by a splicing endonuclease. Science 312(5775):906–910. doi:10.1126/science.1126629, 312/5775/906 [pii]

Zhang J, Rouillon C, Kerou M, Reeks J, Brugger K, Graham S, Reimann J, Cannone G, Liu H, Albers SV, Naismith JH, Spagnolo L, White MF (2012) Structure and Mechanism of the CMR Complex for CRISPR-Mediated Antiviral Immunity. Mol Cell. doi:10.1016/j.molcel.2011.12.013, S1097-2765(11)00957-9 [pii]

Chapter 9
CRISPR-Cas Systems to Probe Ecological Diversity and Host–Viral Interactions

Nicole L. Held, Lauren M. Childs, Michelle Davison, Joshua S. Weitz, Rachel J. Whitaker and Devaki Bhaya

Abstract A key feature of the CRISPR-Cas defense system is the ability of the host to rapidly acquire novel spacers from invasive foreign genetic elements such as plasmids, viruses, or transposons. Consequently, host CRISPR loci have the potential to provide time-resolved information about exposure to foreign genetic elements as well as fine-scale ecological diversity. Furthermore, viral genomes can mutate rapidly, allowing viruses to circumvent the host CRISPR-encoded immunity system, which relies on close matches between spacers and incoming nucleic

N. L. Held · R. J. Whitaker
Department of Microbiology, University of Illinois, 601 S. Goodwin Ave., Urbana, IL 61801, USA
e-mail: nheld2@illinois.edu

R. J. Whitaker
e-mail: rwhitaker@life.illinois.edu

L. M. Childs
School of Biology and School of Mathematics, Georgia Institute of Technology, cAtlanta, GA 30332, USA
e-mail: lauren.childs@biology.gatech.edu

M. Davison · D. Bhaya (✉)
Department of Plant Biology, Carnegie Institution for Science, Stanford, CA 94305, USA
e-mail: dbhaya@stanford.edu

M. Davison
e-mail: shellblu@stanford.edu

J. S. Weitz
School of Biology and School of Physics, Georgia Institute of Technology, Atlanta, GA 30332, USA
e-mail: jsweitz@gatech.edu

M. Davison
Department of Biology, Stanford University, Stanford, CA 94305, USA

R. Barrangou and J. van der Oost (eds.), *CRISPR-Cas Systems*,
DOI: 10.1007/978-3-642-34657-6_9, © Springer-Verlag Berlin Heidelberg 2013

acids. Thus, CRISPR-Cas systems may drive complex, coevolving relationships between bacteria or archaea and viruses. We discuss how ecologically based approaches, in both natural and experimental systems, provide unique insights into host and viral diversity and horizontal gene transfer of CRISPR loci. We critically review recent attempts to model host–viral coevolutionary dynamics in the context of CRISPR loci. Finally, we highlight the future directions in which experimental analyses of host–viral coevolution can be fruitfully combined with theoretical approaches.

Contents

9.1 Introduction

The recent discovery of the CRISPR-Cas system, a novel and widespread adaptive defense mechanism in bacteria and archaea, has stimulated interest in host–viral interactions in the context of unexplored ecological diversity and evolutionary dynamics. Viruses (we adopt the general term 'virus', rather than "bacteriophage" which describes viruses that specifically infect bacteria) are found in most environments and they may outnumber hosts by a factor of ten (Rohwer 2003; Suttle 2005; Lopez-Bueno et al. 2009). Consequently, viruses can also be a driving force in the evolution of microbial communities in many environments, and serve to control both the density and composition of microbial populations (Breitbart and Rohwer 2005; Dinsdale et al. 2008; Rodriguez-Valera et al. 2009; Rohwer and Thurber 2009; Reyes et al. 2010; Rodriguez-Brito et al. 2010; Anderson et al. 2011a, b; Berg Miller et al. 2012). Typically, viruses exhibit high rates of mutation and recombination, which allow them to rapidly evolve and evade host defenses. If host organisms are to survive viral predation, they must have mechanisms of defense and the ability to evolve rapidly to successfully fend off an abundant and rapidly evolving parasite (Hambly and Suttle 2005; Abedon 2009; Labrie et al. 2010; Bikard and Marraffini 2012; Stern et al. 2012). These multifaceted and versatile defense systems operate at various levels and include (1) mechanisms to block adsorption of virus to host receptors, (2) prevention of viral DNA entry, and (3) recognition and degradation of foreign nucleic acids by host restriction modification systems. In addition, there are numerous abortive infection systems which cause host cell death and thereby prevent virus infection (recent reviews on host defense systems include Hyman and Abedon 2010; Labrie et al. 2010; Stern et al. 2012). The recently discovered CRISPR-Cas system, which operates as an intracellular adaptive defense mechanism has now been added to the arsenal of defense systems. One of the unique and powerful aspects of the CRISPR-Cas system is that it is a rapidly evolving defense system which reflects ongoing interactions between virus and host. Thus, it can be used to examine host diversity in an ecological context as well as characterize host–viral coevolution through experimental and theoretical approaches.

Investigation of the CRISPR-Cas system has been facilitated by advances in the ability to rapidly and inexpensively acquire complete or draft genome sequences of microorganisms from diverse environments and that span the microbial tree of life (e.g. Integrated Microbial Genomes at http://img.jgi.doe.gov/cgi-bin/w/main.cgi). Extensive metagenomic information (i.e. a snapshot of the extant genetic diversity) has been acquired from numerous environments with diverse microbial communities, which provide detailed spatial and/or temporal resolution (Riesenfeld et al. 2004; Singh et al. 2009; Willner et al. 2009; Gilbert and Dupont 2011). It is also possible, but not trivial, to acquire extensive genomic information from single bacterial or archaeal cells isolated from the environment (Blainey et al. 2011; Yilmaz and Singh 2012) or to link the presence of certain viruses with single bacterial cells using microfluidic devices (Tadmor et al. 2011). In stark contrast, the availability of annotated genomic information about viruses in most environments is woefully lacking and incomplete (Pignatelli et al. 2008; Rodriguez-Brito et al. 2010; Mokili et al. 2012).

The landmark experimental demonstration of the adaptive defense role of the CRISPR-Cas system (Barrangou et al. 2007) in 2007 was quickly followed by several key advances in the understanding of mechanism of action of the CRISPR-Cas system using a handful of model organisms across all three types (see Chaps. 6 , 7, 8), such as *Escherichia coli, Pseudomonas aeruginosa, Streptococcus thermophilus, Staphylococcus aureus,* and *Pyrococcus furiosus.* By contrast, the diversity, ecological importance, and evolutionary aspects of the CRISPR-Cas systems within natural populations have received less attention. Yet, there is a growing appreciation that the CRISPR-Cas system is very diverse and when studied in natural populations, can provide novel insights into ecological diversity as well as the mechanism and effects of host–viral coevolution. These insights may be hard to acquire if the focus is limited to a few model organisms that have been "domesticated" under prolonged laboratory conditions (i.e. not exposed to their natural predators).

We begin with a short description of features of the CRISPR-Cas system that are particularly relevant in examining ecological diversity and aspects of evolutionary biology. Next, we showcase how data from experimental and natural populations can provide insights into the mechanisms of CRISPR-Cas systems. We examine how CRISPR-Cas loci have been used to analyze host and viral diversity in an ecological context. In addition, we review a limited number of attempts to mathematically model host–viral coevolution mediated by the CRISPR-Cas defense system. Finally, we present scenarios in which experimental evidence and modeling approaches may be combined to understand the importance and role of CRISPR-Cas systems in evolving microbial–viral communities.

9.2 Salient Features of the CRISPR-Cas System in the Context of Natural Populations

In this chapter, the primary focus is on the diversity of the CRISPR loci and spacer content as it pertains to ecological diversity, so only some relevant features of the CRISPR-Cas defense system are described (details of each of these aspects are available in Chaps. 4 through 8).

A functional CRISPR-Cas system requires two components for activity. These have been conceptually described as "two distinct, quasi-independent subsystems", comprising a well-conserved 'information processing' subsystem, and an 'executive' subsystem (Makarova et al. 2011). The first component is the easily identifiable CRISPR locus located on the genome (chromosome or plasmid) which typically contains short repeats that are interspaced with short hypervariable spacers (these are sequences acquired from foreign DNA, either viral or plasmid). Bacteria or archaea often contain a single CRISPR locus (see Chaps. 2 and 6, notably Table 6.3), but there are several species that contain multiple CRISPR loci (see Table 6.3) (Makarova et al. 2011). The ecological or evolutionary advantage of carrying multiple CRISPR-Cas loci on the genome (either on the chromosome or on a plasmid) has not been explored in detail (Diez-Villasenor et al. 2010; Touchon et al. 2011). The second component is a diverse group of *cas* genes located in the vicinity of a CRISPR locus which encode proteins (generically called Cas proteins) that carry out the various enzymatic steps required either for the acquisition of spacers into the CRISPR locus or for expression and interference against invasive genetic elements (Fig. 9.1).

The adaptive immunity process is believed to begin when foreign nucleic acids are recognized as "non-self" and short fragments are incorporated between adjacent CRISPR repeats, although the steps involved in the acquisition process (i.e. how the DNA is recognized, cleaved, and incorporated into the CRISPR locus) are not well characterized. These small DNA fragments or 'spacers' (typically between 26 and 72 bp) are integrated primarily at one end (the 5′ or 'leader' end) of the CRISPR locus. Thus, positional information represents a "time-line"

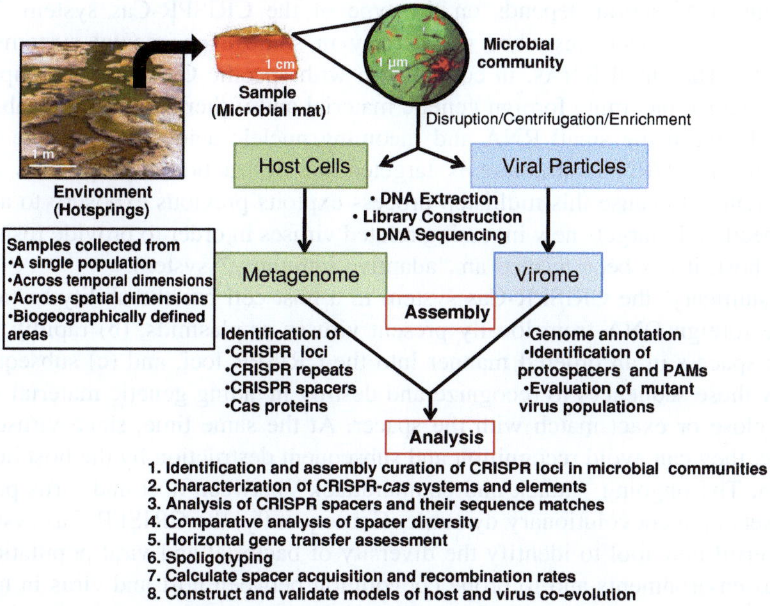

Fig. 9.1 A flow chart illustrating the process by which samples collected from any environment (in this case exemplified by a microbial mat sample from a hot spring) can be used for DNA extraction and library construction followed by DNA sequencing (either next-generation deep sequencing or more traditional methods). The derived sequence information from host cells (metagenome) or from viral particles (virome) can be assembled and analyzed for a number of different purposes (some of which are mentioned here)

of spacer acquisition events, with older spacers typically being closer to the 3′ or 'trailer' end (Fig. 9.1) (Godde and Bickerton 2006; Grissa et al. 2007). This unique feature of the CRISPR-Cas system can be exploited as a tool to examine ecological diversity and the evolutionary dynamics of host–viral interactions.

The vast majority of spacers have no match with any sequences in the available genomic databases, which led to the original designation of these regions as "spacers". This misleading terminology has been widely adopted although we now know that spacers play a critically important role in CRISPR-based immunity. In the rare cases where a spacer matches a sequence on a sequenced viral (or plasmid) genome, the term "proto-spacer" is used to designate the corresponding viral or plasmid sequence (Deveau et al. 2008). In several, but not all cases, a very short stretch of conserved nucleotides on the viral genome, in the immediate vicinity of the proto-spacer, appears to be a recognition motif required for acquisition of the DNA fragment and is referred to as the *p*roto-spacer *a*djacent *m*otif (or PAM) (Mojica et al. 2009; Deveau et al. 2010).

The CRISPR locus is transcribed (either constitutively or in a regulated manner) by host RNA polymerase into a primary transcript called the pre-CRISPR RNA that is precisely processed into small non-coding RNAs, each of which contains a partial repeat and a spacer. This process is complex and the exact mechanism of action depends on the type of the CRISPR-Cas system. These processes have been described extensively in some experimental systems (see Chap. 5). The small RNAs, in conjunction with specific Cas protein complexes, can recognize incoming foreign genetic material and if there is a close or absolute match between the small RNA and incoming nucleic acid sequence, the RNA/DNA (or RNA/RNA) structure is targeted for destruction in a process called interference. Because this multistep process exploits previous exposure to a virus and specifically targets new incoming related viruses in order to provide resistance to the host, it has been termed an "adaptive immunity" system.

In summary, the CRISPR-Cas system in a host cell has the ability to (a) recognize foreign DNA from locally present viruses or plasmids, (b) rapidly incorporate spacers in an ordered manner into the CRISPR loci, and (c) subsequently deploy these sequences to recognize and destroy invading genetic material which has a close or exact match with the spacer. At the same time, since viruses can mutate, they can avoid recognition and subsequent destruction by the host defense system. The ongoing "attack and counterattack" between host and virus populations sets up a coevolutionary dynamic. Consequently, the CRISPR-Cas system is a powerful new tool to identify the diversity of bacterial and viral populations in various environments and to probe coevolution between host and virus in natural environments or in controlled settings. Furthermore, experimental data can be used to develop, test, and refine theoretical models of coevolutionary dynamics.

9.3 Diversity of CRISPR-Cas Loci in Microbial Species

Even before the role or mechanism of action of the CRISPR-Cas system was established, the potential of using the unique hypervariable regions within CRISPR loci for the identification and typing of pathogenic bacteria was recognized by several groups, and is referred to as 'spoligotyping' (see Chap. 2). Subsequently, available genomic and metagenomic databases were used to examine the diversity of CRISPR-Cas loci within (1) populations or at the species level, (2) specific communities, or (3) experimentally defined systems. These studies have revealed a striking degree of diversity in the types of CRISPR-Cas loci as well as in the sequence of spacer arrays within microbial species. Large-scale analyses have provided valuable, new information about the diversity and evolution of the CRISPR-Cas systems, and have also been used to investigate host/viral diversity in various ecological settings (see later and Sect. 9.4). Lastly, this extensive experimental data has been used to examine the coevolution and interaction of host and virus, and to develop and test mathematical models of evolutionary dynamics.

Each CRISPR locus comprises CRISPR repeats and spacers, and in addition, there is an extensive range of Cas proteins associated with these loci. Comparison of CRISPR loci and Cas proteins within and between microbial species has revealed much about the CRISPR-Cas system itself. For instance, CRISPR repeats have been classified into at least 12 groups and there is some correspondence between certain repeats and groups (or subtypes) of Cas proteins associated with them (Kunin et al. 2007; Makarova et al. 2011). In contrast, CRISPR spacer sequences are extraordinarily diverse and unless corresponding viral sequences are available, they are often categorized as being 'unique' (curated databases of CRISPR-Cas loci are available at the CRISPRs Web Server, http://crispr.u-psud.fr/ and at CRISPI http://crispi.genouest.org/) (Grissa et al. 2008; Rousseau et al. 2009). Based on the available information about *cas* gene arrangements and CRISPR repeat sequences, Haft et al. attempted the first classification of CRISPR-Cas systems in 2005 (Haft et al. 2005), which has recently been expanded by Makarova and collaborators (Makarova et al. 2011). Three major types of CRISPR-Cas systems (types I, II, and III) based on a number of criteria (see Chap. 3 for details) have been proposed. The type I system is found in both bacteria and archaea; Type II is exclusively present in bacteria while the type III system appears more commonly in archaea.

The type I CRISPR-Cas system contains the conserved Cas1 and Cas2 proteins and the large 'signature' protein, Cas3 helicase/nuclease, which interacts with the CASCADE protein complex in the "interference" step (Sinkunas et al. 2011). Type I is the most diverse of these systems with six different subtypes (Brouns et al. 2008) (see Chaps. 2 and 6). The type II CRISPR-Cas system is typified by the Cas 9 'signature' protein, which is a large multifunctional protein (Garneau et al. 2010). This type is the simplest of the three types with four genes that comprise the operon (*cas*9, *cas*1, *cas*2, and either *cas*4 or *csn*2). The best studied type II system is that of *S. thermophilus* (see Chap. 7). The type III CRISPR-Cas system, which includes 'signature' proteins such as Cas10 and Cas6, has two subtypes (type IIIA and IIIB). Interestingly, type III systems can target mRNA (in *Pyrococcus furiosus*) (Hale et al. 2009), or DNA (in *Staphylococcus epidermidis* (see Chap. 8) (Marraffini and Sontheimer 2008).

Makarova et al. have tabulated the taxonomic distribution of the three types of CRISPR-Cas types and found that they are present in all major taxonomic groups, albeit with varying distributions (Makarova et al. 2011). Very large metagenomic databases and the sequences of several closely related microbial strains are now available, which represent an untapped resource for CRISPR spacer analyses. On the other hand, other phyla are underrepresented, so generalizations about the presence or absence of CRISPR loci in certain phyla need to be made with caution. For instance, the genomes of marine cyanobacteria sequenced so far appear to lack CRISPR loci, conversely, most freshwater cyanobacterial genomes contain CRISPR loci (Bhaya, unpublished). In this context, the bias in sampling across the tree of life is important to keep in mind and systematic large-scale analysis of the presence of different CRISPR loci types across the microbial tree of life is warranted. It also appears that movement of CRISPR loci can occur across widely diverged lineages by horizontal gene transfer (via plasmids harboring CRISPR-Cas loci or by other gene transfer mechanisms such as transposon activity) (Godde and Bickerton 2006; Heidelberg et al. 2009; Horvath et al. 2009; Portillo and Gonzalez 2009). This raises intriguing questions of what happens following a gene transfer event: e.g., how long do old spacers remain before they are lost and can spacers be used to track horizontal gene transfer events in natural populations?

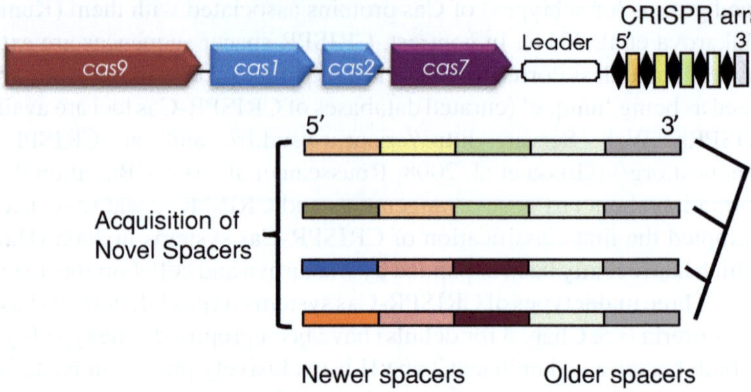

Fig. 9.2 A canonical type II CRISPR-Cas locus in *S. thermophilus* showing the CRISPR-associated (cas) genes (*colored arrows*) in close proximity to a CRISPR array. Each CRISPR array is composed of unique spacers (*colored boxes*), with the most recently acquired spacers at the 5′ end, interspaced between short repeats (*black diamonds*). The cartoon below shows newly acquired spacers at the 5′ end with older, common spacers at the 3′ end; this information can be used to construct a phylogenetic tree of closely related strains (tree shows that the two lower arrays are more closely related to each other than to the top array). See Fig. 9.3 for a more detailed analysis based on data from *Sulfolobus* sp

In one of the first studies of CRISPR spacer diversity in the natural environment, spacer profiles were examined in nearly clonal populations of *Leptospirillum* sp. from acid-mine drainage (AMD) sites (Tyson and Banfield 2008). Comparison of spacers within a single CRISPR locus derived from metagenomic data from two nearby locations in AMD sites demonstrated that spacers near the middle of the locus were population specific, while spacers at the trailer or chronologically 'older' end were shared between the populations. The *Leptospirillum* strains were similar enough that assembly of the metagenomic data, even at the CRISPR array loci, was possible, and enabled the discovery of new spacers at one end of the CRISPR spacer array (Tyson and Banfield 2008). Similar observations have also been reported from several other species where related isolates have been compared, including in the archaeon *Sulfolobus islandicus* (Held et al. 2010), the human pathogen *Yersinia pestis* (Cui et al. 2008), the plant pathogen *Erwinia amylovora* (Rezzonico et al. 2011), and in the extensively studied *Escherichia coli*, in which over a hundred strains are available for analysis (Diez-Villasenor et al. 2010) (Fig. 9.2).

The extent of spacer diversity within a CRISPR array has been used as an estimate of 'activity' level, i.e., how quickly new spacers are being acquired and/or lost. For instance, in *S. thermophilus* strains, which have up to four CRISPR spacer arrays, CRISPR1 (belonging to type II) is by far the most diverse, while CRISPR2 (also belonging to type II) shows very little diversity. Thus, Horvath et al. hypothesized that CRISPR1 is likely the most active and CRISPR2 is suggested to be a degenerate array (Horvath et al. 2008). In studies of other organisms that contain multiple CRISPR spacer arrays, similar results were found, with arrays exhibiting different levels of diversity (Cui et al. 2008; Rezzonico et al. 2011). An alternative explanation for the difference in diversity between loci is that less diverse loci have recently undergone a selective sweep that has purged diversity. This type of detailed analysis is yet to be carried out in many other microorganisms, so it is not yet clear what controls the level of CRISPR activity or whether there are any general rules that define activity levels when a host carries multiple CRISPR arrays.

It has been experimentally demonstrated that spacers are added at the leader end of CRISPR spacer arrays, but arrays can evolve in other ways as well. Comparisons of isolates in *S. thermophilus* (Horvath et al. 2008), *Y. pestis* (Cui et al. 2008), and *S. islandicus* (Held et al. 2010) as well as metagenomic reads in *Leptospirillum* (Tyson and Banfield 2008), and the Global Ocean Survey (GOS) (Sorokin et al. 2010) reveal that spacers can be lost from within CRISPR spacer arrays, possibly by a process involving recombination between CRISPR repeats (Andersson and Banfield 2008). Additionally, duplication of one or more spacers has been discovered in similar CRISPR spacer arrays (Sorokin et al. 2010).

Finally, there is evidence to suggest that the CRISPR-Cas system might play a role which extends beyond its adaptive immunity function against lytic viruses. Some examples are described which suggest that they can protect against integrated mobile elements (Cady et al. 2011) as well as control levels of lysogeny and prophage induction (Edgar and Qimron 2010; Nozawa et al. 2011; Rezzonico et al. 2011). In *S. islandicus*, no spacers match within the genome on which they are found (Held and Whitaker 2009; Held et al. 2010), but there are CRISPR spacers that match integrated elements present in other *S. islandicus* genomes. In *Pseudomonas aeruginosa*, when spacers were compared to virus and plasmid sequences, only matches to integrated elements were found, leading the authors to hypothesize that *P. aeruginosa* CRISPRs play a role in regulating integrated mobile genetic elements rather than conferring resistance to lytic viruses or plasmids. Consistent with this hypothesis, in tested pairs of *P. aeruginosa* and virus, CRISPR spacers did not confer resistance to viruses with identical proto-spacers (Cady et al. 2011). Similarly, in *Erwinia amylovora,* the causal agent of fire blight, there was generally good correlation between the presence of particular spacers and the absence of target plasmids carrying the cognate proto-spacers, indicating their role in preventing plasmid retention. On the other hand, there are also reported cases in which a plasmid is retained even in the presence of spacers that exactly matched sequences on the plasmid. It was hypothesized that this may result from a defective or non-functional CRISPR-Cas system (Rezzonico et al. 2011). *In silico* analysis of all identified CRISPR spacers in 330 sequenced microorganisms identified a low proportion of spacers (0.4 %) that matched host sequences. Although the significance of this finding is unclear, these data along with related observations, might suggest that accidental incorporation of "self DNA" into CRISPRs has a fitness cost (Stern et al. 2010). By examining various strains of the human pathogen, *Streptococcus pyogenes*, Nozawa et al. found that strains which lacked CRISPRs, had significantly more prophages in their genomes and that in some cases this was beneficial to the host, although further experimental demonstration of this is still required (Nozawa et al. 2011).

9.4 Use of CRISPR Spacers to Examine Host Diversity in the Environment

Examining CRISPR-Cas systems in natural environments can shed light on the adaptive defense mechanism as well as reveal selection pressures underlying the evolution of organisms that carry CRISPRs. CRISPR-Cas systems can also help to track relationships between similar species: (1) across short spatial scales or microniches, e.g., different human body locations, (2) across temporal dimensions, (3) across large geographic distances, (4) within closely related strains in a single population, and (5) provide evidence of horizontal gene transfer events. Using CRISPR polymorphism as a measure of host diversity in natural environments and to reconstitute the recent evolutionary history of microorganisms is becoming an increasingly popular method that has been greatly facilitated by next-generation deep sequencing technology coupled with the steady reduction in the cost of sequencing. (See Fig. 9.1 for a general schematic of commonly used methods). In some cases, because they evolve rapidly, CRISPR loci can provide a greater degree of resolution than traditional methods, notably those based on sequencing of rRNA or MLST approaches.

Studies of closely related strains have shown how strain level diversity differs across a single CRISPR spacer array. Because of the polarized addition of spacers to the leader end of an array, the leader end shows more diversity than the trailer end. In the AMD study, carried out in 2008 by Banfield and colleagues, population-specific spacers at the trailer end of the array showed little diversity, while strain-specific spacers at the leader end of the array showed a high amount of diversity (Tyson and Banfield 2008). This analysis is easiest to carry out when comparing strains within a population as in the study of *S. islandicus* (Held et al. 2010) (Figs. 9.2, 9.3), but comparisons of global isolates of *Y. pestis* and *E. amylovora* also show greater leader end diversity than trailer end diversity (Cui et al. 2008; Rezzonico et al. 2011). However, there are exceptions to this rule. For example, in a metagenomic studies of thermophilic cyanobacterial communities in microbial mats, when spacers were compared to one another, 80 % of the ~1,300 spacers identified were found only once, indicating a high level of diversity of *Synechococcus* strains present in the metagenome (Heidelberg et al. 2009). In this case, trends showing spacers preferentially acquired at one end were harder to establish.

CRISPR spacer diversity can be compared across geographic distances as a way to assess microbial diversity. In general, it is observed that when strains from widely separated geographic locations are compared, similarity in spacer content is lower than amongst strains from the same geographic location. In the AMD study, only the population-specific spacers at the trailer end of the array were shared between the metagenomes from the two different locations (Tyson and Banfield 2008). In *S. islandicus*, a comparison of spacers from different geographic locations indicated that the majority of matches occurred between spacers from the same location (Held and Whitaker 2009). Hosts from different locations are therefore distinct from one another in the CRISPR region of their genomes. CRISPR array spacer diversity can also be compared to measures of diversity

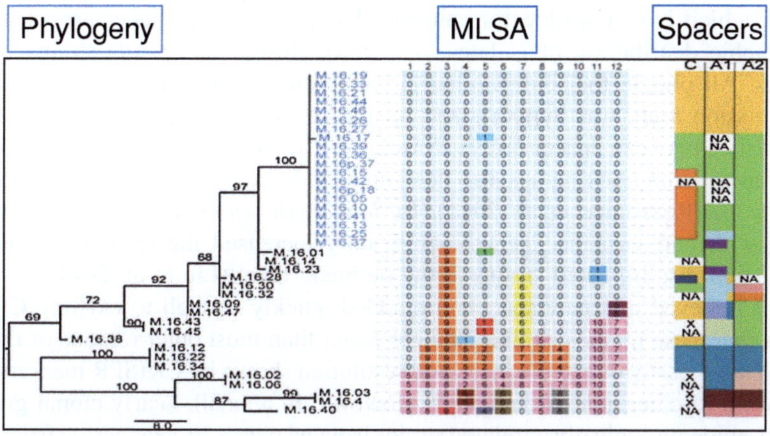

Fig. 9.3 A comparison of core gene phylogeny, MLSA allelic profiles, and CRISPR spacer phylotypes showing that CRISPR loci can differentiate between isolates which phylogenetic profiles and MLSA cannot differentiate. *Left panel* A maximum parsimony phylogeny of a concatenated nucleotide alignment of 12 loci from 39 *S. islandicus* isolates from a single hot spring. Scale bar represents eight nucleotide changes, with bootstrap support from 1,000 replicates. Sequences with nearly identical MLSA profiles are highlighted in *blue*. *Middle panel* The allelic profiles of the 12 MLSA loci show the number of single nucleotide polymorphisms (SNPs) in comparison to strain M.16.19 (*top line*), the background color in each individual cell indicates the allele type for each locus (from 0 to 10). *Right panel* The three colored summary bars to the right of the allelic profiles indicate ancestral groupings of each CRISPR locus by shared spacers. 'X' indicates a CRISPR locus is not present and 'NA' indicates that a locus could not be sequenced. Note that 19 of the isolates (strain names in blue) cannot be distinguished by phylogeny or MLSA but have three distinct CRISPR loci. Modified from Held et al. (2010), with permission

observed in other genomic markers, such as multiple markers from the same genomes in a multi-locus sequence typing (MLST) analysis. In many cases, CRISPR analysis has been found to be a more sensitive approach which is able to resolve closely related genomes, or even clonal populations over time. In *S. islandicus*, MLST indicated the presence of a large clonal group of strains, but analysis of CRISPR sequences showed that the group was actually split into two completely unrelated groups that did not share a single spacer (Held et al. 2010) (Fig. 9.3). A similar analysis was carried out in *E. coli*, where the two different types of CRISPR-Cas systems were mapped onto a phylogenetic tree. When evolutionary distance from the MLST phylogeny was plotted against a measure of spacer relatedness, *E. coli* strains had more spacers in common when they were more closely related phylogenetically (Touchon et al. 2011).

In combination with MLST, CRISPR spacer arrays have also been used to show relatedness between isolates of various pathogens. Relationships of isolates from around the world can easily be determined, and therefore the spread of disease can be tracked. Because analyses of CRISPR spacer arrays provide greater resolution than typical methods of strain typing, the use of CRISPRs is quickly becoming a useful tool in tracking the spread of pathogens. One such pathogen is *Yersinia pestis*, which is responsible for causing plague. Based on spacer arrays and the geographic distribution of isolates, one study clustered *Y. pestis* strains into 12 groups. All plague foci but one had a single cluster of isolates. Additionally, a transmission route was hypothesized for the spread of the *Y. pestis* based on the similarities between spacer arrays in isolates (Cui et al. 2008). Such strain typing and transmission tracking has also been done with the fire blight pathogen *Erwinia amylovora* (Rezzonico et al. 2011). In *Salmonella enterica*, using CRISPR loci along with two virulence genes significantly increased the resolution power of MLST and discriminated strains at the outbreak level (Liu et al. 2011).

It is believed that new spacers are added quickly enough to existing CRISPR arrays that these regions exhibit changes faster than most other regions of the genome. This is partly due to the increased resolution shown by CRISPR markers when compared to other genomic markers. However, in a small, nearly clonal group of *E. coli*, epidemiologically unrelated strains isolated over a 20 year period from vastly different sources had highly similar spacer content, with differences occurring between them due to spacer loss. This was in contrast with analysis of a locus under strong diversifying selection, the O antigen of the *rfb* locus. Each strain had a distinct O antigen allele. Therefore, in these *E. coli* strains, the CRISPRs were not evolving on nearly as fast a timescale as caused by diversifying selection (Touchon et al. 2011). To study the timescale of CRISPR evolution in human subjects, streptococcal strains in human saliva were tracked over the course of a 17 month period. It was found that three spacers were added, and one spacer was deleted from a CRISPR array (Pride et al. 2011). Rho et al. have developed and refined bioinformatics-based methods to identify CRISPR spacers and repeats in the vast array of metagenomic data generated by the Human Microbiome project. This allowed them to track and identify CRISPR spacers across body sites and individuals and provides a detailed view of the dynamics of CRISPR loci (Rho et al. 2012). This study complements the results of Pride et al. and underscores the untapped potential of using CRISPR arrays for the detailed analysis of human microbiome data as well as for environmental samples collected from various locations.

Delaney et al. studied CRISPRs within the pathogenic bacterium *Mycoplasma gallisepticum,* which is associated with poultry birds but has recently moved to infect wild house finches. Although there was high diversity and turnover of CRISPR arrays in poultry strains, after the move of the pathogen to house finches, there was a progressive loss of CRISPR repeat diversity, as well as *cas* genes (Delaney et al. 2012). This study, along with the study of CRISPRs in saliva samples from humans, exemplifies the power of using CRISPR spacers to illuminate the complexity and rapid evolution of host–pathogen/microbe interactions over time.

Analysis of CRISPR loci in microbial strains can also provide strong evidence for horizontal transfer events in various species. In *Y. pestis,* the same CRISPR spacer arrays are found in different parts of the genomes, due to genome rearrangement or possibly horizontal gene transfer of the CRISPR locus (Cui et al. 2008). Comparison of CRISPR-Cas loci in the genomes of two closely related cyanobacterial (*Synechococcus* sp.) isolates from microbial mats showed that one particular isolate contained a CRISPR-Cas locus which closely matched a CRISPR Cas locus (based on CRISPR repeat sequence and Cas operon organization) present in the *Roseiflexus* RS-1 genome (which is also a prominent member of these microbial mats) and in the genome of the Gram-positive thermophile, *Symbiobacterium thermophilum.* The CRISPR-Cas locus was flanked by transposons which could have facilitated DNA transfer of CRISPR-Cas loci across widely separated lineages (Heidelberg et al. 2009). Despite the similarity of the CRISPR locus, none of the spacers at this locus were shared between any of these organisms, which is consistent with the hypothesis that spacer composition reflects the different viruses which attack these hosts. Analysis of the AMD metagenomes demonstrated that the trailer end spacers were shared between the two AMD locations, which could have been the result of horizontal transfer of the locus from one population to the other after these spacers were gained (Tyson and Banfield 2008). In *E. coli*, some very closely related strains were found to have completely different spacer arrays, suggesting transfer of the array from a more distantly related strain (Touchon et al. 2011).

CRISPR loci may also play a role in the horizontal gene transfer and recombination. For instance, in *S. islandicus* and most other species, CRISPR spacers do not have identical matches to sequences in the same genome (Held and Whitaker 2009). In *S. enterica*, spacer sequences matched plasmid and virus sequences from other *S enterica* genomes, indicating that CRISPRs might play a role in controlling horizontal transfer mediated by plasmids and viruses. This is important because plasmids and viruses are associated with virulence in this species, and CRISPRs could be responsible for determining the lifestyle and evolution of *S. enterica* strains (Fricke et al. 2011). This may also be the case in *P. aeruginosa*, as no spacer matches to lytic viruses were found (Cady et al. 2011).

Multiple host strains can have immunity to the same virus through different spacers. In *S. islandicus* isolates from hot springs, comparison of spacer sequences to one another demonstrated that ancestrally unrelated spacers shared similar, partially overlapping sequences, suggesting that there was independent acquisition of spacers from the same virus or plasmid. Additionally, spacers were shown to match different positions on the same viruses. Together, these indicate that isolates from a single hot spring share common, independently acquired immunity. This would result in multiple lineages surviving an epidemic of a single virus and allow a diversity of host strains to survive. CRISPRs were therefore proposed to maintain strain level diversity in this natural system (Held et al. 2010). Independent acquisition of spacers to the same virus was shown experimentally in *S. thermophilus* (Barrangou et al. 2007; Deveau et al. 2008), and in other natural systems such as the AMD *Leptospirillum* metagenomes (Andersson and Banfield 2008).

9.5 Viral Diversity

So far, a search across all available genome and virus-specific databases (e.g. http://www.phantome.org/index.php/philinks/databases/phage) yields very few matches to CRISPR spacers. This serves to highlight a significant deficiency in current genomic databases which have an undersampling of virus and plasmid sequences (the NCBI portal for viruses currently includes 2,756 virus sequences, http://www.ncbi.nlm.nih.gov/genomes/GenomesGroup.cgi?taxid=10239&opt=Virus&sort=genome) The vast majority of viruses, plasmids, and prophages remain uncharacterized (Pignatelli et al. 2008) and furthermore, there is no equivalent to rRNA which has been used extensively to map phylogenetic relatedness. Most viral genomes encode several proteins that cannot be assigned to protein families and are annotated as "unknown" or "hypotheticals". Finally, the high degree of recombination and mutation, diversity in the genomic repertoire, and the lack of marker proteins further complicate the picture (Krupovic et al. 2011). Despite this very limited knowledge of the viral biosphere (Angly et al. 2009; Mokili et al. 2012), the CRISPR-Cas system and the spacer sequences can provide important and unique insights into viral diversity, spatial and temporal patterns of their interactions with host, and the coevolution of host and virus.

For instance, a biogeographic pattern was demonstrated for viruses that infect *S. islandicus* when CRISPR spacers in isolates from three geographically distant locations were compared to all known *Sulfolobus* viruses. Spacers were shown to match at a higher average percent identity to locally present viruses than to foreign viruses. Furthermore, when unique spacers were compared to one another, most of the matches occurred between spacers from genomes from the same geographic location. Since these are spacers independently acquired from the same extra-chromosomal element, this and the higher identity spacer matches to local viruses shows that mobile elements must be present in restricted geographic locations (Held and Whitaker 2009). Similarly, viruses were shown to have a restricted spatial structure in ocean waters. GOS spacers were compared to the entire GOS dataset for similarities to viruses and plasmids and matches between spacers and proto-spacers were more likely to occur when samples were from the same geographic location (Sorokin et al. 2010).

However, not all systems show spatial structuring of viruses. In sludge bioreactors containing *Candidatus* Accumulibacter phosphatis from two geographically distant locations, CRISPR spacers and virus sequences from metagenomic libraries were compared. Two spacers from one location were shown to match virus sequences from the other, which the authors suggest supports geographic dispersal of either the host or virus. However, with only two spacers matching between the two locations, further analysis is needed to support this hypothesis (Kunin et al. 2008). Our knowledge about viral migratory patterns is still very limited but this may be an important area for future study. Examination of metagenomic sequences from cyanobacterial mat communities (Bhaya et al. 2007) indicated that a few spacers were shared between the two different *Synechococcus* strains which were originally isolated from two different temperature locations This suggests that these isolates which are quite closely related, yet occupy different temperature niches, may be infected by common viruses (Breitbart et al. 2004; Heidelberg et al. 2009).

CRISPR spacers have also been used to identify viruses and link host–viral pairs in their natural environment using various different approaches. To date, a few such studies have been carried out so far, and interesting observations have been made. In the AMD metagenomic dataset, reads with sequences matching coexisting spacer sequences were identified. These reads were found to be derived mostly from viruses, though some were from plasmids and transposons. Some of the reads assembled into contigs and scaffolds of partial or complete viral genomes (Andersson and Banfield 2008). Snyder et al. designed a microarray containing CRISPR spacer sequences that were identified in hot spring metagenomes. These microarrays were then used to track viruses (based on their proto-spacer matches to the microarray) and detect changes in the viral community over time and space (Snyder et al. 2010). Another study used a database of CRISPR spacer sequences in order to identify hosts that are infected by a particular viral community. A marine vent virome was queried with all of the spacers (>81,000) from sequenced genomes in NCBI. The spacers that matched the vent community came from a wide range of bacterial and archaeal genomes, and no taxonomic group contained a disproportionate number of matches. Therefore, the authors concluded that marine vent viruses can likely infect a wide range of hosts, although further experimental evidence is required (Anderson et al. 2011a, b).

In yet another approach, about 1,300 spacers from the thermophilic cyanobacterium *Synechococcus* sp., were identified from a microbial mat metagenome and compared to a virome which was derived from the same environment (although from a different year). This yielded several high identity matches between spacers and the viral metagenome (Schoenfeld et al. 2008; Heidelberg et al. 2009). Of these, some proto-spacers were identified within a gene encoding a putative viral lysozyme. A comparison of these proto-spacer sequences to each another revealed the presence of several single nucleotide polymorphisms (SNPs), presumably reflecting individual viral mutants. Interestingly, these SNPs translated into silent or conservative mutations which were unlikely to affect protein function, but could help the virus evade the host CRISPR resistance mechanism (Fig. 9.4). An extension of this analysis with host spacers and viral metagenomes acquired at the same time or from the same environment could provide a unique window into a detailed understanding of how rapidly mutations occur in viral genomes in the context of host acquisition of specific spacers (Davison et al. 2012). The use of next-generation deep sequencing to generate metagenome and viromes is becoming common but often the short reads that are generated are challenging to assemble and annotate, particularly for viral genomes (Rosario et al. 2011; Mokili et al. 2012). To avoid this pitfall, Garcia-Heredia et al. cloned viral DNA, isolated from the low diversity saturated brine environments, into large fosmid clones. This approach combined with the matching of host CRISPR spacers to proto-spacers allowed them to identify several phages that are likely to infect the square archeon *Haloquadratum walsbyi* (Garcia-Heredia et al. 2012).

9.6 Theoretical Models of Host–Virus Interactions

The CRISPR-Cas adaptive defense system has been used to develop and test models that connect resistance mechanisms with the eco-evolutionary dynamics of host-viral systems. In the past, theoretical models have been used in concert with

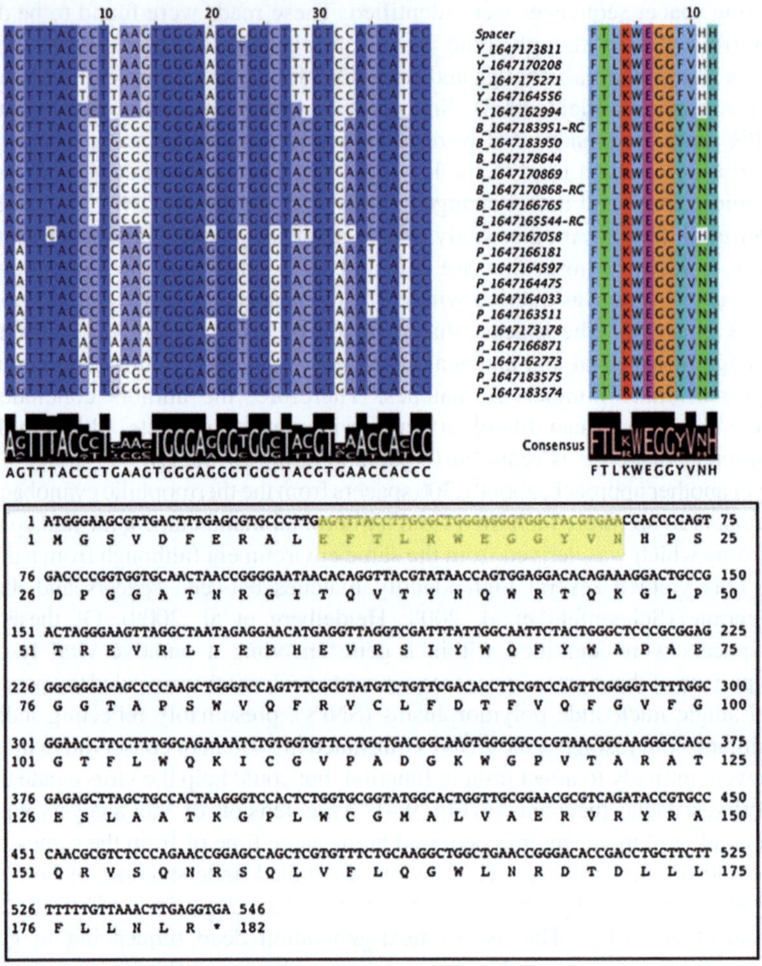

Fig. 9.4 Proto-spacer diversity in a virome for a single CRISPR spacer. *Top left* Alignment of a particular *Synechococcus* sp. spacer (*top line*) with 23 proto-spacers (shown by unique identifiers) identified from a hot spring virome. Varying degrees of nucleotide similarity are shown in *purple*, *light purple*, or *white*. *Top right* The spacer and proto-spacer sequences are translated into amino acids. Note that although there are many differences at the nucleotide level, in many cases the protein sequence has been conserved or there are conservative changes at the amino acid level (e.g. K > R). The consensus sequence of the spacer/proto-spacer and amino acid are shown. *Bottom* The location of the proto-spacer in a viral read that encodes a DUF847 domain (lysozyme-like) protein is shown. Modified from Heidelberg et al. (2009), with permission

experiments to probe the population and evolutionary dynamics of host-viral systems (Abedon 2009). Here, we first introduce the basic elements of host-viral models. Next, we review the limited number of theoretical models that have attempted to integrate molecular principles of CRISPR immune defense with host-viral population and evolutionary dynamics. In doing so, we address, from a theoretical perspective, if and how host–viral coexistence may be mediated by CRISPR immune defense. As we show, every model suggests that coexistence is possible, though (as we discuss) they differ with respect to the mechanisms by which coexistence is maintained, whether CRISPR immunity can drive diversification, and finally, to what extent CRISPR immunity is an active, rather than merely a potential, force in the environment.

Viruses are intracellular parasites and depend on their host for reproduction. Many of the earliest models of host-viral population dynamics considered the case where hosts competed for a limiting resource in a chemostat while subject to viral attack (Levin et al. 1977). These models consider the following basic processes:

nutrient influx	host growth	host lysis by viruses	host washout
$\phi \rightarrow R$	$\epsilon R + N \rightarrow 2N$	$V + N \rightarrow \beta V$	$N \rightarrow \phi$
viral washout	resource washout		
$V \rightarrow \phi$	$R \rightarrow \phi$		

where R are resources, N are hosts, V are viruses, β is the burst size of viruses, ε denotes resource uptake efficiency, and ϕ denotes the absence of a type. This host–viral interaction model is similar to predator–prey models, in that it predicts that viruses and hosts can coexist such that host densities are lower in the presence of viruses (a "top-down" effect) than they would have been when only limited by nutrients (a "bottom-up" effect). Hence, viral-host systems, when modeled strictly ecologically, are predicted to be top-down rather than bottom-up controlled. Chemostats are model experimental apparatuses for some microbial communities, particularly those with continuous flow exchange and relatively little spatial structure. However, natural environments may have greater spatial structure (e.g., as in the case of microbial mats), and greater temporal heterogeneity in resource availability that can lead to switches in physiological states (e.g., nutrient pulsing). Many extensions to these early chemostat models have considered how spatial heterogeneity, physiological differences (e.g., latent period variation or the possibility that viruses integrate into the host genome forming a lysogen), and resource levels influence the number and types of strains that may coexist, e.g., see references in (Abedon 2009). Of course, host and viral lineages do not only change in their relative abundances, they also generate mutants that give rise to new lineages with novel genetic diversity. Hence, models of host-viral evolutionary dynamics have also been proposed to consider how coevolutionary arms races unfold. We illustrate a simplified version of such an arms race process in Fig. 9.5.

A prominent theory to explain the evolutionary dynamics of hosts and viruses is the "kill-the-winner" model in which hosts that reach high abundances are driven to low abundance by viral lysis (Thingstad and Lignell 1997; Thingstad 2000; Winter et al. 2010). Then, the viral strain that infects the previously dominant host strain also drops to low densities, leading to opportunities for the emergence of new and different host–viral pairs. In the kill-the-winner model, a fixed number of host and viral types are assumed to exist in the population. Alternative models (like kill-the-winner) have been proposed that show how host–viral coevolution can lead to

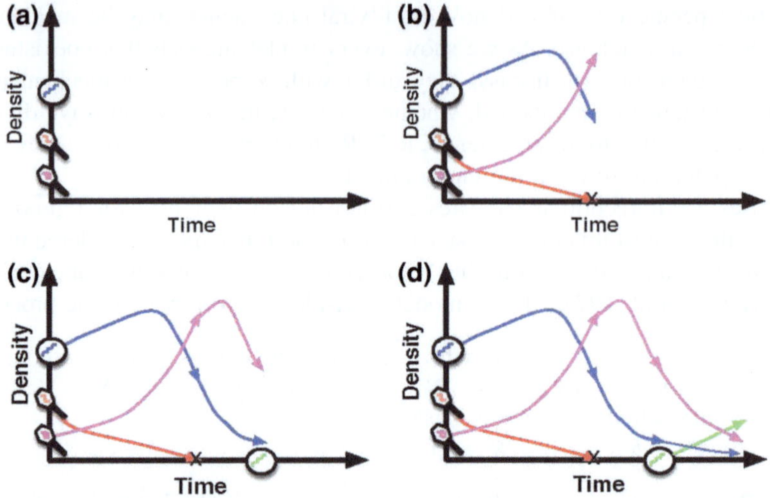

Fig. 9.5 Schematic of the output of a generic evolutionary host-viral model. From *left to right*, the panels depict: **a** initialization of densities of two viruses and one host; **b** population dynamics that lead to the extinction of a strain or; **c** a strain mutation event; **d** continued population dynamics. Together, mechanisms of host–viral interactions can be shown to yield changes in which particular strain may be dominant at a particular moment in time, even as host–viruses coexist over the long-term. Individual models differ with respect to how many strains result during coexistence, whether diversification may result, and to what extent the details of the host–viral defense mechanism alters the coevolutionary arms race

coexistence but also an *increase* in the number of coexisting strains (Weitz et al. 2005). These models extend the chemostat models of Levin et al. (1977) by incorporating evolvable phenotypic traits for hosts and viruses. These traits determine the cross-infection rates of host–viral pairs. Simulations of such coevolutionary models lead to long-term coexistence and diversification of host and viral strains. However, these coevolutionary models lack an explicit genotype–phenotype relationship. Subsequent models have observed that different genotype–phenotype mappings that determine cross-infection rates may play a key role in shaping the extent to which coevolutionary arms races lead to diversification or coexistence of strains (Forde et al. 2008). The CRISPR system provides an opportunity to link molecular mechanisms that determine infection and resistance with ecological rules of interactions. Next, we review a few recently proposed models of this kind.

9.6.1 Immunological Model

He and Deem (2010) explore an ordinary differential equation (ODE) population dynamics model, fashioned similar to established immunological models of viral effects on T-cell function (Nowak and May 2000). Their main goal is to determine whether the interaction and coevolution of hosts and viruses can result in diversity across the CRISPR immunity region over time, where newly added spacers are more diverse than older spacers in the population. The model considers three outcomes of host–viral interaction: host death via virus killing, host survival via immunity coupled with death of the invading virus, or host survival through the acquisition of genetic information from the invading virus which occurs with some low probability. In addition, all hosts can reproduce and all viruses "proliferate" even in the absence of hosts. This exponential growth of viruses is non-standard in ecological models and has the net effect of differentiating viruses in terms of death rates rather than in terms of differential reproduction. Note that in the absence of hosts, this assumption leads to exponential growth of viruses. This unrealistic outcome will not occur so long as hosts are present with spacers matching that of the virus.

In simulations of this model, the hosts evolve through acquisition of viral material at the leader end of the CRISPR locus. Thus, the spacer array is temporally ordered from leader end to trailer end, with recently acquired spacers at the leader end. Additionally, during viral replication there is a small probability that the viral sequence will mutate, introducing novel virus strains into the population. He and Deem (2010) find stable accumulation of a large number of diverse host strains. During comparison of the diversity of spacers by position using an entropy-based metric, they find that newer, leader end spacers are more diverse than older, trailer end spacers. From this result, they hypothesize that the survival of hosts is determined by selection for the spacer arrays that contain spacers that are most effective against the current viral population, which are present as older spacers in the array. They also conclude that spacers that match dominant viruses are more frequently found in CRISPR arrays.

Greater leader end diversity has previously been shown in several natural systems as discussed earlier (Cui et al. 2008; Tyson and Banfield 2008; Held et al. 2010; Rezzonico et al. 2011). To further substantiate this model, one could attempt to observe CRISPR spacer array evolution in the environment or in the laboratory. In order to observe the evolution of CRISPR spacer array diversity in nature, hosts could be tracked temporally, using PCR technology or by analysis of multiple metagenomic datasets. If only a subset of the leader end spacers at the first time point are present in later time points, then this would be consistent with the model proposed by He and Deem (2010). Host spacers along a time-line could be compared to contemporary and past viral sequences in order to test the hypothesis that the less diverse spacers farther from the leader end are those that match dominant viruses in the environment. There are studies that have shown CRISPR spacer matches between hosts and viruses from the same time and location such as in the AMD system (Tyson and Banfield 2008). However, establishing whether dominant viruses in a metagenome are able to infect dominant hosts in a metagenome is not trivial, so a well studied natural system with a low species diversity would be ideal for such an endeavor. This could also be simplified by studying host–viral coevolution in an experimental system in the laboratory, such as in a chemostat.

9.6.2 Ecological Model

Levin (2010) addresses the question of how and why CRISPR systems are maintained in bacterial populations. Levin sets out to examine whether CRISPRs are important factors in defending against invading genetic material by comparing bacterial populations having CRISPR resistance with populations having other forms of resistance. Levin utilizes an ODE population dynamics model framework where types of bacteria, specifically *E. coli*, encounter invading genetic material, both viral and plasmid. The paper presents two models: one model involves comparing CRISPR immunity with envelope resistance during virus invasion and the other involves the comparison of CRISPR immunity in the presence of a deleterious conjugative plasmid. In these models, each host type with some type of potential resistance, CRISPR or otherwise, incurs a fitness cost through decreased growth rate. Levin's models assume that all viruses are identical and hosts come in limited strain types: wild type, immune, CRISPR non-immune, and CRISPR-immune. Hence, this model does not explicitly consider variation in spacer states that would confer different forms of CRISPR-based immunity. Further, although host strains can gain and lose resistance, there is no viral- or plasmid- specific incorporation by host strains. The evolution of viruses or plasmids is not part of the main model (although some possible evolutionary dynamics extensions are discussed). However, what the model lacks in evolutionary complexity it makes up for in terms of ecological complexity. For example, Levin explores the effect of high multiplicity of infection (MOI) on the ability of CRISPR immunity to be maintained.

Levin finds that a wide range of parameters lead to scenarios where CRISPR-immune bacteria can be maintained in populations and can also invade a stable assemblage of bacteria lacking CRISPRs (Levin 2010). The main requirement for the establishment of CRISPR strains in populations that lack CRISPRs is the sustained presence of viral populations that allows the immunity property of CRISPR strains to overcome the reduced growth rate. Although Levin finds parameter sets that allow invasion in both the virus and plasmid models, the plasmid model has a much narrower range of parameters that allows the maintenance of the CRISPR system. Given high MOI, Levin finds that the CRISPR system cannot keep up and is removed from the population. Altogether, Levin cautions that there are certainly scenarios where other types of bacterial resistance could dominate and force CRISPR strains to extinction. This model emphasizes the need to examine the ecological and molecular factors that may determine when CRISPRs are active and important driving forces in natural populations.

Examination of the current CRISPR databases indicates that the CRISPR system is not present in all microbes; it is estimated to be present in ~40 % of bacterial genomes and ~90 % of archaeal genomes (Bhaya et al. 2011). Despite potential sampling biases, the reasons why CRISPR appears in some, but not all, microbes is likely connected to the fitness advantage provided by the CRISPR system relative to other forms of virus resistance present in each organism. In order to test this hypothesis, the "cost" of CRISPR resistance would need to be tested in the laboratory. Then, this can be compared to the "cost" and efficacy of other types of resistance. It may be possible to set up experimental systems to test laboratory strains with different forms of resistance. However, in a natural environment, there are multiple viruses that are constantly evolving as well as multiple types of CRISPR diversity. Therefore, based on the ecological model proposed above, hosts that contain a CRISPR system may fluctuate in frequency with those that lack a CRISPR system. Alternatively, the CRISPR system may be stable or be lost from the host depending on the fitness provided by CRISPR immunity. In order to test such a hypothesis in natural populations, one would need to find a system with CRISPRs and then compare the frequency of CRISPR-based resistance to that of other types of resistance. As natural systems are explored in the future, some of these hypotheses may be tested and this would add valuable new insights into the role and importance of different host resistance mechanisms.

9.6.3 Spatial Model

Haerter et al. (2011), Haerter and Sneppen (2012) introduce a spatial model of host and viral interaction incorporating CRISPR immunity. Their aim is to understand how CRISPRs affect coexistence of hosts and viruses. They explore host and viral populations as they spread on a two-dimensional lattice with periodic boundary conditions. The use of two-dimensional lattices is common in physically based models of living systems and is an important first-step toward representing complex spatial heterogeneity in the environment. In this model, bacterial growth is limited only by available space and not carbon or other limiting nutrients. Virus spread and growth is dependent upon the presence of susceptible bacteria and the "aggressiveness", or ability to infect, of the virus. The speed of spread of both bacteria and viruses is varied in the model and significant consideration is given to the extremes: well-mixed systems and slowly diffusing systems. In addition to the explicit treatment of spatial dynamics, Haerter et al. (2011) extend the model to include dynamic immunity where hosts can evolve via acquisition of resistance to virus types in the system. However, they do not consider evolution of the virus. As an additional feature, they consider CRISPR immunity to be imperfect, so viruses that match a host occasionally survive, reproduce, and kill the host (Haerter et al. 2011).

In the spatial model, Haerter et al. (2011) find that bacterial growth is dependent on the surface area of the growing bacterial cluster. In comparison, virus growth is dependent upon the intersection of the bacterial and virus populations in addition to the surface area of the virus population. In the model with dynamic immunity, Haerter et al. (2011) observe that multiple viruses can coexist as long as a single bacterial strain cannot acquire resistance to all virus strains simultaneously. The amount of resistance a bacterial strain can acquire depends upon the number of CRISPR immunity positions or spacers (a parameter) that are permitted by the model. They demonstrate for bacteria with an arbitrary number of immunity positions (spacers) that coexistence is possible as long as there are more virus types than spacers. Additionally, they find the resultant diversity of host strains present depends upon the speed of spread of populations, the number of immunity positions (or spacers), and in some cases the total number of virus strains present. Thus, their model shows that imperfect CRISPR immunity on a defined spatial structure leads to robust host–viral coexistence when there are more virus strains present than host immunity positions. The mechanism is similar to that of "kill-the-winner" dynamics, however, here the dynamics are realized on spatial mosaics. A follow-up paper by Haerter and Sneppen (2012) extends this earlier model to understand why spacer arrays, particularly in *S. thermophilus,* have an "intermediate" numbers of spacers, i.e., usually dozens, though bioinformatic analysis reveals that the number of spacers can vary significantly between species. This extension presumes that the fitness cost to growth is proportional to the number of spacers; establishing the fitness costs of spacer addition is a key open question. Due to spatial heterogeniety, bacteria only experience a subset of phages in the environment, hence, long spacer arrays do not emerge given that bacteria do not interact with all phage types as they would in a chemostat-based model. Moreover, different spacer compositions within bacteria can emerge due to the local structure of interactions with distinct subpopulations of phages.

Host and virus coexistence can be investigated in natural environments by examining the diversity of viruses that infect a particular host. Andersson et al. have demonstrated that assembling viruses from metagenomes is possible although this may be a challenge in many complex environments (Andersson and Banfield 2008). In a low complexity environment where host–viral pairs are known, deep sequencing of a particular virus might reveal the level of diversity (Davison et al. 2012). If multiple viruses are known to infect a host, then spacer diversity would be predicted to greatly increase in magnitude. A key test of this model would be to evaluate to what extent spatial organization can provide local refuges for hosts to escape viral infection, and likewise, provide a means for propagating fronts of immune hosts. Spatially induced coexistence has been studied in other settings with complex microbial interactions, e.g., in the case of bacteria interacting via colicins (Kerr et al. 2002). Haerter and colleagues suggest comparisons of experimental evolution in well-mixed versus spatially structured environments would be valuable and this is likely to meet with general agreement.

9.6.4 Coevolutionary Ecological Model

Childs et al. (2012) present a stochastically implemented ODE model incorporating both the ecology and evolution of host and viral strains. The objective of this model is to characterize how hosts with CRISPR immune defense evolve in the face of viral infection, and correspondingly, how viruses evolve to avoid host immune defense. Host and viral strains interact through an ODE population dynamics model mediated by CRISPR immunity. When a host encounters a virus, one of the following events occurs: the host is killed via viral release or the virus is eliminated and the host potentially acquires CRISPR immunity. Survival of a host is dependent upon whether it has CRISPR immunity to the invading virus. Immunity is determined based on whether the host has a spacer that matches a proto-spacer of the invading virus. Similar to the model by Haerter et al. (2011) bacterial CRISPR immunity is imperfect and fails with a small probability. During the infection by a virus, the host evolves by acquisition of viral material through the incorporation of a new spacer at the leader end of the spacer array. The model assumes that virus strains may evolve into novel strains via mutation at a proto-spacer. This assumption is a simplification of possible genomic changes in viruses that affect CRISPR function, including changes of PAMs.

Childs et al. (2012) find that low diversity assemblages of both hosts and viruses coevolve to form highly diverse communities. Although this model leads to the maintenance many coexisting strains, individual strains do not permanently coexist but rather emerge, grow, and eventually are replaced. The observed strain dynamics are often similar to (nearly) selective sweeps which rarely reach completion before new strains emerge. Due to the diverse genomic state of the virus as well as the host, the model demonstrates the phenomenon where multiple phenotypically identical but genotypically different strains emerge and co-exist together. Additionally, previously dominant strains sometimes regain dominance as the complex fitness landscape of hosts and viruses evolves. Similar to He and Deem (2010), Childs et al. (2012) observe heightened diversity in the leader end spacers. Their model also leads to the prediction that the leader end spacers match a larger portion of the current viral community and thus are most important to the hosts' CRISPR immunity. This property is consistent with a few experimental observations of preferential deletion of trailer-end spacers (Tyson and Banfield 2008), although more wide-scale analyses are warranted. The extent of diversification and the ways in which strains diversify can then be evaluated as a function of governing ecological and molecular parameters.

To test the predictions of this model, i.e., whether hosts arise due to dominance as (1) a single clone, (2) as coalitions of strains, or (3) both, CRISPR spacer arrays in host populations will have to be temporally tracked. Analysis of metagenomic sequences may allow diversity and the ebb and flow of various host CRISPR spacer arrays to be tracked in a natural environment. If single clone dominance is seen at periodic time points with multiple strains occurring together at others, then this would be consistent with the complex strain dynamics predicted by the model by Childs et al. (2012). In addition, the tenet that leader end spacers have the most matches to the current viruses in the community can be tested by having viral metagenome data acquired from the same place at the same time. This would allow host spacer sequences to be matched to viral proto-spacers and the number of matches for each spacer from leader end to trailer end to be counted. If more matches occur between contemporary host CRISPR spacers at the leader ends of the arrays and viral metagenomic sequences than between those from different time points, then this would suggest that the most recently acquired spacers are indeed the most relevant spacers at any given point in time.

Table 9.1 Summary of features of five alternative CRISPR models of the ecological and coevolutionary dynamics of hosts and viruses

	He and Deem	Levin	Haerter et al.	Childs et al.	Weinberger et al.
Host genomic states (spacers)	Yes	No	Yes	Yes	Yes
Viral genomic states (proto-spacers)	No	No	No	Yes	Yes
Evolution of hosts	Yes	No	Yes	Yes	Yes
Evolution of viruses	Yes	No	No	Yes	Yes
CRISPR immunity	Yes	Yes	Yes	Yes	Yes
Other bacterial immunity	No	Yes	No	No	No
Population dynamic ecology (continuous)	Yes	Yes	No	Yes	No
Population dynamic ecology (discrete)	No	No	Yes	No	Yes
Spatial ecology	No	No	Yes	No	No

Notably, the only component that all models have in common is the inclusion of a submodel to represent CRISPR immunity!

9.6.5 Synthesis of Models

The mathematical models described in the current literature, examine how CRISPR immunity leads to coexistence of diverse host and viral strains. They all assume that the presence of CRISPR immunity, whether explicitly or implicitly modeled, leads to the survival of the host and death of the virus with high probability. In addition, all models permit the population abundances of strains to change with time. Further, all CRISPR immunity is assumed to be equivalent; models assume there is no better or worse viral DNA region to acquire—how valid this is remains to be determined. The way the host and viral interactions are modeled, however, differs. Indeed, the models differ in their inclusion of molecular, physiological, ecological, and evolutionary detail and therefore are able to make different predictions that may be relevant to different types of host-viral systems where CRISPRs are present (see Table 9.1). In addition, Weinberger et al. (2012) have also developed a coevolutionary model of viral–host interactions by extending classical population genetics frameworks to confront the particular challenges of CRISPR immunity. The conceptual innovation of this coevolutionary model is the designation of a population genetics "iteration" in terms of interaction events between hosts and viruses. This model represents a novel approach, complementary to those outlined in this section. Regardless of the implementation details, each model concludes that CRISPR immunity *can be* the driver of such diversity although the means by which the diversity emerges and the specific types of evolutionary dynamics that yield that diversity differ between models.

These models have a number of caveats that should be kept in mind. Specific parameter choices of the models affect the molecular interactions of host and viruses and thus the ability to reach coexistence. All the models which include evolution assume that hosts acquire sequences and incorporate them at the leader end of the spacer array, despite some experimental exceptions to this assumption. In addition, the transfer of CRISPR loci among hosts through horizontal gene transfer or recombination has not been included although it has frequently been observed (Chakraborty et al. 2010; Shah and Garrett 2011). Hence, models may help to address the question of the importance of this aspect of the CRISPR-Cas system. Additionally, viral mutations are all modeled as point mutations; genome level rearrangements are not considered. This raises the question of the importance of viral rearrangements as a means to evade CRISPR-Cas immunity. Finally, only the Levin model considers other types of bacterial immunity and how it will affect the function of CRISPR immunity. Hence, each model may be appropriate for different objectives. We believe it is premature to predict which of the current models will be most useful in guiding and/or interpreting experiments and field studies. In fact, the increasing recognition of different types of CRISPR-Cas systems (Makarova et al. 2011) suggests that multiple types of models will be of use in advancing our understanding of this novel form of coevolutionary dynamics.

9.7 Conclusions and Future Prospects

The current literature clearly shows that analyses of CRISPR-Cas systems can reveal important features about both host and virus communities in natural environments. So far, they have been used to (1) determine whether viruses are globally dispersed or are present in restricted geographic locations, (2) identify and assemble viral reads from metagenomes, (3) track evolving mutations in viral populations, and (4) predict which hosts might coexist with a given viral community. Since little is known about host–viral interactions in natural environments, CRISPRs may turn out to be an essential means to shed light on this important driver of microbial diversity and population trajectories. To facilitate such studies it will be critical to acquire extensive metagenomic and virome data from a variety of environmental niches, across both temporal and spatial scales. This will provide fascinating new information about host–viral interactions and about the unexplored diversity that we can use to explore and model the dynamics of host–viral interactions. Environments which harbor microbial communities of low to moderate complexity may be ideal for such studies. As mentioned, the lack of well-annotated and extensive viral genome databases is also a major obstacle to progress.

To understand the dynamic nature of the CRISPR-Cas system, we advocate for the continued use of models that may (1) stimulate experimental tests and develop model biological systems, in which specific parameters can be measured to validate and test hypotheses, (2) indicate possible shortcomings in understanding of mechanisms thought to be sufficient to describe host–viral coevolution, or (3) help to engage with and make sense of complex data. Given that CRISPR systems are quite diverse, modeling will likely be most useful when assumptions are coupled with specific ecological conditions of interest. The increasing availability of ecological data detailing host–viral coevolution associated with CRISPR-based immunity suggests the potential for greater collaborative work at the interface of theory and data.

A common theme amongst the models currently available is the focus on population dynamics. Therefore, in order to test these models, it is critical to have access to CRISPR population data over different timescales. Low complexity environments would be very useful for investigating such population dynamics, since complex environments still present sampling challenges. Metagenomic data is also useful for investigating all of the models and may be tailored to probe specific questions. So far, very few ecological studies have examined CRISPRs dynamics in populations over short or long timescales. In the future, integrating modeling with data acquired from natural environments will be key to understanding how complex host–viral interactions evolve, and would also further our understanding of the impact of the CRISPR-Cas system on natural populations.

Acknowledgments NLH and RJW acknowledge support from NSF DEB-0816885. DB and MD acknowledge support from the NSF, The Carnegie Institution of Science and Stanford University. JSW acknowledges the support of a grant from the James S. McDonnell Foundation. JSW holds a Career Award at the Scientific Interface from the Burroughs Wellcome Fund.

References

Abedon ST (2009) Phage evolution and ecology. Adv Appl Microbiol 67:1–45

Anderson RE, Brazelton WJ et al (2011a) Is the genetic landscape of the deep subsurface biosphere affected by viruses? Front Microbiol 2:219

Anderson RE, Brazelton WJ et al. (2011b) Using CRISPRs as a metagenomic tool to identify microbial hosts of a diffuse flow hydrothermal vent viral assemblage. FEMS Microbiol Ecol 77(1):120–133

Andersson AF, Banfield JF (2008) Virus population dynamics and acquired virus resistance in natural microbial communities. Science 320(5879):1047–1050

Angly FE, Willner D et al (2009) The GAAS metagenomic tool and its estimations of viral and microbial average genome size in four major biomes. PLoS Comput Biol 5(12):e1000593

Barrangou R, Fremaux C et al (2007) CRISPR provides acquired resistance against viruses in prokaryotes. Science 315(5819):1709–1712

Berg Miller ME, Yeoman CJ et al. (2012) Phage-bacteria relationships and CRISPR elements revealed by a metagenomic survey of the rumen microbiome. Environ Microbiol 14:207–227

Bhaya D, Davison M et al (2011) CRISPR-Cas systems in bacteria and archaea: versatile small RNAs for adaptive defense and regulation. Annu Rev Genet 45:273–297

Bhaya D, Grossman AR et al (2007) Population level functional diversity in a microbial community revealed by comparative genomic and metagenomic analyses. ISME J 1(8):703–713

Bikard D, Marraffini LA (2012) Innate and adaptive immunity in bacteria: mechanisms of programmed genetic variation to fight bacteriophages. Curr Opin Immunol 24:15–20

Blainey PC, Mosier AC et al (2011) Genome of a low-salinity ammonia-oxidizing archaeon determined by single-cell and metagenomic analysis. PLoS One 6(2):e16626

Breitbart M, Rohwer F (2005) Here a virus, there a virus, everywhere the same virus? Trends Microbiol 13(6):278–284

Breitbart M, Wegley L et al (2004) Phage community dynamics in hot springs. Appl Environ Microbiol 70(3):1633–1640

Brouns SJ, Jore MM et al (2008) Small CRISPR RNAs guide antiviral defense in prokaryotes. Science 321(5891):960–964

Cady KC, White AS et al (2011) Prevalence, conservation and functional analysis of *Yersinia* and *Escherichia* CRISPR regions in clinical pseudomonas aeruginosa isolates. Microbiology 157(Pt 2):430–437

Chakraborty S, Snijders AP et al (2010) Comparative network clustering of direct repeats (DRs) and cas genes confirms the possibility of the horizontal transfer of CRISPR locus among bacteria. Mol Phylogenet Evol 56(3):878–887

Childs LM, Held NL et al. (2012) Multi-scale model of CRISPR-induced coevolutionary dynamics: diversification at the interface of Lamarck and Darwin. Evolution 66(7):2015–2029

Cui Y, Li Y et al (2008) Insight into microevolution of *Yersinia pestis* by clustered regularly interspaced short palindromic repeats. PLoS One 3(7):e2652

Davison M, Treangen TJ et al. (2012) Analysis of virome diversity in mixed microbial communities using CRISPR spacers. In preparation

Delaney NF, Balenger S et al (2012) Ultrafast evolution and loss of CRISPRs following a host shift in a novel wildlife pathogen, *Mycoplasma gallisepticum.* PLoS Genet 8(2):e1002511

Deveau H, Barrangou R et al (2008) Phage response to CRISPR-encoded resistance in *Streptococcus thermophilus*. J Bacteriol 190(4):1390–1400

Deveau H, Garneau JE et al (2010) CRISPR/Cas system and its role in phage-bacteria interactions. Annu Rev Microbiol 64:475–493

Diez-Villasenor C, Almendros C et al (2010) Diversity of CRISPR loci in *Escherichia coli*. Microbiology 156(Pt 5):1351–1361

Dinsdale EA, Edwards RA et al (2008) Functional metagenomic profiling of nine biomes. Nature 452(7187):629–632

Edgar R, Qimron U (2010) The *Escherichia coli* CRISPR system protects from lambda lysogenization, lysogens, and prophage induction. J Bacteriol 192(23):6291–6294

Forde SE, Beardmore RE et al (2008) Understanding the limits to generalizability of experimental evolutionary models. Nature 455(7210):220–223

Fricke WF, Mammel MK et al (2011) Comparative genomics of 28 *Salmonella enterica* isolates: evidence for CRISPR-mediated adaptive sublineage evolution. J Bacteriol 193(14):3556–3568

Garcia-Heredia I, Martin-Cuadrado AB et al (2012) Reconstructing viral genomes from the environment using fosmid clones: the case of haloviruses. PLoS One 7(3):e33802

Garneau JE, Dupuis ME et al (2010) The CRISPR/Cas bacterial immune system cleaves bacteriophage and plasmid DNA. Nature 468(7320):67–71

Gilbert JA, Dupont CL (2011) Microbial metagenomics: beyond the genome. Ann Rev Mar Sci 3(1):347–371

Godde JS, Bickerton A (2006) The repetitive DNA elements called CRISPRs and their associated genes: evidence of horizontal transfer among prokaryotes. J Mol Evol 62(6):718–729

Grissa I, Bouchon P et al (2008) On-line resources for bacterial micro-evolution studies using MLVA or CRISPR typing. Biochimie 90(4):660–668

Grissa I, Vergnaud G et al. (2007) CRISPRFinder: a web tool to identify clustered regularly interspaced short palindromic repeats. Nucleic Acids Res 35(Web Server issue): W52-57

Haerter JO, Sneppen K (2012) Spatial Structure and lamarckian adaptation explain extreme genetic diversity at CRISPR locus. MBio 3(4):e00126-12

Haerter JO, Trusina A et al (2011) Targeted bacterial immunity buffers phage diversity. J Virol 85(20):10554–10560

Haft DH, Selengut J et al (2005) A guild of 45 CRISPR-associated (Cas) protein families and multiple CRISPR/Cas subtypes exist in prokaryotic genomes. PLoS Comput Biol 1(6):e60

Hale CR, Zhao P et al (2009) RNA-guided RNA cleavage by a CRISPR RNA-Cas protein complex. Cell 139(5):945–956

Hambly E, Suttle CA (2005) The viriosphere, diversity, and genetic exchange within phage communities. Curr Opin Microbiol 8(4):444–450

He J, Deem MW (2010) Heterogeneous diversity of spacers within CRISPR (clustered regularly interspaced short palindromic repeats). Phys Rev Lett 105(12):128102

Heidelberg JF, Nelson WC et al (2009) Germ warfare in a microbial mat community: CRISPRs provide insights into the co-evolution of host and viral genomes. PLoS One 4(1):e4169

Held NL, Herrera A et al (2010) CRISPR associated diversity within a population of Sulfolobus islandicus. PLoS One 5(9):e12988

Held NL, Whitaker RJ (2009) Viral biogeography revealed by signatures in *Sulfolobus islandicus* genomes. Environ Microbiol 11(2):457–466

Horvath P, Coute-Monvoisin AC et al (2009) Comparative analysis of CRISPR loci in lactic acid bacteria genomes. Int J Food Microbiol 131(1):62–70

Horvath P, Romero DA et al (2008) Diversity, activity, and evolution of CRISPR loci in Streptococcus thermophilus. J Bacteriol 190(4):1401–1412

Hyman P, Abedon ST (2010) Bacteriophage host range and bacterial resistance. Adv Appl Microbiol 70:217–248

Kerr B, Riley MA et al (2002) Local dispersal promotes biodiversity in a real-life game of rock-paper-scissors. Nature 418(6894):171–174

Krupovic M, Prangishvili D et al (2011) Genomics of bacterial and archaeal viruses: dynamics within the prokaryotic virosphere. Microbiol Mol Biol Rev 75(4):610–635

Kunin V, He S et al (2008) A bacterial metapopulation adapts locally to phage predation despite global dispersal. Genome Res 18(2):293–297

Kunin V, Sorek R et al (2007) Evolutionary conservation of sequence and secondary structures in CRISPR repeats. Genome Biol 8(4):R61

Labrie SJ, Samson JE et al (2010) Bacteriophage resistance mechanisms. Nat Rev Microbiol 8(5):317–327

Levin BR (2010) Nasty viruses, costly plasmids, population dynamics, and the conditions for establishing and maintaining CRISPR-mediated adaptive immunity in bacteria. PLoS Genet 6(10):e1001171

Levin BR, Stewart FM et al (1977) Resource-limited growth, competition and predation: a model and experimental studies with bacteria and bacteriophage. Am Nat 111:3–24

Liu F, Barrangou R et al (2011) Novel virulence gene and clustered regularly interspaced short palindromic repeat (CRISPR) multilocus sequence typing scheme for subtyping of the major serovars of *Salmonella* enterica subsp. enterica. Appl Environ Microbiol 77(6):1946–1956

Lopez-Bueno A, Tamames J et al (2009) High diversity of the viral community from an Antarctic lake. Science 326(5954):858–861

Makarova KS, Haft DH et al (2011) Evolution and classification of the CRISPR-Cas systems. Nat Rev Microbiol 9(6):467–477

Marraffini LA, Sontheimer EJ (2008) CRISPR interference limits horizontal gene transfer in staphylococci by targeting DNA. Science 322(5909):1843–1845

Mojica FJ, Diez-Villasenor C et al (2009) Short motif sequences determine the targets of the prokaryotic CRISPR defence system. Microbiology 155(Pt 3):733–740

Mokili JL, Rohwer F et al (2012) Metagenomics and future perspectives in virus discovery. Curr Opin Virol 2(1):63–77

Nowak MA, May RM (2000) Virus Dynamics. Oxford University Press, Oxford

Nozawa T, Furukawa N et al (2011) CRISPR inhibition of prophage acquisition in *Streptococcus pyogenes*. PLoS ONE 6(5):e19543

Pignatelli M, Aparicio G et al (2008) Metagenomics reveals our incomplete knowledge of global diversity. Bioinformatics 24(18):2124–2125

Portillo MC, Gonzalez JM (2009) CRISPR elements in the thermococcales: evidence for associated horizontal gene transfer in *Pyrococcus furiosus*. J Appl Genet 50(4):421–430

Pride DT, Sun CL et al (2011) Analysis of streptococcal CRISPRs from human saliva reveals substantial sequence diversity within and between subjects over time. Genome Res 21(1):126–136

Reyes A, Haynes M et al (2010) Viruses in the faecal microbiota of monozygotic twins and their mothers. Nature 466(7304):334–338

Rezzonico F, Smits TH et al (2011) Diversity, evolution, and functionality of clustered regularly interspaced short palindromic repeat (CRISPR) regions in the fire blight pathogen *Erwinia amylovora*. Appl Environ Microbiol 77(11):3819–3829

Rho M, Wu YW et al (2012) Diverse CRISPRs evolving in human microbiomes. PLoS Genet 8(6):e1002441

Riesenfeld CS, Schloss PD et al (2004) Metagenomics: genomic analysis of microbial communities. Annu Rev Genet 38(1):525–552

Rodriguez-Brito B, Li L et al (2010) Viral and microbial community dynamics in four aquatic environments. ISME J 4(6):739–751

Rodriguez-Valera F, Martin-Cuadrado AB et al (2009) Explaining microbial population genomics through phage predation. Nat Rev Microbiol 7(11):828–836

Rohwer F (2003) Global phage diversity. Cell 113(2):141

Rohwer F, Thurber RV (2009) Viruses manipulate the marine environment. Nature 459(7244):207–212

Rosario K, Marinov M et al (2011) Dragonfly cyclovirus, a novel single-stranded DNA virus discovered in dragonflies (Odonata: Anisoptera). J Gen Virol 92(Pt 6):1302–1308

Rousseau C, Gonnet M et al (2009) CRISPI: a CRISPR interactive database. Bioinformatics 25(24):3317–3318

Schoenfeld T, Patterson M et al (2008) Assembly of viral metagenomes from yellowstone hot springs. Appl Environ Microbiol 74(13):4164–4174

Shah SA, Garrett RA (2011) CRISPR/Cas and Cmr modules, mobility and evolution of adaptive immune systems. Res Microbiol 162(1):27–38

Singh J, Behal A et al (2009) Metagenomics: concept, methodology, ecological inference and recent advances. Biotechnol J 4(4):480–494

Sinkunas T, Gasiunas G et al (2011) Cas3 is a single-stranded DNA nuclease and ATP-dependent helicase in the CRISPR/Cas immune system. EMBO J 30:1335–1342

Snyder JC, Bateson MM et al (2010) Use of cellular CRISPR (clusters of regularly interspaced short palindromic repeats) spacer-based microarrays for detection of viruses in environmental samples. Appl Environ Microbiol 76(21):7251–7258

Sorokin VA, Gelfand MS et al (2010) Evolutionary dynamics of clustered irregularly interspaced short palindromic repeat systems in the ocean metagenome. Appl Environ Microbiol 76(7):2136–2144

Stern A, Keren L et al (2010) Self-targeting by CRISPR: gene regulation or autoimmunity? Trends Genet 26(8):335–340

Stern A, Mick E et al. (2012) CRISPR targeting reveals a reservoir of common phages associated with the human gut microbiome. Genome Res 22(10):1985–1994

Suttle CA (2005) Viruses in the sea. Nature 437(7057):356–361

Tadmor AD, Ottesen EA et al (2011) Probing individual environmental bacteria for viruses by using microfluidic digital PCR. Science 333(6038):58–62

Thingstad TF (2000) Elements of a theory for the mechanisms controlling abundance, diversity, and biogeochemical role of lytic bacterial viruses in aquatic systems. Limnol Oceanogr 45(6):1320–1328

Thingstad TF, Lignell R (1997) Theoretical models for the control of bacterial growth rate, abundance, diversity and carbon demand. Aquat Microb Ecol 13(1):19–27

Touchon M, Charpentier S et al (2011) CRISPR distribution within the *Escherichia coli* species is not suggestive of immune-associated diversifying selection. J Bacteriol 193(10):2460–2467

Tyson GW, Banfield JF (2008) Rapidly evolving CRISPRs implicated in acquired resistance of microorganisms to viruses. Environ Microbiol 10(1):200–207

Weinberger A, Sun C et al. (2012) Persisting viral sequences shape CRISPR-based immunity. PLoS Comput Biol 8(4):e1002475

Weitz JS, Hartman H et al (2005) Coevolutionary arms races between bacteria and bacteriophage. Proc Natl Acad Sci USA 102(27):9535–9540

Willner D, Thurber RV et al (2009) Metagenomic signatures of 86 microbial and viral metagenomes. Environ Microbiol 11(7):1752–1766

Winter C, Bouvier T et al (2010) Trade-offs between competition and defense specialists among unicellular planktonic organisms: the "killing the winner" hypothesis revisited. Microbiol Mol Biol Rev 74(1):42–57

Yilmaz S, Singh AK (2012) Single cell genome sequencing. Curr Opin Biotechnol 23(3):437–443

Chapter 10
Roles of CRISPR in Regulation of Physiological Processes

Gil Amitai and Rotem Sorek

Abstract It is now well established, following numerous independent studies, that the CRISPR-Cas system is an adaptive immune system widespread in bacteria and archaea. The primary role of CRISPR-Cas in protection against foreign DNA is nowadays undisputed. However, there is also strong evidence suggesting that, at least in some organisms, CRISPR-Cas systems have adapted to take nonimmune-related regulatory roles that are hardwired into the core regulatory programs of bacterial physiology. In at least two cases, CRISPR was shown to regulate bacterial social behavior; in other cases, CRISPRs were suggested to directly regulate endogenous gene expression. Curious cases of CRISPR-derived autoimmunity were also described. This chapter discusses major examples of such nonimmune roles of CRISPR-Cas that were described to date, their putative mechanisms of action, and their functional and evolutionary implications.

Contents

G. Amitai · R. Sorek (✉)
Department of Molecular Genetics, Weizmann Institute of Science, 76100 Rehovot, Israel
e-mail: rotem.sorek@weizmann.ac.il

G. Amitai
e-mail: gil.amitai@weizmann.ac.il

R. Barrangou and J. van der Oost (eds.), *CRISPR-Cas Systems*,
DOI: 10.1007/978-3-642-34657-6_10, © Springer-Verlag Berlin Heidelberg 2013

10.1 Introduction

During evolution, it is not uncommon for anti-phage defense systems to gain a second, distinct function in cellular regulation that is independent of phage defense. Such an evolutionary event is called "exaptation" (Brosius and Gould 1992), a term describing the adaptation of a biological structure to acquire a role other than its original one. For example, several restriction/modification (R/M) systems have lost their restriction gene, leaving a methylase that now takes part in epigenetic modifications; such evolutionary events were recorded in the cases of the Dam methylase in *Escherichia coli* and the CcrM methylase in *Caulobacter crescentus* (Marinus and Casadesus 2009), both of which originated from R/M systems. Methylation by Dam was shown to affect important regulatory processes such as replication initiation (via binding of the replication initiation complex to a methylated origin of replication), mismatch repair, and regulation of bacterial pathogenicity (Marinus and Casadesus 2009). The CcrM methylase has been shown to affect the cell cycle in *C. crescentus* (Marinus and Casadesus 2009).

Another kind of defense systems that can acquire regulatory roles are toxin-antitoxin (TA) modules. These modules consist of a toxin (protein) and an antitoxin (protein/RNA) that are tightly bound to each other. In cases of phage attack or other stress insults the antitoxin rapidly degrades, leaving a free toxin that kills the cell or brings it into a dormant state (Gerdes et al. 2005). One of the well studied TA systems, called mazEF, exerts its toxicity by cleaving mRNA molecules in response to different stress signals (Christensen et al. 2003; Pedersen et al. 2002), or phage infection (Hazan and Engelberg-Kulka 2004). However, in *Myxococcus xanthus* the toxin MazF exists without the antitoxin MazE, and it was shown that this toxin mediates programmed cell death during multicellular development of this organism (Nariya and Inouye 2008).

Based on the examples detailed above, it is intriguing but not entirely unexpected to find cases where the CRISPR defense system has adapted to take regulatory roles other than its primary immunity role. The text below presents some of the more studied cases.

10.2 Roles of CRISPR in Modulation of Microbial Social Behavior

Interestingly, involvement of CRISPR in regulating group (social) behavior of bacteria was documented in at least two separate cases, although the exact mechanism of action is still obscure in both cases. Such regulation was described for *Pseudomonas aeruginosa* and *M. xanthus*.

10.2.1 Effect of CRISPR/Prophage Interplay on P. aeruginosa Biofilm Formation

P. aeruginosa is a gram-negative opportunistic pathogen notorious for its ability to colonize the lungs of cystic fibrosis (CF) and immune-suppressed patients (Hoiby et al. 2010), as well as settling on surfaces of man-made medical devices. Such colonization is partly enabled by formation of robust biofilm morphology by this organism, which can resist antibiotic treatment, host immune responses, and biocide treatment (Hermans et al. 1991). To regulate biofilm formation, *P. aeruginosa* uses complex cell-to-cell signaling, enabling behavioral synchronization between cells in the community, which includes swarming motility and cell aggregation (Harmsen et al. 2010; Kirisits and Parsek 2006).

A series of publications from the O'Toole laboratory have described a role for the CRISPR-Cas system in regulation of biofilm formation in *P. aeruginosa* PA14, when a specific prophage is present in its genome (Cady and O'Toole 2011; Cady et al. 2010; Zcgans et al. 2009). Originally, it was shown that when *P. aeruginosa* is infected and lysogenized by phage DMS3 [a temperate Mu-like bacteriophage (Budzik et al. 2004)] biofilm formation is abolished. A transposon random insertion analysis further revealed that biofilm behavior can be restored if the CRISPR locus is mutated (while the prophage is still genomically integrated) (Fig 10.1a).

P. aeruginosa PA14 strain encodes a single CRISPR-Cas locus of the Ypest subtype (Haft et al. 2005) [subtype I-F according to the new nomenclature (Makarova et al. 2011)]. This locus contains six cas genes (*csy*1–4, *cas*1, and *cas*3) and two CRISPR arrays (Fig 10.1b). Disruption of each of the cas genes, except for *cas*1, restored biofilm formation (Zegans et al. 2009), indicating that the interfering activity of CRISPR as a whole, rather than the activity of a single cas gene, is involved in biofilm and swarming repression. Interestingly, deletion of any cas gene or the entire CRISPR locus had no effect on biofilm formation in the absence of the DMS3 phage (Cady and O'Toole 2011), asserting a mutual link between the CRISPR, phage, and *P. aeruginosa* biofilm behavior.

Further refined deletions in the CRISPR array have localized the effect to a single spacer, spacer #1 in array #2 (Fig 10.1b). Mutations in that spacer had abolished CRISPR effect on biofilm formation, so that biofilm was restored,

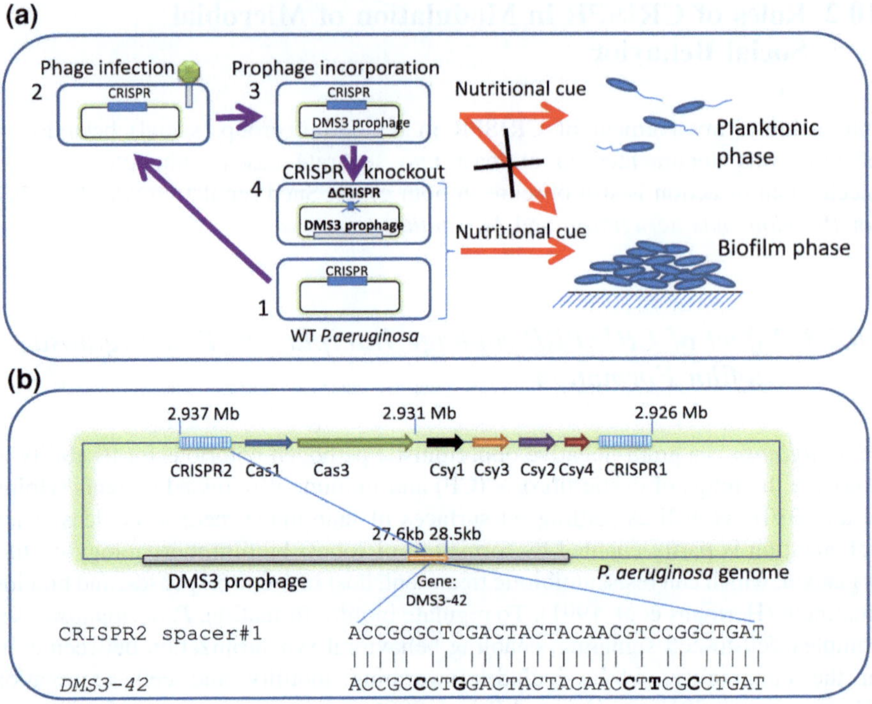

Fig. 10.1 The presence of CRISPR-Cas system and prophage DMS3 triggers a change in *P. aeruginosa* PA14 group behavior, resulting in inhibition of biofilm formation and lack of swarming motility. **a** 1. Upon nutritional cues *P. aeruginosa* PA14 initiates biofilm formation on polyvinylchloride surface (O'Toole et al. 1999; O'Toole and Kolter 1998a, b). 2–3. Infection of *P. aeruginosa* PA14 with the DMS3 temperate phage abolishes the ability of the host to form biofilm. 4. Deletion knockouts of genes in the CRISPR locus in a lysogenized strain restore biofilm formation. **b** *P. aeruginosa* CRISPR and *cas* genes arrangement within the genome (*green rectangle*). The active spacer #1 matches the DMS3-42 gene via sequence complementation of 27 out of 32 bases. Nonaligned bases are in *bold*

while mutations in any other spacer did not show any effect. Spacer #1 shows significant similarity (84 %) to a region in the coding sequence of an essential gene, called DMS3-42, in the prophage (Fig 10.1b). Indeed, synonymous mutations within this gene that altered the proto-spacer sequence eliminated the CRISPR mediated effect (Cady and O'Toole 2011). Furthermore, direct mutations within the spacer could be complemented by reciprocal single base mutations within the proto-spacer, indicating that sequence complementation between the spacer and target is the key mechanism behind target recognition. Interestingly, modification of the spacer to generate 100 % identity with the respective gene in the phage resulted in a typical CRISPR-mediated defense against this phage (Cady et al. 2012).

Two additional spacers, apart from spacer #1, show significant similarity to regions in the prophage genome. Surprisingly, although these two spacers (spacers #17 and #20) match the phage genome by 97 and 100 %, respectively, precise deletions and replacement of these spacers did not have any effect on the CRISPR-mediated biofilm inhibition.

From the mutational analysis it is clear that the complete interfering function of the CRISPR, including the full cascade protein complex, the *Cas*3 effector, and a specific spacer, is needed to cause the CRISPR-mediated effect. However, the exact mechanism governing these intricate interactions between prophage-CRISPR-biofilm is still not known, and remains open to speculations. Several questions remain unanswered: if the CRISPR is active against the phage, how can the phage infect the cell and establish itself as a prophage? How can the prophage become lytic again without facing CRISPR interference? This is especially intriguing since several spacers show high similarity to the phage genome. It is possible that the DMS3 phage carry some specific mechanisms to counteract the primary CRISPR interference action; another possibility is that the CRISPR in the *P. aeruginosa* PA14 strain has evolved to function differently than the canonical CRISPR activity [although a recent study shows that this system still provides defense against temperate phages in *P. aeruginosa* PA14 (Cady et al. 2012)]. Cady and O'Toole proposed that the CRISPR may regulate production and/or stability of the prophage RNA rather than affecting its DNA (Cady and O'Toole 2011). However, this still does not explain the link to biofilm formation. Future studies might shed light on these currently unresolved questions.

10.2.2 CRISPR Regulation of Fruiting Body Formation in M. xanthus

Another example of a CRISPR-mediated regulation of microbial social behavior was described in *M. xanthus*, a gram-negative soil bacterium belonging to the delta proteobacteria lineage. Upon starvation and other environmental or chemical cues, this organism initiates a complex developmental program leading to multicellular organization (Kroos 2007; Rosenberg 1984; Velicer and Vos 2009) (Fig. 10.2a). This developmental program begins by coordinated cell organization into parallel ridges, reminiscent of ripples on a water surface. This step is followed by synchronized gliding movements leading to aggregation of cells into mounds that form fruiting bodies (aggregation step). Subsequently, the rod-shaped cells within the fruiting bodies differentiate into spherical spores (sporulation step), and these spores are eventually released from the fruiting body (Kroos 2007; Mignot and Kirby 2008).

Using random transposon mutagenesis it was shown that mutations in a CRISPR locus within the *M. xanthus* genome abolish cell aggregation and fruiting body formation (Kroos et al. 1990; Thony-Meyer and Kaiser 1993). This organism

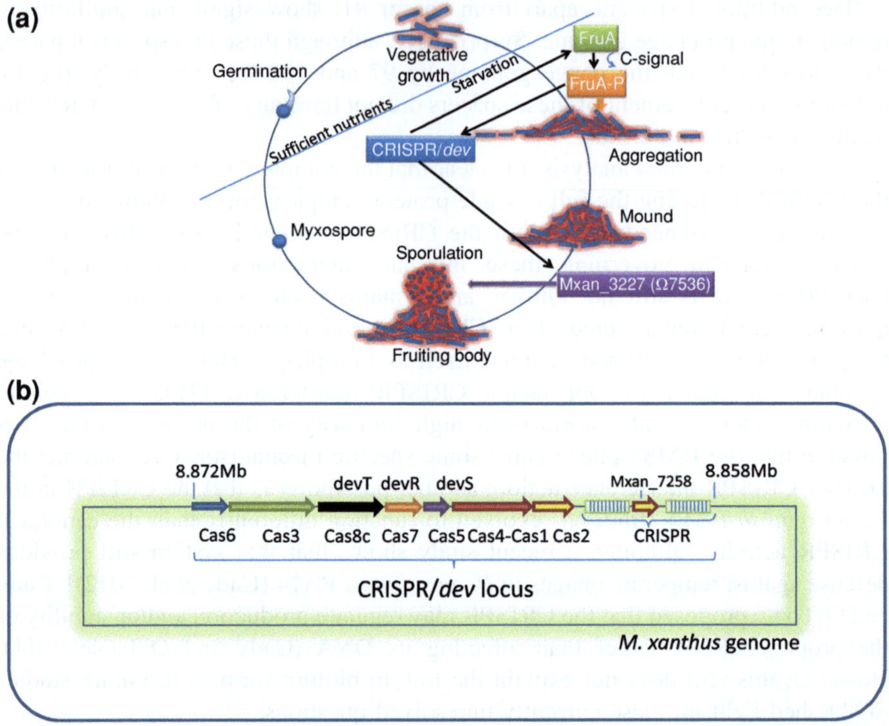

Fig. 10.2 Influence of CRISPR/dev operon components (devR, S, and T) on *M. xanthus* group behavior. **a** The developmental program of *M. xanthus* in response to environmental cues results in cell differentiation and fruiting body formation. The CRISPR/*dev* operon upregulates two genes essential for activating the developmental process of aggregation (FruA) and sporulation (Mxan_3227, also referred to as omega-7536). Phosphorylated FruA was also shown to upregulate the dev locus. **b** Schematic arrangement of the CRISPR/*dev* operon. Gene annotations according to CRISPR nomenclature (Makarova et al. 2011) are shown below the genes, while naming conventions from the early *M. xanthus* literature are shown above the genes

has three CRISPR loci of different subtypes: Dvulg (I-C), Tneap (I-B), and RAMP (III-B). Mutations in specific genes in the Tneap-type CRISPR were those which affected the *M. xanthus* fruiting body formation. Since early *M. xanthus* developmental studies predated the description and classification of CRISPR, the affected genes were named *devR*, *devS,* and *devT*. With current CRISPR classification, it is now clear that these genes correspond to *cas*7, *cas*5, and *cas*8*b*, respectively, in the Tneap cas operon (Fig 10.2b). We therefore refer to this CRISPR operon here as CRISPR/dev.

Attempts to understand the effect of the *dev* (*cas*) genes on fruiting body formation revealed that *devT*-encoded protein stimulates transcription of the *fruA* gene. FruA is a major regulator of the multicellular program in *M. xanthus*, and is essential for fruiting body formation (Ellehauge et al. 1998; Ogawa et al. 1996).

Deletion of *devT* (*cas8b*) has resulted in significant reduction in FruA protein accumulation (Boysen et al. 2002). Deletion of both *devR* and *devS* genes (*cas7* and *cas5*) resulted in elimination of expression of the gene Mxan_3227, which encodes a putative exopolysaccharide transport protein (Licking et al. 2000). Direct deletion of this gene was also shown to abolish formation of functional fruiting body, providing a clear functional link between *devR*/*devS* and fruiting body formation. Interestingly, the expression of the CRISPR/*dev* locus is spatially regulated; using GFP fusion experiments, it was shown that the CRISPR/*dev* operon expression is localized to the center of the fruiting body, but is absent from the peripheral rods (Julien et al. 2000). All in all, these multiple lines of evidence strongly support the involvement of CRISPR/*dev* genes in the *M. xanthus* multicellular developmental program.

It is still unclear how the CRISPR genes regulate this complex developmental process. Since *cas*/*dev* genes were shown to positively regulate FruA and Mxan_3227, direct spacer-mediated targeting of these genes by the CRISPR machinery is unlikely. If such spacer-mediated inhibition indeed occurs, then it is possible that the CRISPR/*dev* locus uses specific spacers to target a negative regulator of FruA and Mxan_3227, and that the effect seen in these two genes is secondary to the primary inhibiting effect of CRISPR. A recent screen for self-targeting spacers has identified a spacer within *M. xanthus* that is 100 % identical to a sequence within a lipoprotein gene in that organism (Stern et al. 2010). However, this spacer is encoded within a different CRISPR array (belonging to the Type I-C subtype), that is separated more than 250 kb from the CRISPR/dev locus. Furthermore, no connection was established between the lipoprotein gene and *M. xanthus* development. Nevertheless, it is possible that other spacers (that have less than 100 % match to self genes) mediate such putative CRISPR-based effects on fruiting body formation. It is also possible that the CRISPR/dev locus presents an extreme example of exaptation, where the cas genes have evolved to conduct a function entirely different from their original one. Finally, perhaps mutations in the cas/dev locus genes have a polar effect, which influences the expression of downstream genes that are more directly involved in *M. xanthus* development. As with many ongoing studies, time might allow bridging the gap between the expected immunity-related functionality of CRISPR and the unexpected noncanonical regulatory phenotypes.

10.3 Additional Evidence for CRISPR-Based Regulation

A number of studies have suggested regulatory roles for crRNA base-pairing with an endogenous gene (with or without involvement of the Cascade complex). Some of these studies are described below.

10.3.1 CRISPR-Based Gene Regulation in Response to Envelope Stress in E. coli

Envelope stress in gram-negative bacteria is manifested in part by accumulation of unfolded proteins in the periplasm (Raivio 2005). Cells lacking essential chaperons are more likely to develop such stress when protein production rates are high. Envelope stress induces the expression of the CRISPR Cascade operon in *E. coli* (Baranova and Nikaido 2002), possibly because such stress can indicate phage invasion. CRISPR induction following envelope stress was shown to be mediated by the BaeSR two component signal transduction system that senses envelope stress. Upon such stress, the BaeSR becomes phosphorylated, binds to the promoter upstream of *cas*A (ygcL), and activates the Cascade operon expression (Baranova and Nikaido 2002).

Serendipity stemming from studies of the twin-arginine translocation (Tat) pathway revealed another possible aspect of CRISPR-based regulation in *E. coli.* The Tat system transfers folded proteins across the lipid membrane to the periplasm, cell-envelope, or the growth medium (Berks et al. 2003, 2005; Palmer et al. 2005), and is dependent on an N-terminal Tat signal sequence in the translocated protein. DeLisa and colleagues have studied the factors required by the Tat system for the export of cytoplasmic folded proteins, using a reporter system where the Tat-dependent TorA signal sequence (ssTorA) was fused to GFP (Perez-Rodriguez et al. 2011). Interestingly, in the absence of the cellular chaperone DnaK, ssTor-GFP mRNA expression was diminished. Furthermore, deletion of genes in the *E. coli* CRISPR operon restored ssTor-GFP expression, pointing to CRISPR involvement in ssTor-GFP expression elimination. Finally, deletion of the BaeSR system had also restored ssTor-GFP mRNA expression, even when the CRISPR locus was intact, linking CRISPR activation to envelope stress.

Three spacers in the *E. coli* MG1655 CRISPR array were suggested to be involved in mediating interference in this system. In all three cases, the hypothesized base pairing between these spacers and the target was very limited (9–14 pairing bases). Nevertheless, synonymous mutations in the ssTor sequence corresponding to the hypothesized base pairing abolished the CRISPR-mediated effect on ssTor-GFP mRNA, implying a possible link between these spacers and CRISPR silencing of ssTor-GFP expression. The very low similarity, though, between these spacers and their suggested targets, raises some doubt as to whether this indeed is the mechanism of CRISPR-mediated regulation.

As in other cases of putative CRISPR-mediated regulation, the exact mechanism of interference is not completely understood. In this case, CRISPR was shown to target the DNA of the plasmid that carries this ssTor-GFP fusion, probably leading to the observed reduction of mRNA product. However, both the ssTor signal peptide sequence and the putatively targeting spacers naturally reside within the host genome, raising the peculiar possibility that the bacterial genomic DNA is targeted by CRISPR in response to envelope stress. More studies are needed in order to establish this hypothesis and the role of *E. coli* CRISPR in regulation of self genes.

10.3.2 CRISPR-Derived Regulatory Noncoding RNA in Listeria monocytogenes EGD-e

L. monocytogenes is a pathogenic gram-positive bacteria causing listeriosis, a lethal human infection with a high mortality rate (Cossart 2007). The pathogenicity of this organism is mediated by an arsenal of virulence genes, as well as a set of regulatory noncoding small RNAs (sRNAs). Intriguingly, although the *L. monocytogenes* EGDe strain completely lacks *cas* genes, one of the sRNAs found in this strain strongly resembles crRNA (Mandin et al. 2007). This sRNA, called rliB, is composed of five 29nt repeats, interspersed by nonrepetitive spacers. Deletion of the rliB affected liver colonization by *L. monocytogenes* EGDe in infected mice, suggesting a role for rliB in controlling virulence (Toledo-Arana et al. 2009). Furthermore, overexpression of rliB in *L. monocytogenes* resulted in elevation of ferrous transport proteins mRNA expression, further suggesting a role for rliB in regulating iron transport.

The repeats in rliB are almost identical (1 base difference) to the repeats in the crRNA of *Listeria seeligeri* serovar 1/2b str. SLCC3954, where adjacent cas operon is present. It is therefore likely that the rliB gene is a "relic" of a past functional CRISPR-Cas system that, in recent evolution, was inactivated in the *L. monocytogenes* genome. Under this hypothesis, the crRNA had acquired a new regulatory role as a sRNA in *L. monocytogenes* that is not dependent on the activity of *cas* genes. Therefore, rliB might be a remarkable example of recent exaptation of components of the CRISPR system into novel functions.

10.3.3 Role of Cas1 in DNA Repair

Cas1 and Cas2 are the only two Cas proteins found universally in all CRISPR-Cas systems (Haft et al. 2005). These two genes were shown to participate in acquisition and integration of proto-spacers into CRISPR cassettes (Datsenko et al. 2012; Yosef et al. 2012). Interestingly, Babu and colleagues have impressively substantiated that the Cas1 protein in *E. coli* is also involved in DNA repair (Babu et al. 2011). This role for Cas1 was demonstrated from multiple angles, providing biochemical, genetic, structural, and functional evidence. First, Cas1 physically co-purifies with RecB and RecC, which are two subunits of the recBCD complex involved in the recombinational repair of double-strand breaks (Kowalczykowski 2000). Cas1 also co-purifies with RuvB, a DNA helicase that binds Holliday junctions, and UvrC, a nuclease of the nucleotide excision repair pathway that is involved in the excision of a small ssDNA fragment flanking a damaged DNA site (Orren et al. 1992).

In vitro studies of the biochemical properties of the *E. coli* Cas1 showed that it is an endonuclease that primarily cleaves linear ssDNA substrates, and to a lesser degree ssRNA, and short (<34nt) dsDNAs (Babu et al. 2011). This protein also cleaves Holliday junctions to produce a nicked Holliday junction duplex, as well as a series of diverse branched substrates such as replication forks, 5′-flap, 3′-flap and splayed arm DNA duplex structures (Babu et al. 2011). In accordance with its biochemical properties, Cas1 knockout strains of *E. coli* showed increased sensitivity to DNA damage induced by UV light or mitomycin C (Babu et al. 2011).

Additional support for the role of Cas1 in DNA repair stems from experiments with a fluorescently tagged Cas1, showing localization of this protein to discrete foci on the nucleoid upon prolonged mitomycin C treatment (Babu et al. 2011). Combined, these data suggest that Cas1 is recruited to loci of DNA damage in the chromosome, where it functions as part of the DNA repair machinery. Nevertheless, its specific functional role within this machinery still remains uncharacterized.

What is the rationale behind the DNA repair/immunity duality in the functions of Cas1? One possible explanation may be that the role of Cas1 in DNA processing is intimately linked to the mechanism of spacer acquisition, and that as part of this role Cas1 interacts with the DNA repair machinery. Nevertheless, the increased sensitivity of ΔCas1 *E. coli* strains to DNA damage suggests that Cas1 may have been adopted by *E. coli* to become an integral part of the DNA repair pathway. Could the natural tendency of Cas1 to localize to unusually organized nucleic acids explain its enigmatic ability of identifying foreign DNA for spacer acquisition? At this stage, the question remains unresolved.

10.4 CRISPR-Driven Autoimmunity and its Evolutionary Implications

One of the challenges faced by any immune system is the necessity to discriminate between self and nonself antigens. To achieve this task, the mammalian adaptive immune system employs a complex, multi organ procedure that "educates" the immune cells and prevents targeting of self antigens (Lio and Hsieh 2011). Nevertheless, instances of autoimmunity, where immune cells and antibodies attack self tissues, can occur. Such instances can lead to severe pathological phenotypes, manifested in disorders such as rheumatoid arthritis, type-I diabetes, multiple sclerosis, and many more (Raman and Mohan 2003).

Being an adaptive immunity system, the CRISPR-Cas system faces similar challenges. Although many of the spacers in a given CRISPR array target phage or plasmid sequences, cases of spacers that fully match regions in the bacterial genome ('self-targeting spacers') were documented (Aklujkar and Lovley 2010; Horvath et al. 2008). A recent study that addressed the question of self targeting found that ~0.2 % of studied spacers show 100 % identity to a gene in the genome of the bacteria in which the spacer resides (Stern et al. 2010). Some of the targeted genes were essential genes, such as in the case of *Lactobacillus acidophilus* NCFM, where one of the spacers targeted the 16S rRNA gene (Stern et al. 2010). Such self-targeting by CRISPR is referred to as CRISPR-driven autoimmunity.

Since most currently studied CRISPR systems are believed to target and cleave the DNA of the invading element, self-targeting by CRISPR is hypothesized to lead to cleavage of self DNA, and thus have deleterious effects on the bacterium. In agreement with this hypothesis, self targeting is frequently associated with loss of CRISPR function, either generally, by mutations inactivating one or more of the *cas* genes, or locally by mutations inactivating the repeats surrounding the self-targeting spacers, or the PAM sequence adjacent to the proto-spacer (Stern et al. 2010). For example, in the case of *L. acidophilus* NCFM, where one of the spacers targeted the 16S rRNA gene, the *cas* operon was completely lost (Stern et al. 2010). None of the identified self-targeting spacers were conserved, suggesting rapid turnover of such spacers. These observations are consistent with the notion of CRISPR-driven autoimmunity, where acquisition of self-targeting spacers leads to negative selective pressure that is mitigated by CRISPR inactivation.

Curiously, archaea seem to have a significantly reduced tendency for acquisition of self spacers: only 3 of the 46 archaea analyzed (6 %) presented evidence for self-targeting spacers, compared to 19 % of bacteria (56 of the 284 analyzed genomes) (Stern et al. 2010). This is linked to the observation that archaea have a less patchy distribution of CRISPR, possibly because they have less tendency to lose the CRISPR system as a result of autoimmunity-mediated CRISPR inactivation. It is therefore possible that archaeal CRISPRs have some kind of protection mechanism against such autoimmunity, or alternatively they are more prone to immediate cell death upon spacer-based autoimmunity.

The causes for acquisition of autoimmune, self-targeting spacers are still unknown. It is possible that such acquisition originates from phages that have acquired bacterial genes from previous rounds of infection (Partridge et al. 2009), leading to CRISPR recognizing these genes as foreign DNA. One might speculate that the tendency of many phages to pack random pieces of host DNA as part of the phage particle has evolved in response to CRISPR immunity. Under this hypothesis, package of host genome sequences reduces CRISPR effectiveness in subsequent rounds of infection, as acquisition of self-targeting spacers will eventually lead to the loss of CRISPR protection. Therefore, phage transduction, which is known to have huge roles in horizontal gene transfer in the microbial world (Liu et al. 2006), and which is crucial for many genetic engineering systems, might have originally evolved as a general anti-CRISPR mechanism. Notably, the concept of autoimmunity mediated by viruses has been also described in mammalian systems (Munz et al. 2009). Such virus-mediated autoimmunity is often triggered by molecular mimicry, i.e., viral antigens that are structurally similar to epitopes found in a self protein. Again, this shows conceptual similarities between the bacterial and mammalian adaptive immune systems.

Faulty incorporation of self-DNA could also occur simply because of CRISPR 'errors', whereas the CRISPR machinery responsible for spacer acquisition uses genomic DNA, instead of foreign DNA, as a substrate. Since the mechanism of spacer incorporation has still not been elucidated (although Cas1 and Cas2 were suggested as the candidate proteins involved in this process (Brouns et al. 2008; Cady and O'Toole 2011; Wiedenheft et al. 2009; Zegans et al. 2009)), the role of spacer acquisition errors in generation of CRISPR-driven autoimmunity is yet to be established.

10.5 Other Possible Roles of CRISPR Yet to be Discovered

The analogy between the prokaryotic CRISPR machinery and the RNAi/miRNA machinery in eukaryotes had led to the suggestion that CRISPRs can also regulate mRNA expression in prokaryotes (Makarova et al. 2006; Sorek et al. 2008). Although little evidence so far points to such regulation, it is intriguing to speculate that such a regulatory role of CRISPR might be established in the future.

In RNAi, the foreign nucleic acids are degraded following full match between the small interfering RNA (siRNA) and the target RNA. Proteins belonging to the RNAi machinery are also participating in regulation by microRNAs (miRNAs), where the base pairing between the miRNA and its target is only partial (Flynt and Lai 2008). Possibly, CRISPR activity could also depend on the extent of base-pairing between the mature crRNA and its target, where partial base-pairing might be leading to regulatory effects other than cleavage and inactivation of the target DNA.

The discovery that the CRISPR RAMP subtype (Type III-B) can target and cleave single-stranded RNA in archaea also raises the possibility that such RAMP type CRISPR-Cas systems have adopted to take a nonimmune role in regulatory degradation of mRNAs of self genes upon a specific signal. Indeed, the Type III-B system in *Pyrococcus furiosus* was shown to cleave an antisense RNA complementary to the CRISPR array, in vivo (Hale et al. 2012). Although this antisense RNA does not code for a protein, it is conceivable that protein coding mRNAs targeted by spacers of this system may be affected in a similar manner.

In cases where a CRISPR system had undergone dramatic changes and evolved to perform gene regulation, one might expect that the structure of the system has also been altered, and possibly differs from the anti-phage CRISPR system as we know it today. Some of the essential genes might have been lost, and others might have been significantly changed. In addition, the organization of the regulatory spacers might be altered, and for example could be composed of only one spacer flanked by partial repeats. Such altered, regulatory CRISPR systems are yet to be discovered.

Another putative role of CRISPR might concern manipulation of beneficial phages and prophages by partial silencing. It is well documented that temperate phages can carry genes that are beneficial to the host; for example, some phages carry antibiotics-resistance genes (Brabban et al. 2005; Muniesa et al. 2004; Witte 2004), while others bring enzymes that increase bacterial fitness (Desiere et al. 2001; Hendrix et al. 2000). Bacteria, therefore, face complex evolutionary considerations, as in some cases it might be beneficial to "allow" phage entry and lysogenization. One may speculate that delicate CRISPR-based regulation can affect the bacterial decision whether or not to tolerate a first cycle insertion of phage and subsequent prophage integration. Under this hypothesis, CRISPRs might have a "sentinel" role in monitoring lysogenized phages against entering the lytic cycle, rather than preventing infection and lysogenization in the first place.

10.6 Conclusions

We have reviewed in detail several key studies where strong evidence for the involvement of CRISPR in regulating nonimmune processes was presented. Such regulation was also alluded to other recent studies, in which spacers targeting self genes were identified (Aklujkar and Lovley 2010; Horvath et al. 2008), strengthening the overall perception of CRISPRs as capable of adopting alternative roles other than immunity.

It is curious that in two different bacterial systems group behavior was shown to be affected by CRISPR. Both biofilm formation in *P. aeruginosa* and fruiting body formation in *M. xanthus* are regulated by complex signals and require concerted activity of multiple cells. Notably, in both organisms the signals that initiate these processes are starvation or other forms of stress. Since cells are more sensitive to phage attacks under stress conditions, and/or phage attacks results in cellular stress, activation of CRISPR is likely to be linked to such stress. This connection might point to the initial steps in CRISPR adaptation into regulating these group behavior processes. Additional connection to stress response was described in the case of *E. coli* CRISPR, which was shown to be activated upon envelope stress (Perez-Rodriguez et al. 2011).

Since CRISPR research is still in its early days, it is likely that currently published studies represent only the tip of the iceberg with respect to our understanding of CRISPR roles in regulation of endogenous, nonimmune processes. As in Eukaryotes, silencing of mRNA expression as opposed to DNA may be much broader than anticipated. Future studies will probably shed more light on this exciting possibility.

Acknowledgments We thank Shany Doron for comments on the manuscript. The Sorek lab is supported, in part, by the ERC-StG program (grant 260432), the Israeli Science Foundation (grant ISF-1303/12), and the Leona M. and Harry B. Helmsley Charitable Trust.

References

Aklujkar M, Lovley DR (2010) Interference with histidyl-tRNA synthetase by a CRISPR spacer sequence as a factor in the evolution of *Pelobacter carbinolicus*. BMC Evol Biol 10:230. doi:10.1186/1471-2148-10-230

Babu M, Beloglazova N, Flick R, Graham C, Skarina T, Nocek B, Gagarinova A, Pogoutse O, Brown G, Binkowski A, Phanse S, Joachimiak A, Koonin EV, Savchenko A, Emili A, Greenblatt J, Edwards AM, Yakunin AF (2011) A dual function of the CRISPR-Cas system in bacterial antivirus immunity and DNA repair. Mol Microbiol 79(2):484–502. doi:10.1111/j.1365-2958.2010.07465.x

Baranova N, Nikaido H (2002) The baeSR two-component regulatory system activates transcription of the yegMNOB (mdtABCD) transporter gene cluster in Escherichia coli and increases its resistance to novobiocin and deoxycholate. J Bacteriol 184(15):4168–4176

Berks BC, Palmer T, Sargent F (2003) The tat protein translocation pathway and its role in microbial physiology. Adv Microb Physiol 47:187–254

Berks BC, Palmer T, Sargent F (2005) Protein targeting by the bacterial twin-arginine translocation (Tat) pathway. Curr Opin Microbiol 8(2):174–181. doi:10.1016/j.mib.2005.02.010

Boysen A, Ellehauge E, Julien B, Sogaard-Andersen L (2002) The DevT protein stimulates synthesis of FruA, a signal transduction protein required for fruiting body morphogenesis in *Myxococcus xanthus*. J Bacteriol 184(6):1540–1546

Brabban AD, Hite E, Callaway TR (2005) Evolution of foodborne pathogens via temperate bacteriophage-mediated gene transfer. Foodborne Pathog Dis 2(4):287–303. doi:10.1089/fpd.2005.2.287

Brosius J, Gould SJ (1992) On "genomenclature": a comprehensive (and respectful) taxonomy for pseudogenes and other "junk DNA". Proc Natl Acad Sci U S A 89(22):10706–10710

Brouns SJ, Jore MM, Lundgren M, Westra ER, Slijkhuis RJ, Snijders AP, Dickman MJ, Makarova KS, Koonin EV, van der Oost J (2008) Small CRISPR RNAs guide antiviral defense in prokaryotes. Science 321(5891):960–964. doi:10.1126/science.1159689

Budzik JM, Rosche WA, Rietsch A, O'Toole GA (2004) Isolation and characterization of a generalized transducing phage for *Pseudomonas aeruginosa* strains PAO1 and PA14. J Bacteriol 186(10):3270–3273

Cady KC, Bondy-Denomy J, Heussler GE, Davidson AR, O'Toole GA (2012) The CRISPR/Cas adaptive immune system of *Pseudomonas aeruginosa* mediates resistance to naturally occurring and engineered phages. J Bacteriol. doi:10.1128/JB.01184-12

Cady KC, O'Toole GA (2011) Non-identity-mediated CRISPR-bacteriophage interaction mediated via the Csy and Cas3 proteins. J Bacteriol 193(14):3433–3445. doi:10.1128/JB.01411-10

Cady KC, White AS, Hammond JH, Abendroth MD, Karthikeyan RS, Lalitha P, Zegans ME, O'Toole GA (2010) Prevalence, conservation and functional analysis of Yersinia and Escherichia CRISPR regions in clinical *Pseudomonas aeruginosa* isolates. Microbiology 157(Pt 2):430–437. doi:10.1099/mic.0.045732-0

Christensen SK, Pedersen K, Hansen FG, Gerdes K (2003) Toxin-antitoxin loci as stress-response-elements: ChpAK/MazF and ChpBK cleave translated RNAs and are counteracted by tmRNA. J Mol Biol 332(4):809–819. doi:S0022283603009227

Cossart P (2007) Listeriology (1926–2007): the rise of a model pathogen. Microbes Infect 9(10):1143–1146. doi:10.1016/j.micinf.2007.05.001

Datsenko KA, Pougach K, Tikhonov A, Wanner BL, Severinov K, Semenova E (2012) Molecular memory of prior infections activates the CRISPR/Cas adaptive bacterial immunity system. Nat Commun 3:945. doi:10.1038/ncomms1937

Desiere F, Mahanivong C, Hillier AJ, Chandry PS, Davidson BE, Brussow H (2001) Comparative genomics of lactococcal phages: insight from the complete genome sequence of lactococcus lactis phage BK5-T. Virology 283(2):240–252. doi:10.1006/viro.2001.0857S0042-6822(01)90857-8

Ellehauge E, Norregaard-Madsen M, Sogaard-Andersen L (1998) The FruA signal transduction protein provides a checkpoint for the temporal co-ordination of intercellular signals in *Myxococcus xanthus* development. Mol Microbiol 30(4):807–817

Flynt AS, Lai EC (2008) Biological principles of microRNA-mediated regulation: shared themes amid diversity. Nat Rev Genet 9(11):831–842. doi:10.1038/nrg2455

Gerdes K, Christensen SK, Lobner-Olesen A (2005) Prokaryotic toxin-antitoxin stress response loci. Nat Rev Microbiol 3(5):371–382. doi:10.1038/nrmicro1147

Haft DH, Selengut J, Mongodin EF, Nelson KE (2005) A guild of 45 CRISPR-associated (Cas) protein families and multiple CRISPR/Cas subtypes exist in prokaryotic genomes. PLoS Comput Biol 1(6):e60. doi:10.1371/journal.pcbi.0010060

Hale CR, Majumdar S, Elmore J, Pfister N, Compton M, Olson S, Resch AM, Glover CV 3rd, Graveley BR, Terns RM, Terns MP (2012) Essential features and rational design of CRISPR RNAs that function with the Cas RAMP module complex to cleave RNAs. Mol Cell 45(3):292–302. doi:10.1016/j.molcel.2011.10.023

Harmsen M, Yang L, Pamp SJ, Tolker-Nielsen T (2010) An update on *Pseudomonas aeruginosa* biofilm formation, tolerance, and dispersal. FEMS Immunol Med Microbiol 59(3):253–268. doi:10.1111/j.1574-695X.2010.00690.x

Hazan R, Engelberg-Kulka H (2004) Escherichia coli mazEF-mediated cell death as a defense mechanism that inhibits the spread of phage P1. Mol Genet Genomics 272(2):227–234. doi:10.1007/s00438-004-1048-y

Hendrix RW, Lawrence JG, Hatfull GF, Casjens S (2000) The origins and ongoing evolution of viruses. Trends Microbiol 8(11):504–508. doi:S0966-842X(00)01863-1

Hermans PW, van Soolingen D, Bik EM, de Haas PE, Dale JW, van Embden JD (1991) Insertion element IS987 from Mycobacterium bovis BCG is located in a hot-spot integration region for insertion elements in *Mycobacterium tuberculosis* complex strains. Infect Immun 59(8):2695–2705

Hoiby N, Ciofu O, Bjarnsholt T (2010) *Pseudomonas aeruginosa* biofilms in cystic fibrosis. Future Microbiol 5(11):1663–1674. doi:10.2217/fmb.10.125

Horvath P, Romero DA, Coute-Monvoisin AC, Richards M, Deveau H, Moineau S, Boyaval P, Fremaux C, Barrangou R (2008) Diversity, activity, and evolution of CRISPR loci in *Streptococcus thermophilus*. J Bacteriol 190(4):1401–1412. doi:10.1128/JB.01415-07

Julien B, Kaiser AD, Garza A (2000) Spatial control of cell differentiation in *Myxococcus xanthus*. Proc Natl Acad Sci U S A 97(16):9098–9103. doi:97/16/9098

Kirisits MJ, Parsek MR (2006) Does *Pseudomonas aeruginosa* use intercellular signalling to build biofilm communities? Cell Microbiol 8(12):1841–1849. doi:10.1111/j.1462-5822.2006.00817.x

Kowalczykowski SC (2000) Initiation of genetic recombination and recombination-dependent replication. Trends Biochem Sci 25(4):156–165. doi:S0968-0004(00)01569-3

Kroos L (2007) The bacillus and myxococcus developmental networks and their transcriptional regulators. Annu Rev Genet 41:13–39. doi:10.1146/annurev.genet.41.110306.130400

Kroos L, Kuspa A, Kaiser D (1990) Defects in fruiting body development caused by Tn5 lac insertions in *Myxococcus xanthus*. J Bacteriol 172(1):484–487

Licking E, Gorski L, Kaiser D (2000) A common step for changing cell shape in fruiting body and starvation-independent sporulation of *Myxococcus xanthus*. J Bacteriol 182(12):3553–3558

Lio CW, Hsieh CS (2011) Becoming self-aware: the thymic education of regulatory T cells. Curr Opin Immunol 23(2):213–219. doi:10.1016/j.coi.2010.11.010

Liu J, Glazko G, Mushegian A (2006) Protein repertoire of double-stranded DNA bacteriophages. Virus Res 117(1):68–80. doi:10.1016/j.virusres.2006.01.015

Makarova KS, Grishin NV, Shabalina SA, Wolf YI, Koonin EV (2006) A putative RNA-interference-based immune system in prokaryotes: computational analysis of the predicted enzymatic machinery, functional analogies with eukaryotic RNAi, and hypothetical mechanisms of action. Biol Direct 1:7. doi:10.1186/1745-6150-1-7

Makarova KS, Haft DH, Barrangou R, Brouns SJ, Charpentier E, Horvath P, Moineau S, Mojica FJ, Wolf YI, Yakunin AF, van der Oost J, Koonin EV (2011) Evolution and classification of the CRISPR-Cas systems. Nat Rev Microbiol 9(6):467–477. doi:10.1038/nrmicro2577

Mandin P, Repoila F, Vergassola M, Geissmann T, Cossart P (2007) Identification of new noncoding RNAs in Listeria monocytogenes and prediction of mRNA targets. Nucleic Acids Res 35(3):962–974. doi:10.1093/nar/gkl1096

Marinus MG, Casadesus J (2009) Roles of DNA adenine methylation in host-pathogen interactions: mismatch repair, transcriptional regulation, and more. FEMS Microbiol Rev 33(3):488–503. doi:10.1111/j.1574-6976.2008.00159.x

Mignot T, Kirby JR (2008) Genetic circuitry controlling motility behaviors of *Myxococcus xanthus*. BioEssays 30(8):733–743. doi:10.1002/bies.20790

Muniesa M, Garcia A, Miro E, Mirelis B, Prats G, Jofre J, Navarro F (2004) Bacteriophages and diffusion of beta-lactamase genes. Emerg Infect Dis 10(6):1134–1137

Munz C, Lunemann JD, Getts MT, Miller SD (2009) Antiviral immune responses: triggers of or triggered by autoimmunity? Nat Rev Immunol 9(4):246–258. doi:10.1038/nri2527

Nariya H, Inouye M (2008) MazF, an mRNA interferase, mediates programmed cell death during multicellular myxococcus development. Cell 132(1):55–66. doi:10.1016/j.cell.2007.11.044

O'Toole GA, Kolter R (1998a) Flagellar and twitching motility are necessary for *Pseudomonas aeruginosa* biofilm development. Mol Microbiol 30(2):295–304

O'Toole GA, Kolter R (1998b) Initiation of biofilm formation in *Pseudomonas fluorescens* WCS365 proceeds via multiple, convergent signalling pathways: a genetic analysis. Mol Microbiol 28(3):449–461

O'Toole GA, Pratt LA, Watnick PI, Newman DK, Weaver VB, Kolter R (1999) Genetic approaches to study of biofilms. Methods Enzymol 310:91–109

Ogawa M, Fujitani S, Mao X, Inouye S, Komano T (1996) FruA, a putative transcription factor essential for the development of *Myxococcus xanthus*. Mol Microbiol 22(4):757–767

Orren DK, Selby CP, Hearst JE, Sancar A (1992) Post-incision steps of nucleotide excision repair in *Escherichia coli*. Disassembly of the UvrBC-DNA complex by helicase II and DNA polymerase I. J Biol Chem 267(2):780–788

Palmer T, Sargent F, Berks BC (2005) Export of complex cofactor-containing proteins by the bacterial tat pathway. Trends Microbiol 13(4):175–180. doi:10.1016/j.tim.2005.02.002

Partridge SR, Tsafnat G, Coiera E, Iredell JR (2009) Gene cassettes and cassette arrays in mobile resistance integrons. FEMS Microbiol Rev 33(4):757–784. doi:10.1111/j.1574-6976.2009.00175.x

Pedersen K, Christensen SK, Gerdes K (2002) Rapid induction and reversal of a bacteriostatic condition by controlled expression of toxins and antitoxins. Mol Microbiol 45(2):501–510

Perez-Rodriguez R, Haitjema C, Huang Q, Nam KH, Bernardis S, Ke A, DeLisa MP (2011) Envelope stress is a trigger of CRISPR RNA-mediated DNA silencing in *Escherichia coli*. Mol Microbiol 79(3):584–599. doi:10.1111/j.1365-2958.2010.07482.x

Raivio TL (2005) Envelope stress responses and gram-negative bacterial pathogenesis. Mol Microbiol 56(5):1119–1128. doi:10.1111/j.1365-2958.2005.04625.x

Raman K, Mohan C (2003) Genetic underpinnings of autoimmunity–lessons from studies in arthritis, diabetes, lupus and multiple sclerosis. Curr Opin Immunol 15(6):651–659 S0952791503001444

Rosenberg E (1984) Myxobacteria: development and cell interactions. Springer, New York

Sorek R, Kunin V, Hugenholtz P (2008) CRISPR–a widespread system that provides acquired resistance against phages in bacteria and archaea. Nat Rev Microbiol 6(3):181–186. doi:10.1038/nrmicro1793

Stern A, Keren L, Wurtzel O, Amitai G, Sorek R (2010) Self-targeting by CRISPR: gene regulation or autoimmunity?

Thony-Meyer L, Kaiser D (1993) devRS, an autoregulated and essential genetic locus for fruiting body development in *Myxococcus xanthus*. J Bacteriol 175(22):7450–7462

Toledo-Arana A, Dussurget O, Nikitas G, Sesto N, Guet-Revillet H, Balestrino D, Loh E, Gripenland J, Tiensuu T, Vaitkevicius K, Barthelemy M, Vergassola M, Nahori MA, Soubigou G, Regnault B, Coppee JY, Lecuit M, Johansson J, Cossart P (2009) The Listeria transcriptional landscape from saprophytism to virulence. Nature 459(7249):950–956. doi:10.1038/nature08080

Velicer GJ, Vos M (2009) Sociobiology of the myxobacteria. Annu Rev Microbiol 63:599–623. doi:10.1146/annurev.micro.091208.073158

Wiedenheft B, Zhou K, Jinek M, Coyle SM, Ma W, Doudna JA (2009) Structural basis for DNase activity of a conserved protein implicated in CRISPR-mediated genome defense. Structure 17(6):904–912. doi:10.1016/j.str.2009.03.019

Witte W (2004) International dissemination of antibiotic resistant strains of bacterial pathogens. Infect Genet Evol 4(3):187–191. doi:10.1016/j.meegid.2003.12.005S1567134804000310

Yosef I, Goren MG, Qimron U (2012) Proteins and DNA elements essential for the CRISPR adaptation process in *Escherichia coli*. Nucleic Acids Res 40(12):5569–5576. doi:10.1093/nar/gks216

Zegans ME, Wagner JC, Cady KC, Murphy DM, Hammond JH, O'Toole GA (2009) Interaction between bacteriophage DMS3 and host CRISPR region inhibits group behaviors of *Pseudomonas aeruginosa*. J Bacteriol 191(1):210–219. doi:10.1128/JB.00797-08

Chapter 11
Applications of the Versatile CRISPR-Cas Systems

Philippe Horvath, Giedrius Gasiunas, Virginijus Siksnys and Rodolphe Barrangou

Abstract CRISPR-Cas systems provide adaptive immunity against viruses and plasmids in bacteria and archaea. Interference is mediated by small non-coding CRISPR RNAs (crRNAs) that guide the Cas machinery towards complementary nucleic acids for sequence-specific cleavage. Several recent studies have shown that CRISPR-encoded immunity can increase the breadth and depth of phage resistance in bacteria, and can provide a barrier to acquisition of undesirable genetic elements, notably plasmid-encoded antibiotic resistance genes. Further, the adaptive and inheritable nature of those idiosyncratic chromosomal loci provide valuable genetic polymorphism which can be leveraged for typing purposes, proprietary strain tagging, ecological surveys, and epidemiological studies. The ability to readily transfer functional CRISPR-Cas systems across even distant bacteria, and re-program their endonuclease activity make them amenable to genetic engineering and useful for genome editing. These features, in combination with recent breakthroughs in unravelling the molecular underpinnings of the CRISPR mechanism of action have paved the way for several applications in a diversity of industrial and biotechnological areas.

P. Horvath (✉)
DuPont Nutrition and Health, Dangé-Saint-Romain 86220, France
e-mail: philippe.horvath@dupont.com

G. Gasiunas · V. Siksnys
Institute of Biotechnology, Vilnius University, LT–02241 Vilnius, Lithuania
e-mail: gasiunas@ibt.lt

V. Siksnys
e-mail: siksnys@ibt.lt

R. Barrangou
DuPont Nutrition and Health, Madison, WI 53716, USA
e-mail: rodolphe.barrangou@dupont.com

R. Barrangou and J. van der Oost (eds.), *CRISPR-Cas Systems*,
DOI: 10.1007/978-3-642-34657-6_11, © Springer-Verlag Berlin Heidelberg 2013

Contents

11.1 Introduction

While warfare is continuously being waged between microbes and their viral counterparts, arguably endlessly, a novel weapon is occasionally discovered in the arsenal, which might be exploited by humans to shift the balance between the conflicting parties. The raging battle between bacteria used as starter cultures in the food industry for the fermentation of milk into appetizing products such as yogurt and cheese, and their predatory bacteriophages has lasted for centuries, ever since the need to store surplus milk has arisen. Recently, CRISPR-Cas systems were shown to provide adaptive resistance against viruses of bacteria and archaea, and numerous studies have documented their functional properties, characterizing the molecular underpinnings of their biochemical mechanism of action. These studies have set the stage for leveraging those versatile molecular systems in a variety of technological applications.

The historical path that the CRISPR field has taken has been discussed in detail previously (see Chap. 1), and the occurrence, distribution, and evolution of those loci outlined (see Chaps. 2 and 3), with approximately 46 % of bacteria and 90 % of archaea carrying CRISPR loci, including many model and industrially relevant organisms. Notwithstanding the various types of CRISPR-Cas systems that have been established in the literature (see Chaps. 3, 5, 6, and 7), there are many elements that are somewhat conserved across CRISPR-Cas systems, both in mechanism of action and in function(s) that set the stage for a wide array of technological applications.

Although the field might be considered by many in its infancy, the CRISPR literature and citation rates reflect both the quantity and quality of the work that has been performed over the past decade. Further, the ability to potentially translate this work into tangible applications can be somewhat measured by the intellectual property activity, as monitored by a number of patent application deposits. To date, 12 patents related to CRISPR uses and applications have been published (see Table 11.1).

These patents span several distinct areas and types of applications, notably the detection and typing of bacterial strains, the development of phage resistance, and the use of CRISPR-Cas systems for interference and cleavage of nucleic acids. We highlight below a number of documented and potential applications of CRISPR-Cas systems.

11.2 Resistance Against Viruses

A variety of roles have been attributed to the diverse CRISPR-Cas systems within the last 10 years, including DNA repair and biofilm inhibition (Babu et al. 2011; Palmer and Whiteley 2011). Nevertheless, it has been quickly and broadly accepted that resistance against bacteriophages (phages), and more generally against viruses, is the primary and most common role of these small RNA-based interference systems. Hypothesized in 2005 (Pourcel et al. 2005; Mojica et al. 2005; Bolotin et al. 2005), the antiviral activity of CRISPR-Cas was demonstrated shortly thereafter with a food-grade bacterium of industrial relevance (Barrangou et al. 2007; see Table 11.1, patent application WO/2007/025097). Indeed, large-scale dairy fermentations using *Streptococcus thermophilus*-containing starter cultures are occasionally impaired by lytic phages, compelling starter cultures companies to constantly devise strategies aimed at controlling phage populations in industrial settings.

11.2.1 "CRISPerization": Phage Resistance Improvement Through Iterative Challenges

Traditionally, phages have been extensively used—intentionally or otherwise—to challenge sensitive bacterial strains, in order to select subpopulations named Bacteriophage-Insensitive Mutants (BIMs) that display increased viral resistance (Labrie et al. 2010). Besides providing naturally improved strains, such approaches have led to the identification of a variety of phage resistance mechanisms, notably those involved in the early steps (phage adsorption onto cell receptor(s), phage DNA injection within the cytoplasm) of the phage–host interaction (Sturino and Klaenhammer 2006). Although relatively easy to generate, BIMs generally show a weak and volatile protection against phages, mainly because phage populations evolve at a faster rate than their hosts. Furthermore, receptor mutations only provide resistance against a narrow spectrum of phages that use a conserved pathway for infection.

Table 11.1 Patent applications related to various uses of CRISPR-Cas systems

Publication number	Title	Inventors	Publication date (priority date)
WO/2006/073445	Detection and typing of bacterial strains	Russell et al.	13.07.2006 (28.04.2004)
WO/2007/025097	Use of CRISPR-associated genes (*cas*)	Horvath et al.	01.03.2007 (26.08.2005)
WO/2007/136815	Tagged microorganisms and methods of tagging	Barrangou et al.	29.11.2007 (19.05.2006)
WO/2008/108989	Cultures with improved phage resistance	Barrangou et al.	12.09.2008 (02.03.2007)
WO/2009/115861	Molecular typing and subtyping of *Salmonella* by identification of the variable nucleotide sequences of the CRISPR loci	Weill et al.	24.09.2009 (28.12.2007)
WO/2010/011961	Prokaryotic RNAi-like system and methods of use	Terns et al.	28.01.2010 (25.07.2008)
US20100076057	Target DNA interference with crRNA	Sontheimer and Marraffini	25.03.2010 (23.09.2008)
WO/2010/054108	Cas6 polypeptides and methods of use	Terns et al.	14.05.2010 (06.11.2008)
WO/2010/054154	Bifidobacteria CRISPR sequences	Romero et al.	14.05.2010 (07.11.2008)
WO/2010/075424	Compositions and methods for downregulating prokaryotic genes	Kunin et al.	01.07.2010 (22.12.2008)
WO/2011/143124	Endoribonuclease compositions and methods of use thereof	Haurwitz et al.	17.11.2011 (10.05.2010)
WO/2012/054726	*Lactococcus* CRISPR-*cas* sequences	Horvath et al.	26.04.2012 (20.10.2010)

Some CRISPR-Cas systems have been shown to be responsive to viral challenge, either naturally (Barrangou et al. 2007; Deveau et al. 2008; van der Ploeg 2009; Mills et al. 2010; Cady et al. 2012; Erdmann and Garrett 2012) or following genetic engineering and priming (Datsenko et al. 2012; Swarts et al. 2012; Yosef et al. 2012). Specifically, in the acquisition stage, small pieces (called proto-spacers) of the viral nucleic acid may be integrated as new spacers in-between new repeats at the leader end of CRISPR array(s), thus providing adaptive immunity. The presence of such additional spacers, subsequently transcribed in order to interfere with any complementary sequence, confers an improved resistance to the surviving host cell.

Based on this CRISPR-Cas adaptive system, "CRISPerization" strategies have been developed to rationally and purposefully generate improved lineages in *S. thermophilus* (see Table 11.1, notably patent application WO/2008/108989, and Fig. 11.1). Provided sufficient—both in number and diversity—virulent phages are available, iterative phage challenges may be performed (endlessly?) to increase the level of resistance of the host strain, leading to a stacking of newly acquired spacers. Furthermore, by selecting genetically diverse and industrially relevant phages, subsequent challenges advantageously broaden the spectrum of resistance of the host strain. Due to the apparent randomness of proto-spacer uptake (though new data suggest the proto-spacer sampling process is not completely random;

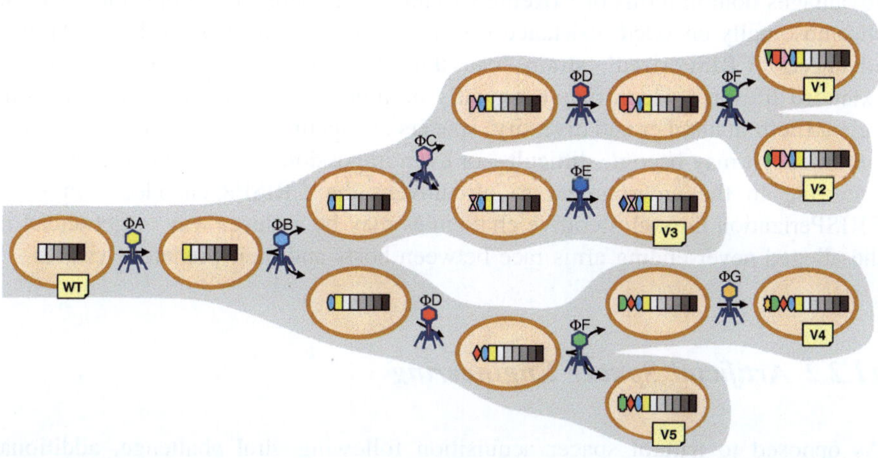

Fig. 11.1 CRISPerization process. The diagram displays the way in which CRISPR BIMs (bacteriophage insensitive mutants) can be selected following iterative exposure to phages (φA through φG), to generate multi-generational variants (V1–V5) that have acquired several new CRISPR spacers, eventually making them resistant to all phages used. *Colored rectangles* and other shapes represent CRISPR spacers newly acquired, with each color corresponding to the phage used in the challenge. WT, wild-type (parental) strain. Note that all phages do not need to be used in each lineage, as some spacers may be efficient against distinct phages sharing common sequences. This strategy can be enhanced by selecting CRISPR BIMs that have acquired spacers in multiple active CRISPR loci, that have acquired multiple spacers in a single round of phage exposure, and by selecting spacers that target highly conserved and/or functional sequences in phage genomes

Datsenko et al. 2012) and the broad reservoir of proto-spacers within each phage genome (typically several hundreds in a 35 kb genome), distinct bacterial lineages with complementary resistances may be generated by using independently the same phages, in the same order or otherwise (see Fig. 11.1).

CRISPerization by iterative challenges holds three major advantages over other phage resistance improvement strategies. First, the resulting variants are "natural microorganisms", a trait which is currently critical to the food industry, notably in Europe. No genetic engineering is involved in the process, which is purely based on the generation, selection, and characterization of surviving subpopulations. Second, all variants that are obtained, whatever the number of iterations they underwent, are isogenic variants of the parental wild-type strain, that have maintained their valuable functional properties. Theoretically they only carry mutations (i.e., additional repeat-spacer units) in their CRISPR array(s), thus maintaining identical physiological and functional properties, another critical trait for the robustness of industrial applications. Obviously, combinations of various isogenic strains in rotation schemes are highly valuable in the dairy manufacturing environment, and provide increased phage resistance both in terms of depth of phage resistance and breadth of the phage resistance spectrum. While the industry historically relies on rotation strategies combining distinct phage resistance mechanisms and phenotypes, CRISPR-mediated phage resistance provides advantages both in terms of isogenic variants' sustainable use, and stability of the chromosomally encoded resistance system, as opposed to plasmid-borne. Highly advanced "CRISPerized" strains can thus be considered as variants with an extended lifespan, which may eventually be immortalized. Finally, provided sufficient (both in number and diversity) spacers are acquired in controlled, laboratory conditions, it may become difficult—or even impossible—for the phages naturally occurring in the environment to circumvent the CRISPR-encoded immunity. CRISPerization through iterative challenges may be a clever way to get ahead in the alleged never-ending arms race between hosts and their predatory viruses.

11.2.2 Artificial Spacer Engineering

As opposed to natural spacer acquisition following viral challenge, additional spacers can be intentionally introduced into CRISPR arrays by using classical genetic engineering approaches. Only a relatively short segment (i.e., the size of a spacer) of the target nucleic acid sequence has to be known in order to build specific immunity against any complementary sequence (provided the associated proto-spacer adjacent motif is taken into account). A conservative, safe strategy is to "copy-paste" naturally occurring spacers (belonging to the same CRISPR-Cas system) between characterized strains. By extension, spacers can be designed entirely de novo prior to their integration between CRISPR repeats, so that it is virtually feasible to confer immunity against nucleic acid sequences that have never been observed yet. Engineered CRISPR arrays can also be an answer to strain improvement when no lytic virus is available for challenge, or when no virus is efficiently stimulating novel spacer acquisition.

We brought the first illustration of this approach in *S. thermophilus* in 2007 (Barrangou et al. 2007). Spacers S1 and S2, simultaneously acquired in the CRISPR1 locus of a BIM following a challenge with phage 858, were cloned from their host strain into a plasmid and transferred to the CRISPR1 locus of another strain, thereby transferring immunity against phage 858. De novo spacer engineering against phage Lambda was also performed in *Escherichia coli* (Brouns et al. 2008; Pougach et al. 2010; Sapranauskas et al. 2011), showing that CRISPR-Cas systems can be specifically engineered to contain particular spacers that target phage sequences and provide resistance against viruses that carry homologous sequences.

The major limit of artificial spacer engineering is the fact that not all sequences constitute efficient CRISPR spacers, due to the need, in some CRISPR-Cas systems, for a proto-spacer-associated motif (PAM) (Deveau et al. 2008; Horvath et al. 2008; Mojica et al. 2009). In such cases, the selection in a target nucleic acid of a sequence to be converted into a CRISPR spacer is constrained by the presence of an adequate PAM sequence.

11.2.3 Transfer Between Microorganisms

The propensity of CRISPR-Cas systems to be subjected to horizontal gene transfer has been documented for a while (Godde and Bickerton 2006; Horvath et al. 2009), and reflects the distribution and evolution of those systems, as discussed in Chaps. 2 and 3, respectively.

Finally, phage resistance may also be obtained through the transfer of a complete CRISPR-Cas system between strains (not necessarily belonging to the same species), as exemplified by Sapranauskas et al. (2011). After cloning of the *S. thermophilus* CRISPR3-Cas system on a plasmid, it was readily transferred into *E. coli,* and could provide resistance against phage and lower plasmid uptake propensity. The next major advance will be to assess whether functional systems can be transferred and/or engineered to provide nucleic acid interference in valuable, important, and model eukaryotic organisms, especially for agricultural, biotechnological, and medical applications (see Table 11.1, patent applications WO/2012/054726 and WO/2011/143124). As the visibility of the field increases, we expect that attempts will be made to engineer CRISPR-encoded interference in yeast, fungi, plants, and perhaps vertebrates.

11.3 Immunity Against Non-Viral Nucleic Acids

Although resistance against viruses is arguably the primary functional role of CRISPR-Cas systems, as it provides immunity against nucleic acids through base-pairing between spacer-derived crRNAs and complementary target sequences,

DNA or RNA molecules other than virus-encoded may be subjected to interference. Indeed, similarity searches of spacer sequences within DNA databases generally show that, besides the large majority of matches with viral sequences, most of the matches correspond to plasmid sequences, followed by a minority of hits to chromosomal sequences (Horvath et al. 2008; Stern et al. 2010). The low occurrence of plasmid- and chromosome-derived spacers in CRISPR arrays may probably be considered as a side effect of the adaptive nature of CRISPR-Cas systems, whereby host genetic material (or the transcription products thereof), are perceived as "foreign" rather than "self" nucleic acid molecules. Thus, CRISPR/Cas systems may be exploited to provide non-viral immunity.

11.3.1 Plasmid Interference

Several reports in the literature document the ability of CRISPR-Cas systems to provide interference against plasmid DNA. It was first established in a milestone and elegant study in *Staphylococcus epidermidis* whereby CRISPR-encoded spacers lowered efficiency of plasmid uptake. This study also established that the primary CRISPR-Cas nucleic acid target is DNA (Marraffini and Sontheimer 2008). Subsequently, several studies showed that spacers can be acquired from plasmid sequences, and interfere with plasmid uptake (Garneau et al. 2010; Sapranauskas et al. 2011; Swarts et al. 2012; Datsenko et al. 2012; Jinek et al. 2012; Gasiunas et al. 2012).

11.3.2 Interference Against Other Mobile Elements

The documented ability of CRISPR-Cas systems to preclude plasmid uptake in *S. epidermidis, S. thermophilus,* and *E. coli* has set the stage for developing CRISPR-based systems that provide interference against mobile genetic elements. Given the elevated concerns about antibiotic resistance marker dissemination, especially in clinically relevant human pathogens, in combination with the circumstantial evidence which indicated a negative correlation between the occurrence of CRISPR-Cas systems and pathogenicity in *Enterococcus* (Palmer and Gilmore 2010) and *Campylobacter* (Schouls et al. 2003; Louwen et al. 2012), there is tremendous potential in leveraging CRISPR-mediated interference against antibiotic resistance genes. A recent report documenting the ability of *S. thermophilus* to naturally acquire spacers that target an antibiotic resistance gene (Garneau et al. 2010), in combination with the ability of the acquired spacers to preclude the uptake of plasmids that carry homologous DNA sequences, sets the stage for vaccination of bacterial strains against antibiotic resistance marker uptake. Similarly, because prophages can also readily mediate the transfer of pathogenic markers, CRISPR-encoded immunity can be used to reduce the pathogenic potential of a microorganism though reduction of its propensity to uptake novel DNA. This is consistent with the reported negative correlation between the occurrence of prophages and CRISPR spacers in *Streptococcus pyogenes* (Nozawa et al. 2011). As such, active CRISPR-Cas systems provide a natural means to select strains that are unlikely to uptake and disseminate antibiotic resistance markers and pathogenic traits. Likewise, the ability to engineer CRISPR-Cas systems with synthetic spacers provides an in vitro means to generate mutants that are refractory to the uptake of undesirable sequences. Indeed, recent reports show that CRISPR can prevent natural transformation and virulence marker acquisition in *Streptococcus pneumoniae* (Bikard et al. 2012), and influence mobilome diversity in *Streptococcus agalactiae* (Lopez-Sanchez et al. 2012).

Prokaryotic genome integrity and stability may be affected by the integration or excision of mobile genetic elements such as transposons and prophages. Spacers designed to target transposons and mobile genetic elements that mediate chromosomal rearrangements and shuffling could be used to increase chromosomal stability and integrity. Likewise, spacers targeting undesirable genes such as those coding for antibiotic resistance, toxins, and virulence factors could be used to generate "safer" strains.

11.4 CRISPR-Based Gene Regulation

Despite the genetic commonalities observed across the three CRISPR-Cas types, and the conservation of several mechanistic steps in various systems, in some Type III systems, at least, CRISPR targets RNA in vitro (Hale et al. 2009; Garrett et al. 2011). Accordingly, there is potential to use CRISPR-Cas systems for the regulation, transcriptional control, or regulation of transcript levels within a cell (Horvath and Barrangou 2010; see Table 11.1, patent application WO/2010/075424). A recent report illustrates the ability of CRISPR spacers to lower transcript levels, showing that a spacer homologous to the histidyl-tRNA synthetase sequence lowers His-tRNA levels (Aklujkar and Lovley 2010). A study documenting several examples of self-targeting spacers shows that this phenomenon may be under-appreciated (Stern et al. 2010). Likewise, several studies have implicated self-targeting CRISPR spacers in *Pseudomonas aeruginosa* lysogeny (Zegans et al. 2009; Cady and O'Toole 2011).

Analogies between CRISPR-mediated interference and RNA interference have been discussed in several reviews, and multiple studies have provided enough circumstantial evidence that crRNA can silence transcripts to pave the way for CRISPR-based mRNA targeting. Further, given the ancillary and emerging roles of CRISPR-Cas systems beyond foreign DNA defensive targeting (see Chap. 10), notably host regulatory and developmental processes, there are several RNA-targeting applications that can be developed. Given the tremendous interest in and the many successes of RNAi in eukaryotic systems, together with the growing importance of non-coding small RNAs in numerous biological functions, there is potential to harness the flexibility and modularity of CRISPR-Cas systems for RNA interference in bacteria and archaea (see Table 11.1, patent applications WO/2010/011961 and WO/2010/054108).

11.5 CRISPR-Based Strain Typing

A historical review of the CRISPR literature over time (see Chap. 1) clearly illustrates the potential of CRISPR loci for genotyping of bacteria. Several early studies that preceded the implication of CRISPR-Cas systems in adaptive immunity, notably spoligotyping in the early 1990s, have actually shown that these loci are both hypervariable, and provide a time-dependent iterative record of the environmental conditions to which a strain has been exposed. A milestone method describing the use of "direct repeat region" DNA sequences in the chromosome of *Mycobacterium tuberculosis* in 1993 (Groenen et al. 1993) had the foresight to observe that there was tremendous polymorphism across a diversity of strains in this particular region, and that sequence content could be digitized to monitor the epidemiology of clinical cases and samples of tuberculosis (Brudey et al. 2006). A similar approach was subsequently used and developed for *Corynebacterium diphtheriae* (Mokrousov et al. 2007, 2009). Undoubtedly, this is an insightful example of the contribution of genomics to the discovery of unknown, uncharacterized, occasionally un- or mis-annotated regions that nonetheless are hypervariable enough to provide a basis for genotyping. Indeed, the literature spans distant industrial or pathogenic bacteria across which CRISPR-based genotyping provides insights, notably *M. tuberculosis* (Abadia et al. 2010; Borile et al. 2011; Brudey et al. 2006; Groenen et al. 1993; Zhang et al. 2010), *Yersinia pestis* (Cui et al. 2008; Pourcel et al. 2005; Riehm et al. 2012; Vergnaud et al. 2007), *C. diphtheriae* (Mokrousov et al. 2007, 2009), *P. aeruginosa* (Cady et al. 2011), *Legionella* (D'Auria et al. 2010; Ginevra et al. 2012), *S. pyogenes* (Hoe et al. 1999; McShan et al. 2008), *S. thermophilus* (Horvath et al. 2008), *Lactobacillus* (see Table 11.1, patent WO/2006/073445), *Propionibacterium acnes* (Brüggemann et al. 2012), *Erwinia amylovora* (Rezzonico et al. 2011; McGhee and Sundin 2012), *Campylobacter* (Tasaki et al. 2012), *Salmonella* (Liu et al. 2011a, b; Fabre et al. 2012; Fricke et al. 2011; see Table 11.1, patent application WO/2009/115861), and pathogenic *E. coli* (Díez-Villaseñor et al. 2010; Delannoy et al. 2012).

Over time, the molecular methods that target CRISPR sequences have evolved. Initially, hybridization-based spoligotyping was developed in *Mycobacterium* and *Corynebacterium*, although results were highly dependent on the reference database, and solely known sequences could be targeted. Later on, Sanger-sequencing of CRISPR PCR amplicons, either completely or partially from the extremities, was developed and implemented for the genotyping of some species (see Fig. 11.2). Alternatives to sequencing were also assessed to compare and contrast CRISPR PCR amplicons, notably restriction fragment length polymorphism (RFLP) assays, capillary electrophoresis analysis, and melting curve analysis

(Price et al. 2007). Nowadays, the ubiquitous and affordable natures of multiple sequencing technologies have rendered such approaches nearly obsolete. In fact, the pace of next-generation sequencing technologies development, in combination with the ever-increasing throughput and rapidly decreasing price, have opened new avenues for deep sequencing analysis of CRISPR amplicons and mixed population metagenomes. Currently, sequencing technologies have out-paced the development of fast, efficient, and convenient bioinformatic tools which provide the reconstruction of CRISPR loci and visualization of their content.

11.6 Bacterial or Viral Strain Tracking

Further, the presence and diversity of CRISPR-Cas systems and their hypervariable spacer sequences in a diversity of industrially relevant bacteria provide a similar basis for genotyping of commercial strains, notably for lactic acid bacteria widely used as starter cultures in the dairy industry (Horvath et al. 2008, 2009; Barrangou and Horvath 2012). Even within a clonal population, active CRISPR loci are hypervariable and adaptive enough to track a strain over time, as shown in *Leptospirillum* isolated from acid mine drainage samples (Andersson and Banfield 2008; Tyson and Banfield 2008). Other metagenomics studies have shown that CRISPR loci can provide critical insights into population diversity and dynamics (Heidelberg et al. 2009; Held and Whitaker 2009; Anderson et al. 2011; Berg et al. 2012; Delaney et al. 2012; Garcia-Heredia et al. 2012; Pride et al. 2011, 2012; Rho et al. 2012; Stern et al. 2012). CRISPR spacer sequences may also be exploited to detect viral sequences or fish out viruses from complex, undefined ecosystems (Snyder et al. 2010). For metagenomic surveys, resolving CRISPR loci for mixed and occasionally complex microbial populations can unravel dynamics and ancestral relationships and occasionally reflect dramatic shifts and events such as selective bottlenecks. Nevertheless, it is important to keep in mind that such loci have variable typing potential across organisms given their broad range of (in-) activity and their highly variable distribution, occurrence, and propensity for horizontal gene transfer. Also, when multiple CRISPR loci are present within a chromosome, it is important to target a universal and polymorphic locus. Accordingly, their epidemiological potential has to be evaluated on a case-by-case basis, preferably using a broad and bio-geographically diverse set of strains and isolates.

11.7 Natural Genetic Tagging

In combination with increased phage resistance, CRISPR-Cas systems provide a tremendous avenue for the development of immortalized industrial workhorses which have highly desirable functional traits for the food supply chain, or that

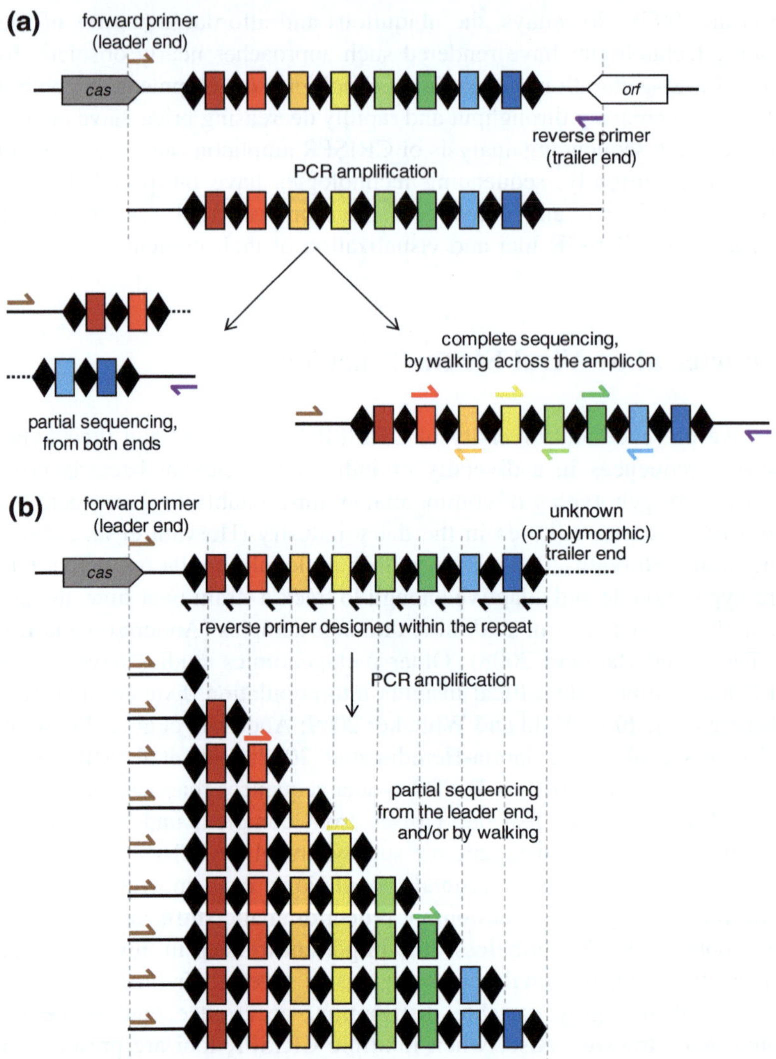

Fig. 11.2 CRISPR-based typing schemes, *Panel A*: sequencing CRISPR arrays from both ends. For strain typing purposes, sequencing from both the leader and trailer (i.e., opposite to the leader) ends should always be preferred, when possible. Ancient spacers at the trailer end allow clustering of distantly related strains, while leader-end spacers, more recently acquired, differentiate closely related strains. In many cases, sequencing the whole CRISPR repeat-spacer array requires significant time and effort but adds only little information. If necessary, sequencing by walking across can be performed by designing primers within non-redundant spacer sequences. *Panel B*: one-sided CRISPR typing. When the sequences surrounding CRISPR arrays are polymorphic or unknown, sequencing is still possible from the conserved leader end, especially when *cas* genes are present. The PCR amplicon mix generated by using a reverse primer designed within the repeat sequence can be sequenced from the leader end and/or from internal spacers

carry valuable biotechnological properties. In a competitive and global environment, although bacteria have been universally used as starter cultures in the food industry for centuries, it is increasingly critical to secure intellectual property and monitor the use of proprietary highly valuable strains.

A broadly used strategy is the deposit of characterized strains in strain banks, notably the culture collections that are official depositories under the Budapest Treaty (Budapest Treaty on the International Recognition of the Deposit of Microorganisms for the Purposes of Patent Procedure, signed on April 28, 1977). However, as strains evolve over time, and the origin and ownership of natural biological entities is difficult to define, it is important to secure intellectual property rights for the use of specific material for particular applications in specific fields. Accordingly, correctly and accurately defining a proprietary strain is critical, and CRISPR provides a unique natural means to generate mutants that have iteratively acquired a unique array of novel spacers in a human-defined order, directed manner, and selected way (see Table 11.1, patent application WO/2007/136815). Thus, iteratively selecting BIMs that have acquired novel CRISPR spacers following exposure to phage(s) (see Fig. 11.1) generates a natural (not genetically engineered) variant with a sequence tag (set of novel CRISPR spacers) which has an extremely remote probability to randomly arise in nature. This unique genetic watermark can subsequently be used to monitor the presence of a proprietary strain in any environment through simple and affordable Sanger sequencing of a CRISPR PCR amplicon.

11.8 Cas Endonuclease Reprogramming and Restriction Enzyme Customization

Two recent reports have shown that Cas-mediated DNA cleavage can be reprogrammed through crRNA design (Gasiunas et al. 2012; Jinek et al. 2012). Jinek et al. showed that both crRNA and tracrRNA direct DNA cleavage in *S. pyogenes*, and that a chimeric RNA can be engineered to redefine cleavage specificity. Gasiunas et al. showed that the *S. thermophilus* Cas9–crRNA ribonucleoprotein complex mediates specific DNA cleavage, and that the Cas9 HNH and RuvC domains nick the complementary and non-complementary DNA strands, respectively, ultimately generating a dsDNA cleavage. This is consistent with previous studies showing that Cas9 cleaves phage and plasmid dsDNA (Garneau et al. 2010; Sapranauskas et al. 2011; Magadán et al. 2012). The ability to nick either or both DNA strand(s) at (re-)programmable locations in a DNA sequence (see Fig. 11.3) opens new avenues for genome editing, stacking, shuffling, and engineering (Barrangou 2012). This essentially adds a new option to the genome engineering toolkit, in addition to zinc finger nucleases (ZFNs) and transcription activator-like effector nucleases (TALENs). Typically, genome engineering relies on site-specific endonucleases that trigger sequence modification by DNA-repair systems at the

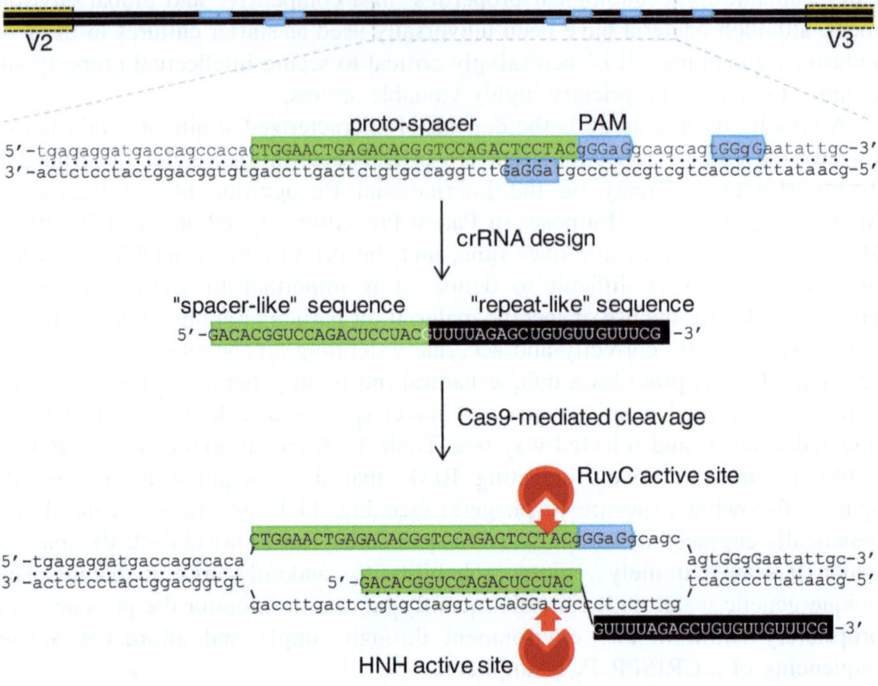

Fig. 11.3 Endonuclease customization. Cas9 endonuclease reprogramming. Any sequence containing at least one appropriate PAM can be cleaved specifically in its vicinity, at a precise location. In the example provided, the aim was to design a cleavage site within the *E. coli* 16S rDNA gene, between the variable regions V2 and V3, using the *S. thermophilus* CRISPR3-Cas system. Eight CRISPR3 PAM sequences (5′-NGNGG-3′, depicted as blue pentagons; Horvath et al. 2008) are found within this 183 bp region. The proper design and use of a chimeric crRNA targeting the proto-spacer (*green rectangle*), in combination with the Cas9 endonuclease, will lead to dsDNA cleavage within the proto-spacer, 3 nt upstream of the PAM (*red arrows*). Furthermore, the use of the wild-type Cas9, or RuvC- or HNH- mutants, lead to a double-stranded, single (+)stranded, or single (−)stranded cleavage, respectively (Gasiunas et al. 2012)

cleavage site. An advantage of Cas-crRNA-mediated cleavage is that specificity can be readily reprogrammed by customizing the crRNA sequence, rather than re-engineering cleavage proteins (ZFNs or TALENs) each time a new sequence has to be targeted (Gasiunas et al. 2012; Hale et al. 2012; Jinek et al. 2012).

11.9 Other Applications of CRISPR-Cas Systems

There are other ancillary and less documented roles and applications of CRISPR-Cas systems that remain to be substantiated and investigated, notably the potential that these loci have for the genesis of "large" amounts of small interfering RNAs (Djordjevic et al. 2012; see Table 11.1, patent application US20100076057), and the ability to generate and select "super phages" that circumvent CRISPR-encoded immunity for advanced biocontrol of microbial populations and phage therapy (see Table 11.1, patent application WO/2008/108989).

11.10 Conclusions and Perspectives

Overall, many intrinsic features of CRISPR-Cas systems provide avenues for applications that cover a broad spectrum, ranging from exploiting genetic hypervariability for typing and epidemiological purposes to increasing viral resistance, immunizing strains against the uptake of undesirable genetic material, through the generation of programmable RNA-guided endonucleases for genome engineering, editing, and stacking. Notwithstanding the tremendous potential of CRISPR loci in bacteria and archaea, it is critical to assess their potential for in vivo activity in eukaryotes to fully assess the potential of CRISPR-Cas systems for white biotechnology and next-generation synthetic biology.

As we reflect upon the past decade of CRISPR research, the impressive quality and quantity of manuscripts that have showcased their many powerful functionalities, in combination with the engaged and collegial CRISPR scientist community that has made the field so enjoyable and productive, it is obvious that the publication and citation rates of CRISPR manuscripts, together with the increased intellectual property activity, highlight the potential that these systems have for a diversity of applications.

Clearly, significant recent advances in phage resistance and strain typing have set the stage for extending the longevity of valuable industrial strains, and new epidemiological frameworks, respectively. For the latter, it is yet to be determined whether CRISPR loci can universally or broadly be used for typing of clinical isolates highly relevant for human health and disease. Nevertheless, we certainly hope the best is yet to come, and that many a talented and creative scientist will come up with innovative ways to harness the beauty and power of CRISPR-Cas systems for valuable and beneficial purposes.

Acknowledgments We are indebted to our DuPont colleagues, notably Patrick Boyaval, Christophe Fremaux, and Dennis Romero, and our many academic collaborators with whom we have had the privilege to share exciting times exploring CRISPR-*cas* systems, notably the Moineau laboratory at Laval, the Banfield laboratory at the University of California Berkeley, the Roberts laboratory at the Pennsylvania State University, the Dudley laboratory at the Pennsylvania State University, the Terns laboratory at the University of Georgia, the Levin laboratory at Emory, the Bhaya laboratory at Stanford University, the VerBerkmoes laboratory at ORNL, Eugene Koonin and Kira Makarova at NCBI, Eric Brown and Marc Allard at FDA, and our dedicated and talented teams. Work in Vilnius was funded by the European Social Fund under Global Grant measure Project R100.

References

Abadia E, Zhang J, dos Vultos T, Ritacco V, Kremer K, Aktas E, Matsumoto T, Refregier G, van Soolingen D, Gicquel B, Sola C (2010) Resolving lineage assignation on *Mycobacterium tuberculosis* clinical isolates classified by spoligotyping with a new high-throughput 3R SNPs based method. Infect Genet Evol 10:1066–1074

Aklujkar M, Lovley DR (2010) Interference with histidyl-tRNA synthetase by a CRISPR spacer sequence as a factor in the evolution of *Pelobacter carbinolicus*. BMC Evol Biol 10:230

Andersson AF, Banfield JF (2008) Virus population dynamics and acquired virus resistance in natural microbial communities. Science 320:1047–1050

Anderson RE, Brazelton WJ, Baross JA (2011) Using CRISPRs as a metagenomic tool to identify microbial hosts of a diffuse flow hydrothermal vent viral assemblage. FEMS Microbiol Ecol 77:120–133

Babu M, Beloglazova N, Flick R, Graham C, Skarina T, Nocek B, Gagarinova A, Pogoutse O, Brown G, Binkowski A, Phanse S, Joachimiak A, Koonin EV, Savchenko A, Emili A, Greenblatt J, Edwards AM, Yakunin AF (2011) A dual function of the CRISPR-Cas system in bacterial antivirus immunity and DNA repair. Mol Microbiol 79:484–502

Barrangou R (2012) RNA-mediated programmable DNA cleavage. Nat Biotechnol 30:836–838

Barrangou R, Horvath P (2012) CRISPR: new horizons in phage resistance and strain identification. Annu Rev Food Sci Technol 3:143–162

Barrangou R, Fremaux C, Deveau H, Richards M, Boyaval P, Moineau S, Romero DA, Horvath P (2007) CRISPR provides acquired resistance against viruses in prokaryotes. Science 315:1709–1712

Berg Miller ME, Yeoman CJ, Chia N, Tringe SG, Angly FE, Edwards RA, Flint HJ, Lamed R, Bayer EA, White BA (2012) Phage-bacteria relationships and CRISPR elements revealed by a metagenomic survey of the rumen microbiome. Environ Microbiol 14:207–227

Bikard D, Hatoum-Aslan A, Mucida D, Marraffini LA (2012) CRISPR interference can prevent natural transformation and virulence acquisition during in vivo bacterial infection. Cell Host Microbe 12:177–186

Bolotin A, Quinquis B, Sorokin A, Ehrlich SD (2005) Clustered regularly interspaced short palindrome repeats (CRISPRs) have spacers of extrachromosomal origin. Microbiology 151:2551–2561

Borile C, Labarre M, Franz S, Sola C, Refrégier G (2011) Using affinity propagation for identifying subspecies among clonal organisms: lessons from *M. tuberculosis*. BMC Bioinf 12:224

Brouns SJ, Jore MM, Lundgren M, Westra ER, Slijkhuis RJ, Snijders AP, Dickman MJ, Makarova KS, Koonin EV, van der Oost J (2008) Small CRISPR RNAs guide antiviral defense in prokaryotes. Science 321:960–964

Brudey K, Driscoll JR, Rigouts L et al (2006) *Mycobacterium tuberculosis* complex genetic diversity: mining the fourth international spoligotyping database (SpolDB4) for classification, population genetics and epidemiology. BMC Microbiol 6:23

Brüggemann H, Lomholt HB, Tettelin H, Kilian M (2012) CRISPR/*cas* loci of type II *Propionibacterium acnes* confer immunity against acquisition of mobile elements present in type I *P. acnes*. PLoS ONE 7:e34171

Cady KC, O'Toole GA (2011) Non-identity-mediated CRISPR-bacteriophage interaction mediated via the Csy and Cas3 proteins. J Bacteriol 193:3433–3445

Cady KC, White AS, Hammond JH, Abendroth MD, Karthikeyan RS, Lalitha P, Zegans ME, O'Toole GA (2011) Prevalence, conservation and functional analysis of *Yersinia* and *Escherichia* CRISPR regions in clinical *Pseudomonas aeruginosa* isolates. Microbiology 157:430–437

Cady KC, Bondy-Denomy J, Heussler GE, Davidson AR, O'Toole GA (2012) The CRISPR/Cas adaptive immune system of *Pseudomonas aeruginosa* mediates resistance to naturally occurring and engineered phages. J Bacteriol 194:5728–5738

Cui Y, Li Y, Gorgé O, Platonov ME, Yan Y, Guo Z, Pourcel C, Dentovskaya SV, Balakhonov SV, Wang X, Song Y, Anisimov AP, Vergnaud G, Yang R (2008) Insight into microevolution of *Yersinia pestis* by clustered regularly interspaced short palindromic repeats. PLoS ONE 3:e2652

Datsenko KA, Pougach K, Tikhonov A, Wanner BL, Severinov K, Semenova E (2012) Molecular memory of prior infections activates the CRISPR/Cas adaptive bacterial immunity system. Nat Commun 3:945

D'Auria G, Jiménez-Hernández N, Peris-Bondia F, Moya A, Latorre A (2010) *Legionella pneumophila* pangenome reveals strain-specific virulence factors. BMC Genomics 11:181

Delaney NF, Balenger S, Bonneaud C, Marx CJ, Hill GE, Ferguson-Noel N, Tsai P, Rodrigo A, Edwards SV (2012) Ultrafast evolution and loss of CRISPRs following a host shift in a novel wildlife pathogen, *Mycoplasma gallisepticum*. PLoS Genet 8:e1002511

Delannoy S, Beutin L, Burgos Y, Fach P (2012) Specific detection of enteroaggregative hemorrhagic *Escherichia coli* O104:H4 strains using the CRISPR locus as target for a diagnostic real-time PCR. J Clin Microbiol 50:3485–3492

Deveau H, Barrangou R, Garneau JE, Labonté J, Fremaux C, Boyaval P, Romero DA, Horvath P, Moineau S (2008) Phage response to CRISPR-encoded resistance in *Streptococcus thermophilus*. J Bacteriol 190:1390–1400

Díez-Villaseñor C, Almendros C, García-Martínez J, Mojica FJ (2010) Diversity of CRISPR loci in *Escherichia coli*. Microbiology 156:1351–1361

Djordjevic M, Djordjevic M, Severinov K (2012) CRISPR transcript processing: a mechanism for generating a large number of small interfering RNAs. Biol Direct 7:24

Erdmann S, Garrett RA (2012) Selective and hyperactive uptake of foreign DNA by adaptive immune systems of an archaeon via two distinct mechanisms. Mol Microbiol 85:1044–1056

Fabre L, Zhang J, Guigon G, Le Hello S, Guibert V, Accou-Demartin M, de Romans S, Lim C, Roux C, Passet V, Diancourt L, Guibourdenche M, Issenhuth-Jeanjean S, Achtman M, Brisse S, Sola C, Weill FX (2012) CRISPR typing and subtyping for improved laboratory surveillance of *Salmonella* infections. PLoS ONE 7:e36995

Fricke WF, Mammel MK, McDermott PF, Tartera C, White DG, Leclerc JE, Ravel J, Cebula TA (2011) Comparative genomics of 28 *Salmonella enterica* isolates: evidence for CRISPR-mediated adaptive sublineage evolution. J Bacteriol 193:3556–3568

Garcia-Heredia I, Martin-Cuadrado AB, Mojica FJ, Santos F, Mira A, Antón J, Rodriguez-Valera F (2012) Reconstructing viral genomes from the environment using fosmid clones: the case of haloviruses. PLoS ONE 7:e33802

Garneau JE, Dupuis MÈ, Villion M, Romero DA, Barrangou R, Boyaval P, Fremaux C, Horvath P, Magadán AH, Moineau S (2010) The CRISPR/Cas bacterial immune system cleaves bacteriophage and plasmid DNA. Nature 468:67–71

Garrett RA, Shah SA, Vestergaard G, Deng L, Gudbergsdottir S, Kenchappa CS, Erdmann S, She Q (2011) CRISPR-based immune systems of the Sulfolobales: complexity and diversity. Biochem Soc Trans 39:51–57

Gasiunas G, Barrangou R, Horvath P, Siksnys V (2012) Cas9-crRNA ribonucleoprotein complex mediates specific DNA cleavage for adaptive immunity in bacteria. Proc Natl Acad Sci U S A 109:E2579–E2586

Ginevra C, Jacotin N, Diancourt L, Guigon G, Arquilliere R, Meugnier H, Descours G, Vandenesch F, Etienne J, Lina G, Caro V, Jarraud S (2012) *Legionella pneumophila* sequence type 1/Paris pulsotype subtyping by spoligotyping. J Clin Microbiol 50:696–701

Godde JS, Bickerton A (2006) The repetitive DNA elements called CRISPRs and their associated genes: evidence of horizontal transfer among prokaryotes. J Mol Evol 62:718–729

Groenen PM, Bunschoten AE, van Soolingen D, van Embden JD (1993) Nature of DNA polymorphism in the direct repeat cluster of *Mycobacterium tuberculosis*; application for strain differentiation by a novel typing method. Mol Microbiol 10:1057–1065

Hale CR, Zhao P, Olson S, Duff MO, Graveley BR, Wells L, Terns RM, Terns MP (2009) RNA-guided RNA cleavage by a CRISPR RNA-Cas protein complex. Cell 139:945–956

Hale CR, Majumdar S, Elmore J, Pfister N, Compton M, Olson S, Resch AM, Glover CV 3rd, Graveley BR, Terns RM, Terns MP (2012) Essential features and rational design of CRISPR RNAs that function with the Cas RAMP module complex to cleave RNAs. Mol Cell 45:292–302

Heidelberg JF, Nelson WC, Schoenfeld T, Bhaya D (2009) Germ warfare in a microbial mat community: CRISPRs provide insights into the co-evolution of host and viral genomes. PLoS ONE 4:e4169

Held NL, Whitaker RJ (2009) Viral biogeography revealed by signatures in *Sulfolobus islandicus* genomes. Environ Microbiol 11:457–466

Hoe N, Nakashima K, Grigsby D, Pan X, Dou SJ, Naidich S, Garcia M, Kahn E, Bergmire-Sweat D, Musser JM (1999) Rapid molecular genetic subtyping of serotype M1 group A *Streptococcus* strains. Emerg Infect Dis 5:254–263

Horvath P, Barrangou R (2010) CRISPR/Cas, the immune system of bacteria and archaea. Science 327:167–170

Horvath P, Romero DA, Coûté-Monvoisin AC, Richards M, Deveau H, Moineau S, Boyaval P, Fremaux C, Barrangou R (2008) Diversity, activity, and evolution of CRISPR loci in *Streptococcus thermophilus*. J Bacteriol 190:1401–1412

Horvath P, Coûté-Monvoisin AC, Romero DA, Boyaval P, Fremaux C, Barrangou R (2009) Comparative analysis of CRISPR loci in lactic acid bacteria genomes. Int J Food Microbiol 131:62–70

Jinek M, Chylinski K, Fonfara I, Hauer M, Doudna JA, Charpentier E (2012) A programmable dual-RNA-guided DNA endonuclease in adaptive bacterial immunity. Science 337:816–821

Labrie SJ, Samson JE, Moineau S (2010) Bacteriophage resistance mechanisms. Nat Rev Microbiol 8:317–327

Liu F, Barrangou R, Gerner-Smidt P, Ribot EM, Knabel SJ, Dudley EG (2011a) Novel virulence gene and clustered regularly interspaced short palindromic repeat (CRISPR) multilocus sequence typing scheme for subtyping of the major serovars of *Salmonella enterica* subsp. *enterica*. Appl Environ Microbiol 77:1946–1956

Liu F, Kariyawasam S, Jayarao BM, Barrangou R, Gerner-Smidt P, Ribot EM, Knabel SJ, Dudley EG (2011b) Subtyping *Salmonella enterica* serovar enteritidis isolates from different sources by using sequence typing based on virulence genes and clustered regularly interspaced short palindromic repeats (CRISPRs). Appl Environ Microbiol 77:4520–4526

Lopez-Sanchez MJ, Sauvage E, Da Cunha V, Clermont D, Ratsima Hariniaina E, Gonzalez-Zorn B, Poyart C, Rosinski-Chupin I, Glaser P (2012) The highly dynamic CRISPR1 system of *Streptococcus agalactiae* controls the diversity of its mobilome. Mol Microbiol 85:1057–1071

Louwen R, Horst-Kreft D, de Boer AG, van der Graaf L, de Knegt G, Hamersma M, Heikema AP, Timms AR, Jacobs BC, Wagenaar JA, Endtz HP, van der Oost J, Wells JM, Nieuwenhuis EE, van Vliet AH, Willemsen PT, van Baarlen P, van Belkum A (2012) A novel link between *Campylobacter jejuni* bacteriophage defence, virulence and Guillain-Barré syndrome. Eur J Clin Microbiol Infect Dis. doi:10.1007/s10096-012-1733-4

Magadán AH, Dupuis MÈ, Villion M, Moineau S (2012) Cleavage of Phage DNA by the *Streptococcus thermophilus* CRISPR3-Cas System. PLoS ONE 7:e40913

Marraffini LA, Sontheimer EJ (2008) CRISPR interference limits horizontal gene transfer in staphylococci by targeting DNA. Science 322:1843–1845

McGhee GC, Sundin GW (2012) Erwinia amylovora CRISPR elements provide new tools for evaluating strain diversity and for microbial source tracking. PLoS ONE 7:e41706

McShan WM, Ferretti JJ, Karasawa T, Suvorov AN, Lin S, Qin B, Jia H, Kenton S, Najar F, Wu H, Scott J, Roe BA, Savic DJ (2008) Genome sequence of a nephritogenic and highly transformable M49 strain of *Streptococcus pyogenes*. J Bacteriol 190:7773–7785

Mills S, Griffin C, Coffey A, Meijer WC, Hafkamp B, Ross RP (2010) CRISPR analysis of bacteriophage-insensitive mutants (BIMs) of industrial *Streptococcus thermophilus*—implications for starter design. J Appl Microbiol 108:945–955

Mojica FJ, Díez-Villaseñor C, García-Martínez J, Soria E (2005) Intervening sequences of regularly spaced prokaryotic repeats derive from foreign genetic elements. J Mol Evol 60:174–182

Mojica FJ, Díez-Villaseñor C, García-Martínez J, Almendros C (2009) Short motif sequences determine the targets of the prokaryotic CRISPR defence system. Microbiology 155:733–740

Mokrousov I, Limeschenko E, Vyazovaya A, Narvskaya O (2007) *Corynebacterium diphtheriae* spoligotyping based on combined use of two CRISPR loci. Biotechnol J 2:901–906

Mokrousov I, Vyazovaya A, Kolodkina V, Limeschenko E, Titov L, Narvskaya O (2009) Novel macroarray-based method of *Corynebacterium diphtheriae* genotyping: evaluation in a field study in Belarus. Eur J Clin Microbiol Infect Dis 28:701–703

Nozawa T, Furukawa N, Aikawa C, Watanabe T, Haobam B, Kurokawa K, Mauyama F, Nakagawa I (2011) CRISPR inhibition of prophage acquisition in *Streptococcus pyogenes*. PLoS ONE 6:e19543

Palmer KL, Gilmore MS (2010) Multidrug-resistant enterococci lack CRISPR-cas. MBio 1:e00227–e00310

Palmer KL, Whiteley M (2011) DMS3-42: the secret to CRISPR-dependent biofilm inhibition in *Pseudomonas aeruginosa*. J Bacteriol 193:3431–3432

Pougach K, Semenova E, Bogdanova E, Datsenko KA, Djordjevic M, Wanner BL, Severinov K (2010) Transcription, processing and function of CRISPR cassettes in *Escherichia coli*. Mol Microbiol 77:1367–1379

Pourcel C, Salvignol G, Vergnaud G (2005) CRISPR elements in *Yersinia pestis* acquire new repeats by preferential uptake of bacteriophage DNA, and provide additional tools for evolutionary studies. Microbiology 151:653–663

Price EP, Smith H, Huygens F, Giffard PM (2007) High-resolution DNA melt curve analysis of the clustered, regularly interspaced short-palindromic-repeat locus of *Campylobacter jejuni*. Appl Environ Microbiol 73:3431–3436

Pride DT, Sun CL, Salzman J, Rao N, Loomer P, Armitage GC, Banfield JF, Relman DA (2011) Analysis of streptococcal CRISPRs from human saliva reveals substantial sequence diversity within and between subjects over time. Genome Res 21:126–136

Pride DT, Salzman J, Relman DA (2012) Comparisons of clustered regularly interspaced short palindromic repeats and viromes in human saliva reveal bacterial adaptations to salivary viruses. Environ Microbiol 14:2564–2576

Rezzonico F, Smits TH, Duffy B (2011) Diversity, evolution, and functionality of clustered regularly interspaced short palindromic repeat (CRISPR) regions in the fire blight pathogen *Erwinia amylovora*. Appl Environ Microbiol 77:3819–3829

Rho M, Wu YW, Tang H, Doak TG, Ye Y (2012) Diverse CRISPRs evolving in human microbiomes. PLoS Genet 8:e1002441

Riehm JM, Vergnaud G, Kiefer D, Damdindorj T, Dashdavaa O, Khurelsukh T, Zöller L, Wölfel R, Le Flèche P, Scholz HC (2012) *Yersinia pestis* lineages in Mongolia. PLoS ONE 7:e30624

Sapranauskas R, Gasiunas G, Fremaux C, Barrangou R, Horvath P, Siksnys V (2011) The *Streptococcus thermophilus* CRISPR/Cas system provides immunity in *Escherichia coli*. Nucleic Acids Res 39:9275–9282

Schouls LM, Reulen S, Duim B, Wagenaar JA, Willems RJ, Dingle KE, Colles FM, Van Embden JD (2003) Comparative genotyping of *Campylobacter jejuni* by amplified fragment length polymorphism, multilocus sequence typing, and short repeat sequencing: strain diversity, host range, and recombination. J Clin Microbiol 41:15–26

Snyder JC, Bateson MM, Lavin M, Young MJ (2010) Use of cellular CRISPR (clusters of regularly interspaced short palindromic repeats) spacer-based microarrays for detection of viruses in environmental samples. Appl Environ Microbiol 76:7251–7258

Stern A, Keren L, Wurtzel O, Amitai G, Sorek R (2010) Self-targeting by CRISPR: gene regulation or autoimmunity? Trends Genet 26:335–340

Stern A, Mick E, Tirosh I, Sagy O, Sorek R (2012) CRISPR targeting reveals a reservoir of common phages associated with the human gut microbiome. Genome Res 22:1985–1994

Sturino JM, Klaenhammer TR (2006) Engineered bacteriophage-defence systems in bioprocessing. Nat Rev Microbiol 4:395–404

Swarts DC, Mosterd C, van Passel MW, Brouns SJ (2012) CRISPR interference directs strand specific spacer acquisition. PLoS ONE 7:e35888

Tasaki E, Hirayama J, Tazumi A, Hayashi K, Hara Y, Ueno H, Moore JE, Millar BC, Matsuda M (2012) Molecular identification and characterization of clustered regularly interspaced short palindromic repeats (CRISPRs) in a urease-positive thermophilic *Campylobacter* sp. (UPTC). World J Microbiol Biotechnol 28:713–720

Tyson GW, Banfield JF (2008) Rapidly evolving CRISPRs implicated in acquired resistance of microorganisms to viruses. Environ Microbiol 10:200–207

van der Ploeg JR (2009) Analysis of CRISPR in *Streptococcus mutans* suggests frequent occurrence of acquired immunity against infection by M102-like bacteriophages. Microbiology 155:1966–1976

Vergnaud G, Li Y, Gorgé O, Cui Y, Song Y, Zhou D, Grissa I, Dentovskaya SV, Platonov ME, Rakin A, Balakhonov SV, Neubauer H, Pourcel C, Anisimov AP, Yang R (2007) Analysis of the three *Yersinia pestis* CRISPR loci provides new tools for phylogenetic studies and possibly for the investigation of ancient DNA. Adv Exp Med Biol 603:327–338

Yosef I, Goren MG, Qimron U (2012) Proteins and DNA elements essential for the CRISPR adaptation process in *Escherichia coli*. Nucleic Acids Res 40:5569–5576

Zegans ME, Wagner JC, Cady KC, Murphy DM, Hammond JH, O'Toole GA (2009) Interaction between bacteriophage DMS3 and host CRISPR region inhibits group behaviors of *Pseudomonas aeruginosa*. J Bacteriol 191:210–219

Zhang J, Abadia E, Refregier G, Tafaj S, Boschiroli ML, Guillard B, Andremont A, Ruimy R, Sola C (2010) *Mycobacterium tuberculosis* complex CRISPR genotyping: improving efficiency, throughput and discriminative power of 'spoligotyping' with new spacers and a microbead-based hybridization assay. J Med Microbiol 59:285–294

Chapter 12
CRISPRs in the Microbial Community Context

Jillian F. Banfield

Contents

It is hard to remember the exact date when CRISPR-related repeat-spacer regions became of interest to us (probably late 2005 or early 2006), but I can remember why we knew to sit up and take notice. Our work at the time focused on natural acid mine drainage (AMD) microbial biofilm populated mostly by uncultivated organisms. We had begun to recover genomes for the dominant members of these biofilm in 2002, and by early 2004 they had published our first paper on the topic (Tyson et al. 2004). What was not appreciated by many readers is that these genome reconstructions did not employ reference isolate sequences, but were reconstructed de novo from short DNA sequences, each of which derived from a different but coexisting cell in a natural population. The fact that we were not sampling clonal populations posed a challenge for genome reconstruction, beyond one familiar isolated genome assemblies associated with repetitive regions: loci where individual genotypes differ significantly will confound assembly. Strain variation creates branch points, which are of vital interest scientifically because they tell us about rapid evolutionary processes that generate diversity within populations (e.g., Simmons et al. 2008). Our attention initially focused on transposon, which often occupy different genomic locations in closely related individuals due to rapid transposon relocation. The CRISPR locus changed our perspective on 'rapid', and on bacterial populations dynamics and diversity overall.

J. F. Banfield (✉)
Department of Earth and Planetary Sciences, University of California,
Berkeley, CA 94720, USA
e-mail: jbanfield@berkeley.edu

R. Barrangou and J. van der Oost (eds.), *CRISPR-Cas Systems*,
DOI: 10.1007/978-3-642-34657-6_12, © Springer-Verlag Berlin Heidelberg 2013

Gene Tyson, then a Ph.D. student, found the first CRISPR region in one of our AMD bacterial genomes the good old-fashioned way: he spotted the strange stripes in a nucleotide sequence. Such regions are places where assemblies fail. In fact, we had heard of these repeat regions previously, so it was not their existence (or even their potential role in phage immunity) that caught our attention. It was this: our sequencing reads came from a natural bacterial population with relatively little sequence variation genome-wide, yet this region was extremely polymorphic. High levels of CRISPR spacer variation made sense if the sequences were involved in immunizing against a very rapidly evolving target population. By this time, phage/viral targeting had been suggested by others (see Chap. 1). The matching of spacers to coexisting phage was comparatively easy to verify because metagenomic methods simultaneously access host and phage/plasmid populations. Thus, despite the still they were widely prevailing view of bacterial populations as essentially clonal, evidence suggested that this region was evolving so fast that, potentially, most individuals had different genotypes. We wrote up a paper for Science later in 2006, and it was rejected. They had a better CRISPR paper in their hands.

Barrangou et al. (2007) was a landmark in biological research. Beyond the vital proof of the function suspected by others (see other chapters in this volume), the experimental studies described behaviors highly comparable to those evident in natural system studies (e.g., Tyson and Banfield 2007). Specifically, the laboratory studies showed, in real time, the incredible (daily) rate at which CRISPR loci can diversify. Other findings with parallels in natural and laboratory systems included unidirectional locus expansion, targeting of phage and plasmids (in the natural system, many different types), loss of spacer-repeat units (in the natural system, from the older end), and transposon interruption of the locus. The natural system studies had one additional lesson: the locus in *Leptospirillum* group II resides on a genome fragment that had apparently moved from one species to another in a recent lateral transfer event. This finding of locus mobility is in concordance with other studies that have underlined the lack of phylogenetic congruity (from the perspective of *Cas* genes and repeats), as well as the finding by many groups (e.g., Andersson and Banfield 2008) that the plasmid pool represents a rich and still underappreciated reservoir of CRISPR systems.

The Cas proteins of the CRISPR locus first came into our view for a different reason. A common finding is that genomes of archaea and bacteria contain numerous blocks encoding proteins of unknown function, but the level of importance of these enigmatic regions in organism metabolism was uncertain. By late 2004 we were using our newly obtained genomic data from AMD biofilm to enable mass spectrometry-based identification of proteins in microbial community samples. One of our goals was to evaluate the representation of proteins of unknown function in the proteome of uncultivated natural microbial communities. The measurements revealed that one unusually large block of hypothetical proteins was highly expressed, and some proteins encoded in this region were among the most abundant in the proteome (Ram et al. 2005). Later, of course, we realized that these were the Cas proteins associated with the CRISPR locus. Thus, we learned that for bacteria and archaea that rely upon CRISPR-Cas defense in our system, response to phage/viral/plasmid challenge is both a high priority and metabolically demanding. This has been recently reinforced by several reports indicating that Cas proteins are present at high levels in the proteomes of bacteria with active CRISPR loci (Young et al. 2012), and that crRNAs are amongst the dominant ncRNAs in the cell (Deltcheva et al. 2011).

Across our (now many) AMD biofilm community datasets, CRISPR loci turned up in almost every organism, and many showed population-level diversity in spacer content. This primed us to believe that the CRISPR/Cas system is almost everywhere, a finding that is apparently incorrect. The fraction of bacteria and archaea with these loci is often stated to be ~46 and ~85 %, respectively (based on the CRISPR database, see details in Chap. 2). These numbers could reflect bias arising from the currently sampling of genomic diversity in bacteria and archaea, and the shallow or non-existent data for many phylogenetic lineages (i.e., the over-representation of pathogenic and commonly studied laboratory species). Further, such analyses overlook the finding that, within a genome, loci can come and go. In fact, even in a single population of AMD bacteria and archaea, loci lengths can range dramatically, and some cells may have no locus at all (thus sequencing of a single isolated genotype would give unrepresentative information). In fact, recently we returned to the question of loci in lineage that represented the major exception to the rule that AMD biofilm microorganisms all have CRISPR loci: the deep branching ARMAN nanoarchaea. Interestingly, a CRISPR locus has now turned up in a new ARMAN genome. The locus encodes Cas1, Cas2, Cas4, and a large protein with a putative Cas9/Csn1 domain (possibly a Type IIb system as defined by Makarova et al., 2011).

Despite dramatic progress, it is not yet possible to determine the phylogenetic distribution of CRISPR/Cas systems in bacterial and archaeal lineages because there are massive gaps in genomic coverage of these domains. For example, currently there are 32 bacterial phyla between zero genomes (compared to over 1,489 for the *Proteobacteria*). Six phyla have only a single genome sequence, and some of these (e.g., TM7) are incomplete. However, genome sequences for novel lineages are beginning to appear, some as the result of single cell sequencing (e.g., TM7 and 270 kb of sequence from an OP11 cell) and some from community metagenomic reconstruction (Wrighton et al. 2012). An initial survey of our currently as yet unpublished genomes from several bacterial phyla indicates that CRISPR/Cas systems are recognizable in a small subset of organisms from each lineage. It will be interesting to see if patterns emerge, and new types of systems are recognized as more genomes become available. Given the magnitude of what we do not know, it seems probable that new phage/viral defense systems will come to light. The story of the CRISPR/Cas locus serves as a case study, illustrating what perhaps awaits us.

Rodolphe Barrangou, Mark Young, and I hosted the first CRISPR meeting in 2008 at UC Berkeley. This has become an annual tradition (the fifth meeting was held in June, 2012). With the ever-expanding group of researchers, we have seen the field bloom and diversify over the years, with scientific interest and research pace seemingly increasing over time. Notwithstanding the diversity in paths that have lead the attendees to join the CRISPR field, much of the excitement has related to discoveries of the biochemical and molecular underpinnings of the mechanism of action (see Chaps. 5, 6, 7, 8). The perhaps equally important topic of the operation of the CRISPR/Cas system in the context of microbial communities has received less attention. In fact, arguably, the CRISPR system only makes sense when understood in the context of diversifying host and phage/viral populations. Indeed, several modeling studies have now shown that CRISPR-Cas systems play a critical role in host-virus population dynamics, and we anticipate that ecological surveys of microbial populations will further substantiate the important role that CRISPR loci play in the host-virus coexistence and evolution in natural habitats (see Chap. 9).

There are many questions that come to mind when contemplating phage/viruses and host population interaction dynamics. When viewed from the laboratory experiment perspective, it seems remarkable that one party (e.g., bacteria or phage) does not eventually loose out. This may prompt us to ask what processes in natural systems ensure stable (or at least sustainable) co-existence? What is the role of migration/immigration? Natural population studies clearly point out the occurrence of major host bottlenecks, probably linked to phage/viral predation, but is diversity only lost locally? How, and how often, are loci laterally transferred, and how does this process impact lineage divergence? Clearly, many different phage/virus populations target the same hosts, and some phage/viruses target multiple hosts. What are the interactions and feedbacks that establish these patterns?

Some of the most important consequences of dramatic changes in phage/virus infectivity and host resistance may be for evolution. The cell that "gets lucky" through immunization (e.g., following appearance of a new phage/virus or after a phage/virus mutates to high virulence) should proliferate, potentially carrying fixation to any novel genomic traits. When contemplating this possibility, it is important to underline that natural populations are rarely clonal so that such events, even if rare, could alter the genetic characteristics of a lineage. Iplasma, an archaeon, may tell such a story. This organism has a CRISPR locus that is almost clonal, except for spacer's right at the leader end. These match a virus that is not targeted by the older, clonal spacers. Such a pattern could suggest a recent population bottleneck caused when the first Iplasma cell acquired immunity to a new virus. Interestingly, the genomic region carrying the Iplasma locus is rich in transposons and novel proteins, thus it may have been introduced on mobile element. If so, an alternative explanation is that the bottleneck was associated with locus acquisition. A similar story was inferred in *Leptospirillum* Group II (Tyson and Banfield 2007), when locus transfer apparently led a selective sweep that eliminated genotypes lacking the recombinant block.

My hope in contributing this brief final perspective is to turn some attention toward the broadest, and as yet largely unknown, implications of heritable adaptive immunity. It is also my intention to frame virus-host interaction dynamics in the context of microbial community structure and functioning, and to underscore the arguably unparalleled importance of such processes in the world around us.

References

Andersson A, Banfield JF (2008) Virus population dynamics and acquired virus resistance in natural microbial communities. Sci 230:1047–1050

Barrangou R, Fremaux C, Deveau H, Richards M, Boyaval P, Moineau S, Romero DA, Horvath P (2007) CRISPR provides acquired resistance against viruses in prokaryotes. Science 315:1709–1712

Deltcheva E, Chylinski K, Sharma CM, Gonzales K, Chao Y et al (2011) CRISPR RNA maturation by trans-encoded small RNA and host factor RNase III. Nature 471:602–607

Makarova KS, Haft DH, Barrangou R, Brouns SJJ, Charpentier E, Horvath P, Moineau W, Mojica FJM, Wolf YI, Yakunin AF, van der Oost J, Koonin EV (2011) Evolution and classification of the CRISPR–Cas systems. Nature Reviews Microbiology 9:467–477

Ram RJ, VerBerkmoes NC, Thelen MP, Tyson GW, Baker BJ, Blake RC II, Shah M, Hettich RL, Banfield JF (2005) Community proteomics of a natural microbial biofilm. Sci 308:1915–1920

Simmons SL, DiBartolo G, Denef VJ, Aliaga Goltsman D, Thelen MP, Banfield JF (2008) Population genomic analysis of strain variation in *Leptospirillum* group II bacteria involved in acid mine drainage formation. *PLoS Biol*, 6(e177):1427–1442

Tyson GW, Chapman J, Hugenholtz P, Allen E, Ram RJ, Richardson P, Solovyev V, Rubin E, Rokhsar D, Banfield JF (2004) Insights into microbial community structure and metabolism by reconstruction of genomes from a natural environment. Nature 428:37–43

Tyson GW, Banfield JF (2007) Rapidly evolving CRISPRs implicated in acquired resistance of microorganisms to viruses. Environ Microbiol 10:200–207

Wrighton KC, Thomas BC, Sharon I, Miller CS, Castelle CJ, VerBerkmoes NC, Wilkins MJ, Hettich RL, Lipton MS, Williams KH, Long PE, Banfield JF (2012) Fermentation, hydrogen and sulfur metabolism in multiple uncultivated bacterial phyla. Science. 337:1661–1665

Young J, Dill BD, Pan C, Hettich RL, Banfield JF, Shah M, Fremaux C, Horvath P, Barrangou R, Verberkmoes NC (2012) Phage-induced expression of CRISPR-associated proteins is revealed by shotgun proteomics in Streptococcus thermophiles. PLoS One, 7, DOI:10.1371/journal.pone.0038077

Glossary

Acquisition: process by which a new spacer is integrated into the CRISPR locus. The ability of CRISPR loci to acquire novel spacers derived from invasive nucleic acids drives the adaptive nature of this immune system.

***Cas*: CRISPR-associated gene.** These *cas* genes encode a functionally diverse set of Cas proteins that are directly involved in one or more stages of the CRISPR mechanism of action: spacer acquisition, CRISPR locus expression, and target interference. These genes often reside in operons and are typically genetically linked with CRISPR repeat-spacer arrays.

Cascade: CRISPR-associated complex for antiviral defense. Multi-subunit Cas protein complex required for crRNA processing and maturation, and CRISPR-mediated interference. Homologs from the archaetype Cascade complex from *E. coli* are found in all Type-I CRISPR-Cas subtypes. The other CRISPR (sub) types have distinct ribonucleoprotein complexes (Type-II, a multi-domain protein Cas9; Type-III, a multi-subunit RAMP complex).

CRISPR: Clustered regularly interspaced short palindromic repeats. Genetic loci that contain arrays of homologous direct DNA repeats separated by short variable sequences called spacers, although not all repeat sequences, are actually palindromic. This hallmark of CRISPR-Cas systems encodes the CRISPR transcript, which is processed into small interfering RNAs.

CRISPR-Cas: immune system comprising a CRISPR repeat-spacer array and accompanying *cas* genes. There are generally three distinct types of CRISPR-*cas* systems, namely Type-I, Type-II, and Type-III, as defined by the content and sequences of their elements, notably *cas* genes.

crRNA: CRISPR RNA. Mature small non-coding CRISPR RNA generated by cleavage and processing of a precursor CRISPR transcript (pre-crRNA), which guides the *Cas* interference machinery towards homologous invading nucleic acids.

R. Barrangou and J. van der Oost (eds.), *CRISPR-Cas Systems*,
DOI: 10.1007/978-3-642-34657-6, © Springer-Verlag Berlin Heidelberg 2013

Interference: process by which invasive DNA or RNA is targeted by crRNA-loaded Cas proteins. This process relies on sequence homology and complementarity between the crRNA and the target nucleic acid. In some cases, ancillary elements are necessary for target interference, such as tracrRNA and PAMs, and for preventing auto-immunity.

Leader: AT-rich sequence located upstream of the first CRISPR repeat. This sequence serves as a promoter for the transcription of the repeat-spacer array. Also, this sequence defines CRISPR locus orientation for transcription and polarized spacer acquisition.

PAM: Protospacer Adjacent Motif. Short signature sequences (typically 2–5 nt) flanking a protospacer, which is necessary for the interference step in most Type I and Type II DNA-targeting systems.

Pre-crRNA: pre-CRISPR RNA. Full length transcript generated by the CRISPR repeat-spacer array, which serves as the precursor for crRNA biogenesis via one or more processing and maturation steps.

Proto-spacer: spacer precursor in invasive nucleic acid. Precursor sequence of CRISPR spacers in the DNA of invasive elements that will be sampled by the CRISPR-Cas immune system as part of the acquisition process and subsequently targeted by crRNA as part of the interference process.

RAMP: repeat-associated mysterious proteins. Subset of *cas* proteins with an RNA recognition motif (RRM fold). Some RAMPs have endoribonuclease activity involved in the maturation and processing of crRNA.

R-loop: section of DNA associated with RNA forming a loop. Structure in which RNA hybridizes with double-stranded DNA. The RNA **base pairs with a complementary sequence in one of the strands of** a DNA molecule, causing the displaced strand to form a loop.

Repeats: short sequence repeated within a CRISPR array, separated by spacers. Highly similar sequences that form direct repeats spaced by short variable sequences of conserved length. The CRISPR repeat sequence is critical for crRNA maturation and processing, and is functionally coupled with *cas* genes to form a functional CRISPR-Cas system. Only a subset of all repeat types are palindromic.

RNAi: RNA interference. Process in eukaryotes by which small non-coding RNA molecules guide enzymatic cleavage of complementary mRNAs, through the RNA-induced silencing complex (RISC).

Seed sequence: short sequence within the crRNA which requires perfect base pairing with the target sequence. Short stretch of nucleotides (7–9 nt) which enthalpically drives hybridization between the interfering crRNA and the complementary target strand in the vicinity of the PAM site, which **supports R-loop formation and generally results in interference.**

Signature gene: key *cas* gene which defines a CRISPR- *cas* Type. Idiosyncratic *cas* gene(s) that define the type of a particular CRISPR-*cas* system, which are *cas*3, *cas*9, and *cas*10 for Type-I, Type-II, and Type-III, respectively.

Spacer: small variable nucleotide sequence within a CRISPR array, flanked by repeats. A spacer is the sequence derived from invasive genetic elements (notably viruses and plasmids) which is integrated within the CRISPR locus.

Spoligotyping: SPacer OLIGOnucleotide TYPING. Strain typing method based on the detection of specific CRISPR spacers, typically through hybridization of labeled CRISPR PCR amplicons.

TracrRNA: trans-encoded CRISPR RNA. Small non-coding RNA partially homologous to CRISPR repeat sequences, which is involved in the crRNA maturation process in Type II systems.

Index

R. Barrangou and J. van der Oost (eds.), *CRISPR-Cas Systems*,
DOI: 10.1007/978-3-642-34657-6, © Springer-Verlag Berlin Heidelberg 2013

MIX
Papier aus verantwortungsvollen Quellen
Paper from responsible sources
FSC® C105338

If you have any concerns about our products,
you can contact us on
ProductSafety@springernature.com

In case Publisher is established outside the EU,
the EU authorized representative is:
Springer Nature Customer Service Center GmbH
Europaplatz 3, 69115 Heidelberg, Germany

Printed by Libri Plureos GmbH
in Hamburg, Germany